Animal Biology: Taxonomy, Anatomy and Physiology

Animal Biology: Taxonomy, Anatomy and Physiology

Edited by Adalina Woodbury

SYRAWOOD
PUBLISHING HOUSE

New York

Published by Syrawood Publishing House,
750 Third Avenue, 9th Floor,
New York, NY 10017, USA
www.syrawoodpublishinghouse.com

Animal Biology: Taxonomy, Anatomy and Physiology
Edited by Adalina Woodbury

© 2019 Syrawood Publishing House

International Standard Book Number: 978-1-68286-837-9 (Hardback)

Cataloging-in-Publication Data

Animal biology : taxonomy, anatomy and physiology / edited by Adalina Woodbury.
    p. cm.
Includes bibliographical references and index.
ISBN 978-1-68286-837-9
1. Zoology. 2. Animals--Classification. 3. Veterinary anatomy. 4. Physiology.
5. Veterinary physiology. I. Woodbury, Adalina.
QL45.2 .A55 2019
590--dc23

TABLE OF CONTENTS

Permissions

List of Contributors

Index

PREFACE

The study of the animal kingdom that comprises of an analysis of the structure, evolution, embryology, classification, habits and distribution of animals is under the scope of animal biology or zoology. It incorporates the disciplines of comparative anatomy, animal physiology, taxonomy, zoography, vertebrate and invertebrate zoology, etc. Animals are classified into distinct groups based on shared characteristics. The branch of science concerned with the identification, description, nomenclature and classification of animals is known as taxonomy. Anatomy deals with the structural organization of all animals. The focus of physiology is to understand how the different structures of the organism such as cells, biomolecules, organs and organ systems execute the various physical and chemical functions essential to the organism. This book provides comprehensive insights into the field of animal biology. It unfolds the innovative aspects of the study of taxonomy, anatomy and physiology, which will be crucial for the progress of this field in the future. It will serve as a valuable source of reference for graduate and post graduate students as well as experts.

The information contained in this book is the result of intensive hard work done by researchers in this field. All due efforts have been made to make this book serve as a complete guiding source for students and researchers. The topics in this book have been comprehensively explained to help readers understand the growing trends in the field.

I would like to thank the entire group of writers who made sincere efforts in this book and my family who supported me in my efforts of working on this book. I take this opportunity to thank all those who have been a guiding force throughout my life.

Editor

Influence of vegetation physiognomy, elevation and fire frequency on medium and large mammals in two protected areas of the Espinhaço Range

Fernando Ferreira de Pinho[1], Guilherme Braga Ferreira[1], Adriano Pereira Paglia[2]

[1]Instituto Biotrópicos. Praça JK 25, 39100-000 Diamantina, MG, Brazil.
[2]Programa de Pós-Graduação em Ecologia, Conservação e Manejo da Vida Silvestre, Departamento de Biologia Geral, Universidade Federal de Minas Gerais. Avenida Antonio Carlos 6627, 31270-901 Belo Horizonte, MG, Brazil.
Corresponding author: Fernando Ferreira de Pinho (fernandopinho@biotropicos.org.br)

http://zoobank.org/5C7A9255-0D5C-4B83-8258-E4E338ACB878

ABSTRACT. The objectives of this study were to determine the richness of medium and large mammal species in two protected areas of the Espinhaço Mountain Range, state of Minas Gerais, Brazil; and to investigate the factors affecting the occurrence of those species. To accomplish that we placed 49 camera traps activated by heat and motion at Rio Preto State Park (RPSP) and 48 at Sempre Vivas National Park (SVNP). We also collected data on three environmental variables: vegetation physiognomy, elevation and wildfire frequency, to evaluate the influence of these factors on species richness and use intensity (inferred from camera trap detection rate) by large mammals. We recorded 23 large mammal species in the two parks combined. The lowest species richness was found at the rupestrian habitat of RPSP, and in the open grasslands of SVNP. The forest and savannah physiognomies were used more intensively by large mammals. Species richness was higher and use was greater at lower elevations of RPSP. In SVNP, fire frequency did not affect species richness or use intensity. The savannah habitat had very similar richness compared to the forests of the two protected areas. The high species richness and use intensity observed in these forest habitats highlights the importance of riparian environments in the Cerrado biome. The highest species richness and use intensity observed at low elevation follows patterns found in the literature, probably due to variation in the vegetation, which results in greater resource availability. Although rupestrian habitats at high elevations of the Espinhaço Range are known to have a high degree of endemism for some taxa, large mammal richness and use were not high in this habitat. These results indicate that the protection of native vegetation at lower elevations is crucial for the long-term conservation of large mammals in the Espinhaço Range.

KEY WORDS. Cerrado, campo rupestre, species richness, use of habitat, wildfire.

INTRODUCTION

Understanding which factors affect species richness is a challenge for ecologists. Resource availability in the ecosystem, the degree of specialization of species and the coexistence of species that share the same resources are key determinants of local species richness (MacArthur 1972). Generally, heterogeneous habitats provide the conditions for the establishment of a large number of species (Kerr and Packer 1997, Kreft and Jetz 2007, Stein et al. 2014). In fact, a review of studies on environmental heterogeneity and diversity found that 85% of the publications arrived at a positive correlation between heterogeneous ecosystems and species richness (Tews et al. 2004), including mammals (e.g., Southwell et al. 1999, Williams et al. 2002). Among the exceptions to this is the work of August (1983), who did not find a positive relationship between habitat heterogeneity (as defined by the horizontal variation within a habitat) and mammal species richness in the Venezuelan llanos, even though habitat complexity (as defined by the vertical stratification of the habitat) in his data was positively correlated with richness.

Habitat heterogeneity also allows the coexistence of competitor species, therefore contributing to local species richness. For example, Schuette et al. (2013) observed that carnivore mammals in Africa responded in different ways to environmental factors, and noted that this variation could be responsible for the high local diversity of carnivores. Similarly, environmental heterogeneity is regarded as one of the main factors allowing the coexistence of two large Neotropical predators, the jaguar, *Panthera onca* (Linnaeus, 1758), and the puma, *Puma concolor* (Linnaeus, 1771) (Sollmann et al. 2012). To limit the negative effects of competition on species' fitness, natural selection favors certain adaptations in morphology, behavior and natural history (Hardin 1960, Creel et al. 2001, Pfennig and Pfennig 2005). These adaptations are evident in several large terrestrial mammals, where differences in morphology (such as body mass) and ecology (diet, habitat selection and activity patterns) allow species coexistence and resource partitioning (Karanth and Sunquist 1995, Fedriani et al. 2000, Sinclair et al. 2003, Hayward and Slotow 2009, Oliveira-Santos et al. 2012, Macandza et al. 2012, Schuette et al. 2013, Ferreguetti et al. 2015).

Several studies have investigated how biotic and abiotic features affect mammal distribution (e.g., Kinnaird and O'Brien 2012, Linkie et al. 2007, Sunarto et al. 2012). Likewise, habitat use in sympatric species is also frequently studied in tropical communities (e.g., Kinnaird and O'Brien 2012, Sollmann et al. 2012, Ahumada et al. 2013, Schuette et al. 2013). For example, Kinnaird and O'Brien (2012) found that environments with greater degrees of restriction on human activities favor the occurrence of medium and large mammals in Kenya. In Costa Rica, Ahumada et al. (2013) assessed how environmental factors such as altitude and canopy height influence the occurrence of mammals and found a variety of responses among the species evaluated. Sunarto et al. (2012) studied populations of tigers in Sumatra and showed that vegetation cover is essential for the presence of the species.

In the Brazilian Cerrado, gallery forests are known to play an important role in mammal distribution, as they provide unique resources within the ecosystem (Redford and Fonseca 1986, Johnson et al. 1999). Nevertheless, in a local or regional scale there is a lack of understanding about the environmental determinants of large mammal occurrence and species richness in the Cerrado. This biome is the second largest ecosystem in Brazil and is formed by a mosaic of grasslands, savannas and forests (Ribeiro and Walter 2008). Due to its large extension, contact with four major Brazilian ecosystems and environmental heterogeneity, the Cerrado harbors a large number of species (Silva et al. 2006), and is recognized as the most biodiverse savannah in the world (Klink and Machado 2005). Regarding mammals, 251 species have been recorded in this ecosystem (Paglia et al. 2012), of which 42 are medium and large species (body mass > 1 kg; Marinho-Filho et al. 2002).

Considering the high environmental heterogeneity found in the Cerrado and the relevance of protected areas for the conservation of large mammals, the objectives of this study were to determine large mammal species richness in two protected areas (PAs) of the Espinhaço Mountain Range and to investigate the factors affecting the occurrence of these species. We hypothesized that large mammal species richness and intensity of use would be higher in more complex habitats, in lower elevation, and in areas that are less frequently affected by fires.

MATERIAL AND METHODS

The Espinhaço Mountain Range is one of the most important Brazilian biogeographic regions. It extends over

parts of three major ecosystems: Caatinga, Cerrado and Atlantic Rainforest, the last two being recognized as global biodiversity hotspots (Myers et al. 2000, Mittermeier et al. 2005). This mountain range is one of the most important centers for species endemism in South America, with several new species of different taxonomic groups described recently (Azevedo and Silveira 2005, Oliveira and Sano 2009, Freitas et al. 2012, Barata et al. 2013, Pardiñas et al. 2014) and its biological relevance has been highlighted in prioritization studies for biodiversity conservation (e.g., Drumond et al. 2005, MMA 2007, Silva et al. 2008).

The Espinhaço Mosaic (Mosaico do Espinhaço: Alto Jequitinhonha-Serra do Cabral, as it is called in Portuguese) is composed of seven strict PAs (IUCN categories I–IV) and five multiple-use PAs (Area de Proteção Ambiental; IUCN category V) located in the southern portion of the Espinhaço Range. Most of this mosaic is within the Cerrado ecosystem, whereas the Atlantic Forest covers its eastern-most portion. Due to its irreplaceability, the Espinhaço Mosaic has been recognized as a priority area for conservation (Silva et al. 2008). This study focused on two strict PAs in this region: Rio Preto State Park (RPSP) and Sempre-Vivas National Park (SVNP).

Established in 1994, RPSP encompasses an area of 121 km² covered mostly by open vegetation such as 'campos rupestres' (rocky outcrops covered by scattered herbs, grasses and shrubs) and savannah, but also with gallery forests bordering water courses. There is a broad elevational gradient in the park, ranging from 750 to 1,800 m asl. Due to the fact that this reserve was established two decades ago, and that it is very effectively managed (WWF 2016), the impact of recent anthropogenic factors on the mammals that inhabit the area is assumed to be negligible.

Sempre-Vivas National Park was established in 2002 and covers an area of 1,241 km² The main vegetation physiognomies in SVNP are campos rupestres and open grasslands, but there are also portions of savannah, veredas (a humid grassland dominated by the palm species *Mauritia flexuosa* L. f.), dry forests and gallery forests. The elevation ranges from 650 to 1,525 m asl. However, due to SVNP's large area and logistical constraints, sampling was limited to between 1,000 and 1,400 m asl, including most of the park's vegetation physiognomies, but not veredas and dry forests normally associated with lower elevation. Within the limits of SVNP there are several land-use issues that result in anthropogenic impacts (such as poaching, wildfire and fuelwood collection). Wildfires are frequently detected in this PA (INPE 2015), but there is no reliable measure of poaching and fuelwood collection intensity inside the park. Both are certainly more intensive than at RPSP, but likely to be less common than in adjacent unprotected natural lands.

We used camera traps activated by heat and motion (Bushnell Trophycam) to study the large mammal community. The sampling design followed recommendations from Team Network (2008) and O'Brien (2010) with minor adaptations. Potential sampling sites (location where camera traps would be installed) were plotted at a density of 1 site per 2 km², representing a distance of 1.5 km between sites. At RPSP potential sites were plotted throughout the entire park and in a small private reserve adjacent to the park, whereas in the much larger SPNP potential sites were established in the southern portion, with the park's lodge roughly in the centre.

In the field, we used a GPS unit to navigate to the potential sites and establish the camera trap within a 100 m radius from the predetermined location, choosing the spots with the highest probability of recording medium and large mammal species. Due to difficult access conditions, in four occasions at RPSP and in two occasions at SVNP camera trap were established out of the 100 m radius, but within a 250 m radius. Additionally, we had to relocate two potential sites at RPSP, because we were unable to reach them (e.g. cliffs). As a result, these two sites were established at a distance of 1.2 km of the nearest neighboring site.

In total we set up 51 camera trap sites at RPSP (including three sites at an adjacent private reserve) and 55 at SVNP, however, due to malfunctioning of some camera trap units, the final number of sites used in the analysis was 49 at RPSP and 48 at SVNP (Fig. 1). We conducted the surveys only during the dry season, from June–September 2013 at RPSP (average sampling days per site: 57.18, range: 7–108) and May–June 2014 at SVNP (average sampling days per sites: 39.75, range: 9–50). Surveys happened during the dry season to minimize camera malfunction from the accumulation of humidity inside the equipment. We assumed this would not have an important effect on our data since large Neotropical mammals do not perform long-distance seasonal migrations. Conducting camera trapping in the dry season has been adopted in several tropical regions of the world by the Team Network (2008). No lure was used to attract animals in either of the PAs.

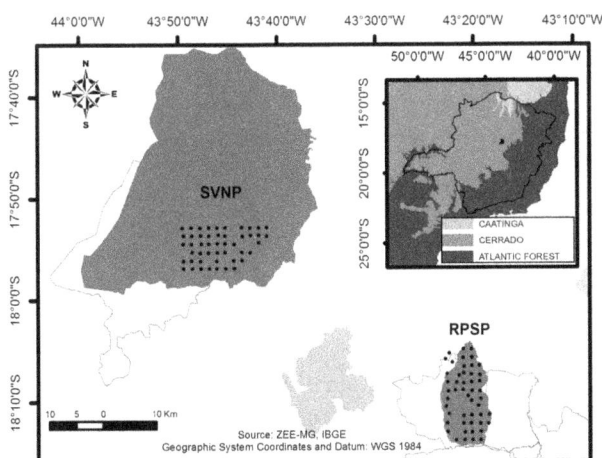

Figure 1. Camera trap sites (black circles) surveyed at Rio Preto State Park and Sempre Vivas National Park. Areas of integral protection in gray and areas of sustainable use in hollow poligones.

Considering our hypothesis, we collected data on three environmental variables: (1) vegetation physiognomy, (2) elevation and (3) wildfire intensity. Physiognomy was assigned to four possible categories: forest (gallery forest or capão forest), savannah (small trees and shrubs with a herbaceous layer), open grassland (with or without scattered trees) and rupestrian habitat (campos rupestres or rock outcrops with scattered vegetation). Elevation was measured in the field using a GPS unit. Wildfire frequency was extracted from a Kernel Map provided by PA management team, which was produced using data available from INPE (2015) on heat spots density within the limits of SVNP, between 1999 and 2012. In this map, the pixel value of the camera trap site was used as a covariate, where smaller values indicated low fire frequency and larger values indicated high fire frequency. We used ArcGIS 10 to produce the map and to extract values for the analysis.

We built a matrix of independent camera trap records of large mammals for each PA. When there were two records of the same species in a single sampling site we assumed them to be independent if the observations were at least 24hs apart. We used the software EstimateS 9.1.0 (Cowell 2013) to randomize the sampling and obtain all species richness estimates presented here. To be able to compare species richness between the two parks, we applied the rarefaction function to data from SVNP, as it had a smaller sampling effort than RPSP. The parameter "S(Est)", which represents the average number of recorded species as a function of sampling effort, and its 95% CI were used to build species accumulation curves for each park.

To assess the effect of the Cerrado physiognomies on species richness, we grouped the records from each vegetation physiognomy (forest, savannah, grassland, rupestrian). Since there were large differences in sampling effort between physiognomies, the rarefaction function provided estimates with very broad CIs. For this reason, we decided to perform the comparison, controlling for the physiognomy with the smallest sampling effort (forest in both parks). Therefore, we obtained the average species richness in each physiognomy based on 452 and 420 sampling days/physiognomy at RPSP and SVNP, respectively. We checked for overlap between the 95% CI for each estimate to infer statistical significance. Additionally, as an overall measure of large mammal use, we grouped the independent records from all large mammal species within a physiognomy (overall photographic rate) and compared them through an analysis of variance (ANOVA). Although we cannot consider the overall photographic rate as a measure of abundance, we used this rate as a measure of the intensity of use by large mammals, in which a large number of independent records represents high use intensity of a given area. This analysis was performed for each park independently.

The effect of elevation on species richness was only assessed for RPSP, since the elevation variation of the sampling sites at SVNP only ranged from 1,041 to 1,369 m asl. In RPSP the difference between the lowest and highest sampling site was almost 1,000 m (range: 800–1,720 m asl). Following a well-established classification used by the park's management team, we divided the RPSP into two regions according to elevation: below 1,000 m asl and above 1,200 m asl. This division reflects markedly different environments, where low elevation sites are dominated by savannah and forest habitats and high elevation sites are normally located in grasslands or rupestrian habitats. We thus grouped data from all sites at low (<1,000 m) and high elevation (>1,200 m), eliminating those sites established between 1,000–1,200 m asl. Species richness estimates with 95% CI was obtained for each region controlling for differences in sampling effort. To evaluate the intensity of use by large mammals we grouped the independent records from all species in low and high altitudes and compared them through independent T-tests.

Finally, we used linear regression to assess the effect of wildfire frequency on species richness and use intensity (defined as average records of all species of medium and large mammals) at SVNP. We excluded RPSP from this analysis because wildfires are rare in this park and there have been no occurrences of it in at least the last five years. To represent wildfire frequency in the sampling site, we extracted the value derived from the Kernel Map (see Data collection) for the pixel where the camera trap had been set. The average species richness and use intensity at the site level, controlling for differences in sampling effort, were obtained for all sites surveyed for more than 30 days at SVNP.

RESULTS

We recorded 23 large mammal species in the two parks combined, 19 at RPSP after 2,865 sampling days and 18 at SVNP after 2,010 sampling days (Table 1). Observed species richness was very similar among parks, with almost a complete overlap of estimates (Fig. 2). Despite the similarity in species richness, five species were recorded exclusively in RPSP: naked-tailed armadillo, *Cabassous* sp.; crab-eating fox, *Cerdocyon thous* (Linnaeus, 1766); puma, *P. concolor*; crab-eating raccoon, *Procyon cancrivorus* (G.[Baron] Cuvier, 1798); and agouti, *Dasyprocta* sp.; and four exclusively in SVNP: lowland tapir, *Tapirus terrestris* (Linnaeus, 1758); hoary-fox, *Lycalopex vetulus* (Lund, 1842); collared pecary; *Pecari tajacu* (Linnaeus, 1758); and capybara, *Hydrochoerus hydrochaeris* (Linnaeus, 1766).

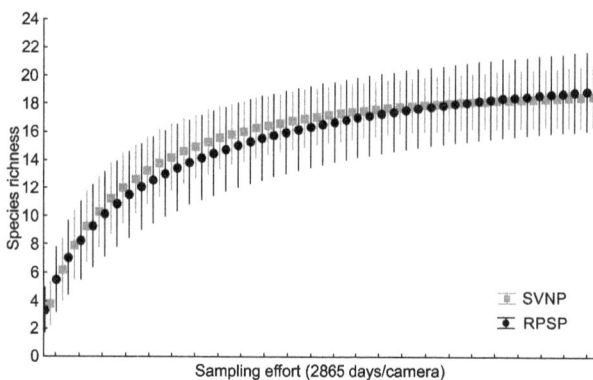

Figure 2. Species accumulation curve (richness x effort) and confidence interval (95%) in Rio Preto State Park (RPSP) and Sempre Vivas National Park (SVNP).

Table 1. Mammal species recorded at Rio Preto State Park (RPSP) and Sempre Vivas National Park (SVNP).

Taxonomic group	Popular name	RPSP	SVNP
Pilosa			
Myrmecophagidae			
Tamandua tetradactyla (Linnaeus, 1758)	Southern-anteater	x	x
Myrmecophaga tridactyla Linnaeus, 1758	Giant-anteater	x	x
Cingulata			
Dasypodidae			
Dasypus sp.	Armadillo	x	x
Cabassous unicinctus (Linnaeus, 1758)	Naked-tailed armadillo	x	
Euphractus sexcinctus (Linnaeus, 1758)	Yellow-armadillo	x	x
Priodontes maximus (Kerr, 1792)	Giant-armadillo	x	x
Carnivora			
Canidae			
Cerdocyon thous (Linnaeus, 1766)	Crab-eating fox	x	
Chrysocyon brachyurus (Illiger, 1815)	Maned-wolf	x	x
Lycalopex vetulus (Lund, 1842)	Hoary-fox		x
Mephitidae			
Conepatus semistriatus (Boddaert, 1785)	Striped hog-nosed skunk	x	x
Mustelidae			
Eira barbara (Linnaeus, 1758)	Tayra	x	x
Felidae			
Puma yagouaroundi (É. Geoffroy, 1803)	Jaguarundi	x	x
Leopardus pardalis (Linnaeus, 1758)	Ocelot	x	x
Leopardus tigrinus (Schreber, 1775)	Oncilla	x	x
Puma concolor (Linnaeus, 1771)	Puma	x	
Procionidae			
Procyon cancrivorus (G. Cuvier, 1798)	Crab-eating raccoon	x	
Nasua nasua (Linnaeus, 1766)	Coati	x	x
Perissodactyla			
Tapiridae			
Tapirus terrestris (Linnaeus, 1758)	Tapir		x
Artiodactyla			
Cervidae			
Mazama gouazoubira (G. Fischer, 1814)	Gray-brocket	x	x
Tayassuidae			
Pecari tajacu (Linnaeus, 1758)	Collared-peccary		x
Rodentia			
Cuniculidae			
Cuniculus paca (Linnaeus, 1766)	Spotted-paca	x	x
Dasyproctidae			
Dasyprocta sp.	Agouti	x	
Caviidae			
Hydrochoerus hydrochaeris (Linnaeus, 1766)	Capybara		x

The observed species richness was statistically lower in rupestrian habitats than other vegetation physiognomies in RPSP (Fig. 3) and in SVNP the species richness in grassland was statistically lower than forest and savannah (Fig. 4). The use intensity (records rate of all species grouped) showed significant variation among physiognomies (RPSP: $F_{3;2819} = 18.9$, $p < 0.001$; SVNP: $F_{3;1959} = 17.3$, $p < 0.001$). The forest in RPSP was clearly the most intensively used habitat, whereas the rupestrian habitat had a very low level of use (Fig. 5). Similarly, forest in SVNP, was the most frequently used habitat (although the savannah showed similar levels), while the grasslands and rupestrian habitats had much lower use intensity (Fig. 6).

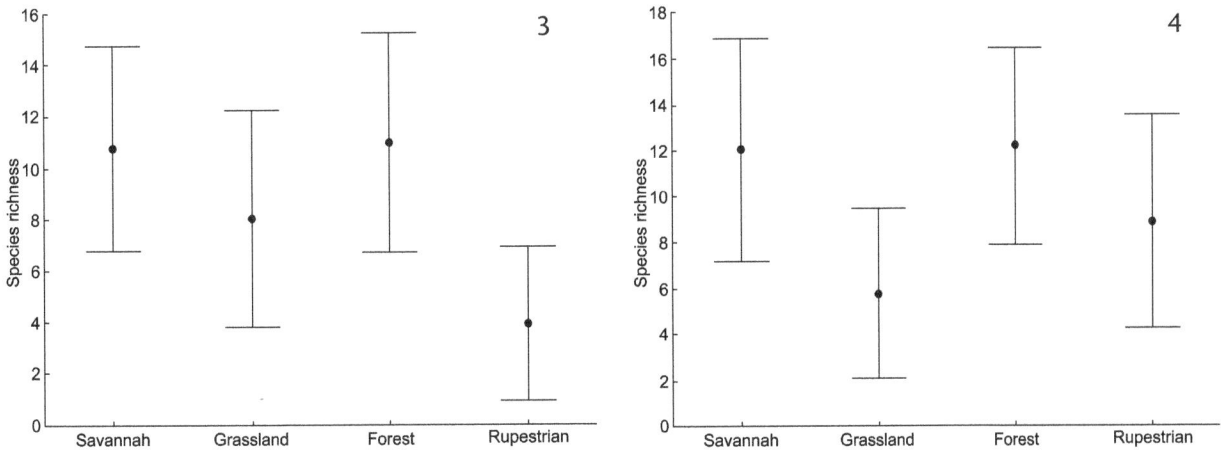

Figures 3–4. Observed species richness and confidence interval (95%) by physiognomy in (3) Rio Preto State Park and (4) Sempre Vivas National Park.

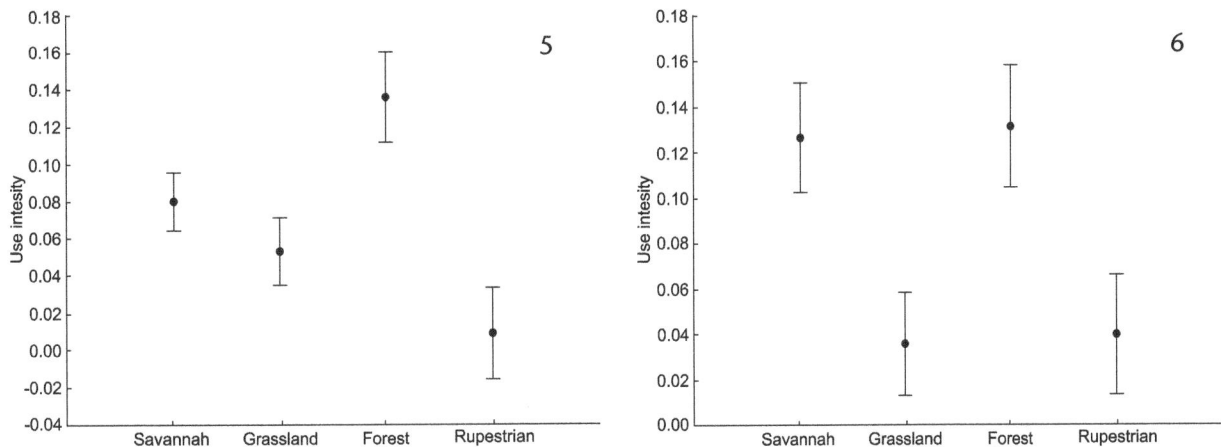

Figures 5–6. Medium and large mammals use intensity and confidence interval (95%) by physiognomy in (5) Rio Preto State Park and (6) Sempre Vivas National Park.

At lower elevations, species richness (Fig. 7) and use intensity ($F_{(2178)}$ = 3.79, p < 0.001) (Fig. 8) were statistically higher in RPSP. The wildfire frequency did not show a significant effect on species richness ($R_{(1, 38)}$ = 0.073, p = 0.65) or use intensity ($R_{(1, 38)}$ = 0.13, p = 0.43) in SVNP.

DISCUSSION

Total species richness recorded here represents 70% and 55% of all large mammals known to occur at the Espinhaço Range (Lessa et al. 2008) and the Brazilian Cerrado (Marinho-Filho et al. 2002), respectively. While the trend of species accumulation curves' to stabilize indicates that a large proportion of the target community has been recorded, some species known to occur in these parks were not detected. For instance, the Neotropical otter, *Lontra longicaudis* (Olfers, 1818), and the lesser grison, *Galictis cuja* (Molina, 1782)

have been recorded in RPSP (Lessa et al. 2008), whereas *P. concolor* and *P. onca* have been recently recorded by camera traps in SVNP (Instituto Biotropicos Archive). Considering these four species and the species we recorded in this study, RPSP and SVNP together harbor virtually all large mammal species known to currently occur in the Espinhaço Range. The presence of some species recorded exclusively in one of the parks (such as puma and crab-eating raccoon in RPSP or capybara in SVNP) can be explained by the relatively short duration of the surveys. Nevertheless, unpublished data from RPSP and information from experienced park rangers suggest that the tapir and the collared peccary are likely restricted to SVNP. Poachers frequently target these two species, and therefore their numbers may decline in the presence of anthropogenic pressure (Cullen-Jr et al. 2000). It is possible that they have become locally extinct at RPSP (which is smaller and easier to access than SVNP) before the PA was established.

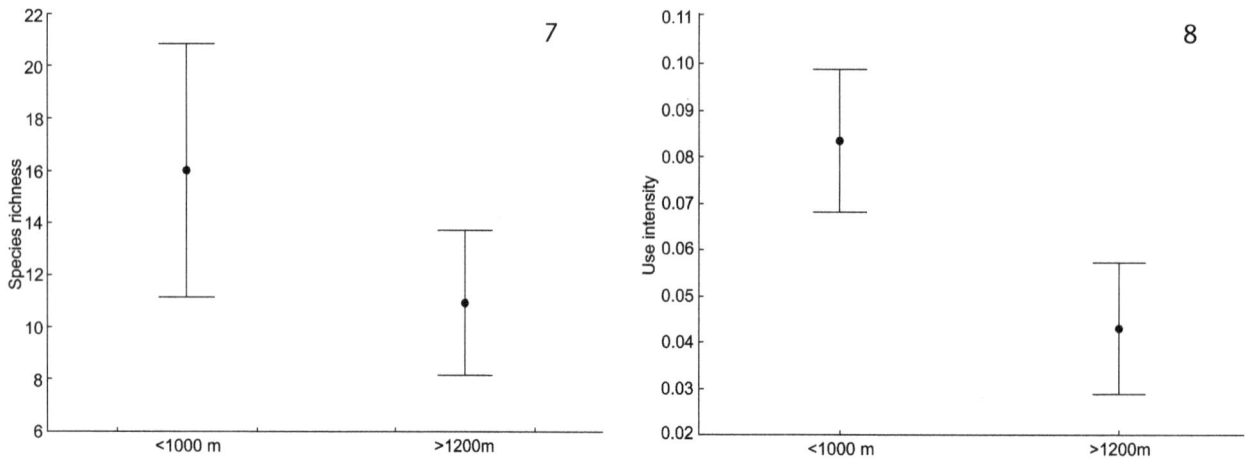

Figures 7–8. (7) Species richness and confidence interval (95%) in low (<1000 m) and high elevation (>1200 m) in Rio Preto State Park. (8) Use intensity by medium and large mammals and confidence interval (95%) in low (<1000 m) and high elevation (>1200 m) in Rio Preto State Park.

The high species richness and use intensity observed in forest habitats highlight the importance of riparian environments in the Cerrado, which may provide essential resources for mammals, especially during the dry season (Johnson et al. 1999, Redford and Fonseca 1986). Forest habitats cover less than 10% of the study areas, but a large proportion of total species richness, 58 and 67% in RPSP and SVNP, respectively, occurred in forests. Higher large mammal species richness in forest environments was also observed in another PA in the Espinhaço Range (Serra do Cipó National Park; Oliveira et al. 2009). As these habitats are associated with watercourses in the region, we can infer that at least part of the home range of most large mammal species must encompass forest habitats.

The species richness in savannah and forest habitats was very similar in the two PAs studied. Some studies show that the mammal community of riparian forests appears to be more similar to the savannah than to any other physiognomy of the Cerrado (Alho et al. 1986, Johnson et al. 1999). Although the composition varied between these two habitats in each PA, only one species – the southern anteater, *Tamandua tetradactyla* (Linnaeus, 1758) – had been registered in the savannah and not in the forest, whereas all the species registered in the forest were also found in the savannah. Grasslands also represented an important part of the richness found in the two PAs (11 in RPSP and 7 in SVNP), with a total of 15 species registered in this vegetation physiognomy. This species richness is higher than the richness of medium and large mammals found by Oliveira et al. (2009) in the open areas of Serra do Cipó National Park.

The species richness found in rupestrian habitats of RPSP was much lower than in the other vegetation physiognomies, whereas in SVNP species richness in this habitat was reasonably high – higher than in the grasslands and not varying significantly in relation to forest and savannah. This variation may be due

to the specific characteristics of the PAs. Due to the greater elevational gradient of RPSP the rupestrian environments in this park are, in general, steeper, which can make it difficult for the animals to move. In SVNP the elevational variation between rupestrian environments and other vegetation physiognomies are milder and less steep (at least in the survey area), facilitating displacement in these environments in the PA. The variations in species richness between physiognomies (especially in RPSP) could be attributed to differences in productivity, habitat complexity, and protection against predators among them. Areas that are more productive tend to have higher species richness (Waide et al. 1999, Jetz et al. 2009, Sandom et al. 2013), which could explain the lower species richness in the rupestrian habitat, a less productive environment due to its soil characteristics (Rodela 1998). The majority of the species recorded in this environment probably only use it occasionally, since it is difficult to maneuver in it and there are few resources and shelter for large animals.

The frequency of records between physiognomies indicated clear variations in the use of different environments. The high amount of records in forests suggests that these environments are more intensely used by wild mammals. Forest habitats play an important role in mammal diversity in the Cerrado, as they provide refuge for several species and, generally, have more food and water available (Redford and Fonseca 1986). August (1983) pointed out that forest habitats provide a variety of food types that are rarely available in other habitats. It is important to highlight that, although the two PAs are dominated by rupestrian habitats, the rate of independent records in this environment was almost always lower than those of the forest and savannah physiognomies. To our knowledge, no study has evaluated the use of rupestrian habitats by medium and large sized mammals. Although several studies have been carried out in PAs that have this predominant vegetation physiognomy, in many cases (e.g., Leal et al. 2008, Lessa et al. 2008, Oliveira et al. 2009), the use of this environment by large mammals has not been evaluated.

The higher species richness and intensity of use in lower elevations was expected. The decline in richness due to increasing elevation is widely accepted as a general standard (Rahbek 1995, 2005). Rahbek (1995) highlighted two main models of species distribution along elevational gradients: (1) the monotonic model, in which there is a linear decrease in the number of species with increasing altitude, observed in plants (Stevens 1992), bats (Patterson et al. 1996) and birds (Terborgh 1971, 1977, Rahbek 1997) and (2) the dome-shaped model, with higher species richness in the intermediate ranges of the elevation gradient, observed in non-flying mammals (Geise et al. 2004), insects (Mccoy 1990, Fleishman et al. 1998) and birds (Rajão and Cerqueira 2006).

Although the maximum elevation of RPSP (not exceeding 2,000 m) is insufficient for direct effects on the medium and large mammal fauna, it is certainly sufficient to change the vegetation structure and can thus indirectly influence their distribution. The elevational gradient has traditionally been used as a substitute for productivity, with the assumption that at higher elevation the environmental conditions contribute to lower productivity (MacArthur 1969, Terborgh 1971, Navarro 1992, Lee et al. 2004, Rahbek 2005). In fact, in RPSP, the more complex and productive habitats are in the lower part of the PA, where there are riparian forests along the Preto River (and other streams) and savannah areas with denser vegetation cover. These features may act as an indirect variable on the composition of medium and large mammals. The upper part of the park has a large area of rupestrian habitats and grasslands, with just a few forest patches that have a more abrupt transition to open vegetation, minimizing the productivity and niche availability and consequently species richness. The area of open grasslands in the upper part of RPSP is unfavorable to the determining factors of species richness, such as complexity (August 1983), habitat heterogeneity (Southwell et al. 1999, Williamset al. 2002) and primary productivity (Waide et al. 1999, Jetz et al. 2009, Sandomet al. 2013).

The absence of effects of wildfire frequency on species richness and use intensity is likely a result of the adaptation of Cerrado plant species to this characteristic of the biome. Felfili et al. (2000) pointed out that the Cerrado is highly fire resilient due to the characteristics of the vegetation, which prevent high mortality of trees and a rapid regrowth of the herbaceous layer in those situations. Although the immediate effect of a fire is a drastic change in the landscape, the mobility of medium and large mammals would allow most of them to escape and seek refuge in adjacent areas (Frizzo et al. 2011). In these cases, gallery forests are important refuge areas for these species (Silveira et al. 1999, Prada and Marinho-Filho 2004). Due to the rapid regrowth of the vegetation, soon after the fire, these highly resilient environments may be able to sustain populations of medium and large mammals again.

Two main aspects need to be considered in order to understand the influence of fire on the fauna: frequency and intensity. Infrequent fires result in a large accumulation of biomass. When the area eventually catches on fire, it is much more intense, with larger flames, higher temperatures and lasts longer (Miranda et al. 1993). This type of fire can have catastrophic effects on the local fauna (Ramos-Neto and Pivello 2000). At Emas National Park, Silveira et al. (1999) observed that a high-intensity fire caused negative effects on mammal populations. A second fire, in the following year with less biomass to burn, was less intense, and its effects were reduced. The recent history of annual fires at SVNP will likely result in less intense future fires and may minimize the impacts on the mammal fauna.

Even though the Espinhaço range is recognized as an important center for species endemism in Brazil (Silva and Bates 2002, Simon and Proença 2000), only 2.6% of its area is covered by strict PAs (Silva et al. 2008). Additionally, only a small portion of these areas is large enough (>500 km²) to safeguard relevant populations of large mammal species (Chiarello 2000a, b). In fact, many of the PAs in the Brazilian Cerrado are not effectively managed and are suffering from moderate or high anthropogenic pressure (WWF-Brasil and ICMbio 2012), which can result in local extinctions.

A global analysis revealed that PAs networks are generally established on higher elevations, steeper slopes and greater distance from urban centers, limiting their ability to avoid natural habitat conversion (Joppa and Pfaff 2009). This is true for the great majority of PAs in the Espinhaço Range, which are located at higher elevations and on steeper slopes. Nevertheless, in this specific case, the bias is biologically reasonable due to the large number of endemic species in rupestrian habitats, a habitat associated with higher elevations in this mountain range (e.g., Azevedo and Silveira 2005, Oliveira and Sano 2009, Freitas et al. 2012, Barata et al. 2013, Pardiñas et al. 2014). Our data, however, showed that this habitat does not seem to be relevant for the large mammal community in the Espinhaço. In fact, we observed higher levels of large mammal diversity in habitats usually found in lower elevations, forest and savannah. Therefore, the protection of native vegetation at lower elevations is crucial for long-term large mammal conservation in the Espinhaço Range. This could be achieved with strategic expansion of the PA network, as well as through the establishment of biodiversity corridors using riparian forests to connect existing PAs. The establishment of new PAs in the Espinhaço Range has been recommended before, as the present size of the PAs is considered insufficient to promote mammal conservation (Rocha et al. 2005).

In this study we found that the evaluated PAs are home to a significant portion of the known medium and large mammals of the Espinaco Range. The SVNP, however, is one of the PAs of the Espinhaço that still suffers from anthropogenic pressures. Because this national park is the largest PA in the Espinhaço Mosaic and the second largest in the Espinhaço Range, it is essential to eliminate these pressures in order to maintain the high diversity of mammals in the region. In recent years some groups have joined efforts for the re-categorization (downgrading) of SVNP from a strict PA to a multiple-use PA. From a conservation point of view, this would be a major setback for the biodiversity of the Espinhaço Range and we emphasize that priority must be given to effective regulation and implementation of the strict PAs for the conservation of numerous threatened species.

ACKNOWLEDGMENTS

Funding was provided by CNPQ (SISBIOTA Program, process 563134/2010-0 and PPBIO Program, process 457434/2012-0). FAPEMIG provides a scholarship to FFP and CNPQ to GBF (process 207195/2014-5). FFP and APP are also affiliated to Programa de Pós Graduação em Ecologia de Biomas Tropicais at Universidade Federal de Ouro Preto. GBF is also affiliated to the Centre for Biodiversity and Environment Research at University College London and the Institute of Zoology at the Zoological Society of London. Marcell Soares Pinheiro assisted with fieldwork. We are grateful to staff of RPSP and SVNP for helping with field logistics, and IEF-MG and ICMBio for the research permit.

REFERENCES

Ahumada JA, Hurtado J, Lizcano D (2013) Monitoring the Status and Trends of Tropical Forest Terrestrial Vertebrate Communities from Camera Trap Data: A Tool for Conservation. PLoS ONE 8: e73707. https://doi.org/10.1371/journal.pone.0073707

Alho CJR, Pereira LA, Paula AC (1986) Patterns of habitat utilization by small mammal populations in cerrado biome of central Brazil. Mammalia 50: 447–460. https://doi.org/10.1515/mamm.1986.50.4.447

August PV (1983) The role of habitat complexity and heterogeneity in structuring tropical mammal communities. Ecology 64: 1495–1507. https://doi.org/10.2307/1937504

Azevedo AA, Silveira FA (2005) Two new species of Centris (Trachina) Klug, 1807 (Hymenoptera: Apidae) from the state of Minas Gerais, Brazil, with a note on Centrispachysoma Cockerell, 1919. Lundiana 6(Suppl.): 41–48.

Barata IM, Santos MT, Letite FS, Garcia PC (2013) A new species of Crossodactylodes (Anura: Leptodactylidae) from Minas Gerais, Brazil: first record of genus within the Espinhaço Mountain Range. Zootaxa 3731: 552–560. https://doi.org/10.11646/zootaxa.3731.4.7

Chiarello AG (2000a) Conservation value of a native forest fragment in a region of extensive agriculture. Revista Brasileira de Biologia 60: 237–247. https://doi.org/10.1590/S0034-71082000000200007

Chiarello AG (2000b) Density and population size of mammals in remnants of brazilian atlantic forest. Conservation

Biology 14: 1649–1657. https://doi.org/10.1111/j.1523-1739.2000.99071.x

Cowell RK (2013) EstimateS: Statistical estimation of species richness and shared species from samples. Version 9. Available online at: http://purl.oclc.org/estimates [Accessed: 16/03/2016]

Creel S, Spong G, Creel NM (2001) Interspecific competition and the population biology of extinction-prone carnivores. In: Gittleman JL, Funk SM, Macdonald D, Wayne RK (Eds) Carnivore Conservation. Cambridge, Cambridge University Press, 35–60.

Cullen-Jr L, Bodmer RE, Pádua CV (2000) Efects of hunting in habitat fragments of the Atlantic forests. Brazil. Biological Conservation 95: 49–56. https://doi.org/10.1016/S0006-3207(00)00011-2

Drumond GM, Martins CS, Machado ABM, Sebaio FA, Antonini Y (2005) Biodiversidade em Minas Gerais: um atlas para sua conservação. Belo Horizonte, Fundação Biodiversitas, 222pp.

Fedriani JM, Fuller TK, Sauvajot RM, York EC (2000) Competition and intraguild predation among three sympatric carnivores. Oecologia 125: 258–270. https://doi.org/10.1007/s004420000448

Felfili JM, Rezende AV, Silva Junior MC, Silva MA (2000) Changes in the floristic composition of cerrado sensu stricto in Brazil over a nine-year period. Journal of Tropical Ecology 16: 579–590. https://doi.org/10.1017/S0266467400001589

Ferreguetti AC, Tomás WM, Bergallo HG (2015) Density, occupancy, and activity pattern of two sympatric deer (Mazama americana and M. gouazoubira) in the Atlantic Forest. Journal of Mammalogy 96: 1245–1254. https://doi.org/10.1093/jmammal/gyv132

Fleishman E, Austin GT, Weiss AD (1998) An empirical test of Rapoport's rule: elevational gradients in montane buttery communities. Ecology 79: 2482–2493. https://doi.org/10.2307/176837

Freitas GH, Chaves AV, Costa LM, Santos FR, Rodrigues M (2012) A new species of Cinclodes from the Espinhaço Range, southeastern Brazil: insights into the biogeographical history of the South American highlands. The International Journal of Avian Science 154: 738–755. https://doi.org/10.1111/j.1474-919X.2012.01268.x

Frizzo TLM, Bonizário C, Borges MP, Vasconcelos HL (2011) Revisão dos efeitos do fogo sobre a fauna de formações savânicas do Brasil. Oecologia Australis 15: 365–379. https://doi.org/10.4257/oeco.2011.1502.13

Geise L, Pereira LG, Bossi DE, Bergallo HG (2004) Patterns of elevational distribution and richness of nonvolant mammals in Itatiaia National Park and surroundings, in Southeastern Brazil. Brazilian Journal of Biology 64: 1–15. https://doi.org/10.1590/S1519-69842004000400007

Hardin G (1960) The competitive exclusion principle. Science 131: 1292–1297. https://doi.org/10.1126/science.131.3409.1292

Hayward MW, Slotow R (2009) Temporal partitioning of activity in large African carnivores: tests of multiple hypotheses. South African Journal of Wildlife Research 39: 109–125. https://doi.org/10.3957/056.039.0207

INPE (2015) Portal do Monitoramento de Queimadas e Incêndios. Brasília, Instituto Nacional de Pesquisas Espaciais, available online at: http://www.inpe.br/queimadas [Accessed: 16/03/2016]

Jetz W, Kreft H, Ceballos G, Mutke J (2009) Global associations between terrestrial producer and vertebrate consumer diversity. Proceedings of the Royal Society B 276: 269–278. https://doi.org/10.1098/rspb.2008.1005

Johnson MA, Saraiva PM, Coelho D (1999) The role of gallery forests in the distribution of cerrado mammals. Revista Brasileira de Biologia 59: 421–427. https://doi.org/10.1590/S0034-71081999000300006

Joppa LN, Pfaff A (2009) High and Far: Biases in the Location of Protected Areas. PLoS ONE 4: e8273. https://doi.org/10.1371/journal.pone.0008273

Karanth KU, Sunquist ME (1995) Prey selection by tiger, leopard and dhole in tropical forests. Journal of Animal Ecology 64: 439–450. https://doi.org/10.2307/5647

Kerr JT, Packer L (1997) Habitat heterogeneity as a determinant of mammal species richness in high-energy regions. Nature 385: 252–254. https://doi.org/10.1038/385252a0

Kinnaird MF, O'Brien TG (2012) Effects of Private-Land Use, Livestock Management, and Human Tolerance on Diversity, Distribution, and Abundance of Large African Mammals. Conservation Biology 26: 1026–1039. https://doi.org/10.1111/j.1523-1739.2012.01942.x

Klink CA, Machado RB (2005) Conservation of the Brazilian Cerrado. Conservation Biology 19: 707–713. https://doi.org/10.1111/j.1523-1739.2005.00702.x

Kreft H, Jetz W (2007) Global patterns and determinants of vascular plant diversity. Proceedings of the National Academy of Sciences of the United States of America 104: 5925–5930. https://doi.org/10.1073/pnas.0608361104

Leal KPG, Batista IR, Santiago FL, Costa CG, Câmara EM (2008) Mamíferos registrados em três unidades de conservação na Serra do Espinhaço: Parque Nacional da Serra do Cipó, Parque Nacional das Sempre Vivas e Parque Estadual da Serra do Rola Moça. Sinapse Ambiental (SI), 40–50.

Lee P, Ding T, Hsu F, Geng S (2004) Breeding bird species richness in Taiwan: distribution on gradients of elevation, primary productivity and urbanization. Journal of Biogeography 31: 307–314. https://doi.org/10.1046/j.0305-0270.2003.00988.x

Lessa LG, Costa BMA, Rossoni DM, Tavares VC, Dias LG, Júnior EAM, Silva JA (2008) Mamíferos da Cadeia do Espinhaço: riqueza, ameaças e estratégias para conservação. Megadiversidade 4: 241–254.

Linkie M, Dinata Y, Nugroho A, Haidir IA (2007) Estimating occupancy of a data deficient mammalian species living in tropical rainforests: Sun bears in the Kerinci Seblat region, Sumatra. Biological Conservation 137: 20–27. https://doi.org/10.1016/j.biocon.2007.01.016

Macandza VA, Owen-Smith N, Cain JW (2012) Habitat and resource partitioning between abundant and relatively rare grazing ungulates. Journal of Zoology 287: 175–185. https://doi.org/10.1111/j.1469-7998.2012.00900.x

MacArthur RH (1969) Patterns of communities in the tropics. Biological Journal of the Linnean Society 1: 19–30. https://doi.org/10.1111/j.1095-8312.1969.tb01809.x

MacArthur RH (1972) Geographical ecology: patterns in the distributions of species. New York, Harper and Row, 269 pp.

Marinho-Filho J, Rodriguez FHG, Juarez KM (2002) The Cerrado mammals: diversity, ecology and natural history. In: Oliveira PS, Marquis RJ (Eds) The Cerrados of Brazil. Columbia University Press, New York, 266–284. https://doi.org/10.7312/oliv12042-013

McCoy ED (1990) The distribution of insects along elevational gradients. Oikos 58: 313–322. https://doi.org/10.2307/3545222

MMA (2007) Projeto de conservação e uso sustentável da diversidade biológica brasileira (PROBIO) – Mapeamento de cobertura vegetal do bioma cerrado. Brasília, Ministério do Meio Ambiente, 33 pp.

Miranda AC, Miranda HS, Dias IDO, Dias BFD (1993) Soil and air temperatures during prescribed cerrado fires in Central Brazil. Journal of Tropical Ecology 9: 313–320. https://doi.org/10.1017/S0266467400007367

Mittermeier RA, Gil RP, Hoffman M, Pilgrim J, BrooksT, Mittermeier CG, Lamoreux J, Fonseca GAB (2005) Hotspots revisited: Earth's biologically richest and most endangered terrestrial ecoregions. Boston, University of Chicago Press, 392 pp.

Myers N, Mittermeier RA, Mittermeier CG, Fonseca GAB, Kent J (2000) Biodiversity hotspots for conservation priorities. Nature 403: 853–858. https://doi.org/10.1038/35002501

Navarro AGS (1992) Altitudinal distribution of birds in the Sierra Madre del Sur, Guerrero, Mexico. Condor 94: 29–39. https://doi.org/10.2307/1368793

O'Brien T (2010) Wildlife Picture Index: Implementation Manual Version 1.0. New York, Wildlife Conservation Society, Working Papers 39.

Oliveira RS, Sano PT (2009) Two new species Habranthus (Amaryllidaceae) from the Espinhaço Range, Brazil. Kew Bulletin 64: 537–541. https://doi.org/10.1007/s12225-009-9144-0

Oliveira VB, Camara EM, liveira LC (2009) Large and medium sized mammals from Parque Nacional da Serra do Cipó, Mina Gearais, Brazil. Mastozoología Neotropical 16: 355–364.

Oliveira-Santos LGR, Graipel ME, Tortato MA, Zucco CA, Cáceres NC, Goulart FVB (2012) Abundance changes and activity flexibility of the oncilla, Leopardus tigrinus (Carnivora: Felidae), appear to reflect avoidance of conflict. Zoologia 29: 115–120. https://doi.org/10.1590/S1984-46702012000200003

Paglia AP, Fonseca, GAB, Rylands AB, Herrmann G, Aguiar LMS, Chiarello AG, Leite YLR, Costa LP, Siciliano S, Kierulff MCM, Mendes SL, Tavares VC, Mittermeier RA, Patton JL (2012) Lista Anotada dos Mamíferos do Brasil. Conservation International (2nd edn), Arlington. Occasional Papers in Conservation Biology 6, 76 pp.

Pardiñas UF, Lessa G, Teta P, Salazar-Bravo J, Câmara EM (2014) A new genus of sigmodontine rodent from eastern Brazil and the origin of the tribe Phyllotini. Journal of Mammalogy 95: 201–215. https://doi.org/10.1644/13-MAMM-A-208

Patterson BD, Pacheco V, Solari S (1996) Distributions of bats along an elevational gradient in the Andes of southeastern Peru. Journal of Zoology 240: 637–658. https://doi.org/10.1111/j.1469-7998.1996.tb05313.x

Pfennig KS, Pfennig DW (2005) Character displacement as the 'best of a bad situation': fitness trade-offs resulting from selection to minimize resource and mate competition. Evolution 59: 2200–2208.

Prada M, Marinho-Filho J (2004) Effects of fire on the abundance of Xenarthrans in Mato Grosso, Brazil. Austral Ecology 29: 568–573. https://doi.org/10.1111/j.1442-9993.2004.01391.x

Rahbek C (1995) The elevational gradient of species richness: a uniform pattern? Ecography 18: 200–205. https://doi.org/10.1111/j.1600-0587.1995.tb00341.x

Rahbek C (1997) The relationship among area, elevation, and regional species richness in neotropical birds. The American Naturalist 149: 875–902. https://doi.org/10.1086/286028

Rahbek C (2005) The role of spatial scale and the perception of large-scale species-richness patterns. Ecology Letters 8: 224–239. https://doi.org/10.1111/j.1461-0248.2004.00701.x

Rajão H, Cerqueira R (2006) Distribuição altitudinal e simpatria das aves do gênero Drymophila Swainson (Passeriformes, Thamnophilidae) na Mata Atlântica. Revista Brasileira de Zoologia 23: 597–607. https://doi.org/10.1590/S0101-81752006000300002

Ramos-Neto MB, Pivello VR (2000) Lightning fires in a Brazilian Savanna National Park: Rethinking management strategies. Environmental Management 26: 675–684. https://doi.org/10.1007/s002670010124

Redford KH, Fonseca GAB (1986) The role of Gallery Forests in the Zoogeography of the Cerrado's non-volant Mammalian Fauna. Biotropica 18: 126–135. https://doi.org/10.2307/2388755

Ribeiro JF, Walter BMT (2008) As principais fitofisionomias do Bioma Cerrado. In: Sano SM, Almeida SP, Ribeiro JF (Eds) Cerrado: ecologia e flora. Embrapa Cerrados, Planaltina, 151–212.

Rocha WJ, Juncá FA, ChavesJM, Funch L (2005) Considerações finais e recomendações para conservação. In: Jucá FA, Funch L, Rocha W (Eds) Biodiversidade e conservação da Chapada Diamantina. Ministério do Meio Ambiente, Brasília, 411–435.

Rodela GL (1998) Cerrados de altitude e campos rupestres do Parque Estadual Do Ibitipoca, dudeste de Minas Gerais: distribuição florística por subfisionomias da vegetação. Revista do Departamento de Geografia 12: 163–189.

Sandom C, Dalby L, Flojgaard C, Kissling WD, Lenoir J, Sandel B (2013) Mammal predator and prey species richness are strongly linked at macroscales. Ecology 94: 1112–1122. https://doi.org/10.1890/12-1342.1

Schuette P, Wagner AP, Wagner ME, Creel S (2013) Occupancy patterns and niche partitioning within a diverse carnivore community exposed to anthropogenic pressures. Biological Conservation 158: 301–312. https://doi.org/10.1016/j.biocon.2012.08.008

Silva JA, Machado RB, Azevedo AA, Drumond GM, Fonseca RL, Goulart MF, Júnior EA, Martins CC, Neto MBR (2008) Identificação de áreas insubstituíveis para conservação da Cadeia do Espinhaço, estados de Minas Gerais e Bahia, Brasil. Megadivesidade 4: 248–269.

Silva JF, Fariñas MR, Felfili JM, Klink CA (2006) Spatial heterogeneity, land use and conservation in the Cerrado region of Brazil. Journal of Biogeography 33: 536–548. https://doi.org/10.1111/j.1365-2699.2005.01422.x

Silva JMC, Bates JM (2002) Biogeographic patterns and conservation in the South American Cerrado: a tropical savanna hotspot. Bioscience 52: 225–233. https://doi.org/10.1641/0006-3568(2002)052[0225:BPACIT]2.0.CO;2

Silveira L, Rodrigues FH, Jacomo AT, Diniz JAF (1999) Impact of wildfires on the megafauna of Emas National Park, central Brazil. Oryx 33: 108–114. https://doi.org/10.1046/j.1365-3008.1999.00039.x

Simon MF, Proença C (2000) Phytogeographic patterns of Mimosa (Mimosoideae, Leguminosae) in the Cerrado biome of Brazil: an indicator genus of high-altitude centers of endemism? Biological Conservation 96: 279–296. https://doi.org/10.1016/S0006-3207(00)00085-9

Sinclair ARE, Mduma SAR, Brashares JS (2003) Patterns of predation in a diverse predator-prey system. Nature 425: 288–290. https://doi.org/10.1038/nature01934

Sollmann R, Furtado MM, Hofer H, Jácomo AT, Tôrres NM, Silveira L (2012) Using occupancy models to investigate space partitioning between two sympatric large predators, the jaguar and puma in central Brazil. Mammalian Biology 77: 41–46. https://doi.org/10.1016/j.mambio.2011.06.011

Southwell CJ, Cairns SC, Pople AR, Delaney R (1999) Gradient analysis of macropod distribution in open forest and woodland of eastern Australia. Australian Journal of Ecology 24: 132–143. https://doi.org/10.1046/j.1442-9993.1999.241954.x

Stein A, Gerstner K, Kreft H (2014). Environmental heterogeneity as a universal driver of species richness across taxa, biomes and spatial scales. Ecology Letters 17: 866–880. https://doi.org/10.1111/ele.12277

Stevens GC (1992) The elevational gradient in altitudinal range: an extension of Rapoport's latitudinal rule to altitude. The American Naturalist 140: 893–911. https://doi.org/10.1086/285447

Sunarto S, Kelly MJ, Parakkasi K, Klenzendorf S, Septayuda E, Kurniawan H (2012) Tiger need cover: Multi-scale occupancy study of the big cat in Sumatran forest and plantation landscapes. Plos One 7: e30859. https://doi.org/10.1371/journal.pone.0030859

Team Network (2008) Terrestrial Vertebrate Protocol Implementation Manual, v. 3.0. Arlington, Tropical Ecology, Assessment and Monitoring Network, Center for Applied Biodiversity Science, Conservation International.

Terborgh J (1971) Distribution on environmental gradients: theory and a preliminary interpretation of distributional patterns in the avifauna of the Cordillera Vilcabamba, Peru. Ecology 52: 23–40. https://doi.org/10.2307/1934735

Terborgh J (1977) Birds species diversity on an Andean elevational gradient. Ecology 58: 1007–1019. https://doi.org/10.2307/1936921

Tews J, Brose U, Grimm V, Tielborger K, Wichmann MC, Schwager M, Jeltsch F (2004) Animal species diversity driven by habitat heterogeneity/diversity: the importance of keystone structures. Journal of Biogeography 31: 79–92. https://doi.org/10.1046/j.0305-0270.2003.00994.x

Waide RB, Willig MR, Steiner CF, Mittelbach G, Gough L, Dodson SI, Juday GP, Parmenter R (1999) The relationship between productivity and species richness. Annual Review of Ecology and Systematics 30: 257–300. https://doi.org/10.1146/annurev.ecolsys.30.1.257

Williams SE, Marsh H, Winter J (2002) Spatial scale, species diversity, and habitat structure: Small mammals in Australian tropical rain forest. Ecology 83: 1317–1329. https://doi.org/10.1890/0012-9658(2002)083[1317:SSSDAH]2.0.CO;2

WWF-Brasil, ICMBio (2012) Efetividade de gestão das unidades de conservação federais do Brasil: Resultados de 2010. Brasília, World Wide Fund for Nature Brasil, Ministério do Meio Ambiente, 67 pp.

WWF-Brasil (2016) Implementação da Avaliação Rápida e Priorização da Gestão de Unidades de Conservação (RAPPAM) em Unidades de Conservação estaduais de Minas Gerais. Brasília, World Wide Fund for Nature Brasil, 102 pp.

Author Contributions: FFP, GBF and APP contributed equally for the manuscript.

Competing Interests: The authors have declared that no competing interests exist.

Influence of artificial lights on the orientation of hatchlings of *Eretmochelys imbricata* in Pernambuco, Brazil

Thyara Noely Simões[1], Arley Candido da Silva[2], Carina Carneiro de Melo Moura[3]

[1]*Universidade Estadual de Santa Gruz, Programa de Pós-graduação em Ecologia e Conservação da Biodiversidade. Rodovia Jorge Amado, km 16, Salobrinho, 45662-900 Ilhéus, BA, Brazil.*
[2]*Ecoassociados NGO – Conservação de tartarugas marinhas, baobás e recifes de corais. Rua das Caraúnas, Porto de Galinhas, 55590-000 Ipojuca, PE, Brazil.*
[3]*Department of Biology, Institute of Pharmacy and Molecular Biotechnology, Heidelberg University. Heidelberg, Im Neuenheimer Feld 364, 69120, Heidelberg, Germany.*
Corresponding author: Thyara Noely Simões (thyara.noely@gmail.com)

http://zoobank.org/B1E05B5F-628B-43F0-8645-33C3BE373B4C

ABSTRACT. Sea turtle hatchlings, in natural abiotic conditions, emerge from their nests at night and go directly to the sea, following the moonlight's reflection in the ocean. Increased human activities such as tourism and artificial lights on the coasts, however, have interfered with the ability of sea turtle neonates to find their correct destination, negatively affecting their survival rates. Here we endeavored to assess the influence of artificial lights on the hatchlings of the sea turtle *Eretmochelys imbricata* (Linnaeus, 1766) in the south coast of the state of Pernambuco, Brazil. To that end, 10 experiments were conducted with 15 hatchlings/test subjects. Five experiments took place in artificially illuminated areas and five in non-illuminated areas. Circles with a 2 m radius were drawn on the sand a small 2–3 cm depression was made at the center of each circles. The neonates were then placed in the depressions to simulate their coming from a nest. After the neonates crossed the edge of the circles, their tracks were photographed and drawn on a diagram. To ascertain if the trajectories of the neonates differed between the two groups (hatchlings from illuminated versus non-illuminated nests), the Rayleigh test was used. The significance of those differences was tested using ANOVA. To evaluate similarities and significance of clusters, a Multi-Dimensional Scaling was used. The tracks of 86.67% (N = 65) of the hatchlings from nests at illuminated areas departed from their correct trajectory. The distribution of trajectories was considered random (V = 19.4895, p > 0.05) only for tracks originating from artificially illuminated areas. The movement patterns of hatchlings from illuminated and non-illuminated areas differed significantly (F < 0.0001, p < 0.01). Consistent with this, two distinct groups were identified, one from illuminated and one from non-illuminated areas. Therefore, we conclude that artificial illumination impacts the orientation of hawksbill hatchlings. This suggests that in order to protect this species it is necessary to safeguard its nesting areas from artificial lights.

KEY WORDS. Anthropogenic impacts, cheloniidae, conservation, hawksbill turtle, light pollution.

INTRODUCTION

Sea turtle hatchlings, after emerging from their nest at night, are immediately oriented to the ocean in environments without light pollution. The main signs for their orientation are visual, and are primarily associated with light intensity and relief elevation (Salmon et al. 1992, Witherington and Martin 2003). The hatchlings move towards the brightest regions, usually following the reflection of the moonlight over the ocean, a mechanism known as phototaxy (Mrosovsky and Shettleworth 1968, Mrosovsky 1970, 1972, Van Rhijn and Van Gorkom 1983, Lohmann and Lohmann 1996).

An increase in tourist activities and the development of small towns along the Brazilian coast have contributed to an increase in the amount of artificial light pollution on beaches (Witherington and Bjorndal 1991b). This has altered the ability of hatchlings to follow the natural light from the moon, which is what orients them toward the sea (Witherington 2000, Silman et al. 2002, Salmon 2003, Tuxbury and Salmon 2005, Deem et al. 2007).

According to Verheijen (1960), artificial lights are stronger stimuli than natural light, although they are less intense than celestial sources, they have more glare. In order to mitigate the impact of artificial lights on beaches, the Ordinance #11 of Instituto Brasileiro do Meio Ambiente e dos Recursos Naturais Renováveis (IBAMA), implemented in Brazil on January 30, 1995, prohibits the use of artificial lights on beaches where sea turtles nest. Despite this ordinance, the use of artificial lighting is still common and is responsible for the death of thousands of hatchlings of hawksbill turtles, *Eretmochelys imbricata* (Linnaeus, 1766), green turtles, *Chelonia mydas* (Linnaeus, 1758), and loggerhead turtles, *Caretta caretta*, every year (Witherington 1997).

The presence of artificial lights causes several problems to turtle hatchlings, for instance dehydration, and increased predation and mortality (Limpus 1971, Philibosian 1976, Mann 1978, Mortimer 1979, Peters and Verhoeven 1994, Witherington and Martin 2003). Even when the hatchlings are able to reach the ocean despite the stimuli from artificial lights, they might be already significantly weakened by then, since the energy be required for their first few hours of swimming is reduced (Kraemer and Bennett 1981).

These impacts, associated with other anthropogenic interferences such as fishing activities (Gallo et al. 2006, Marcovaldi et al. 2009), pollution and degradation of nesting environment, have contributed to a reduction in the numbers of sea turtles (Hamann et al. 2010). As a consequence, all sea turtle species are, to some degree, under threat of extinction. The hawksbill, which will be addressed in this study, is classified by the International Union for Conservation of Nature (IUCN 2016) as "critically endangered".

In order to contribute useful information for the proper management and conservation of the hawksbill turtle at the coast of the state of Pernambuco, this study aimed to verify the influence of artificial light sources and moon phases on the ability of hatchlings to orient themselves.

MATERIAL AND METHODS

The study area is located in Ipojuca, 57 km from Recife, with geographic coordinates 08°24'06"S, 35°03'45"W. It comprises 32 km of coastal area. Sea turtles nestings are recorded by the Ecoassociados Non-governmental organization (institution working in the monitoring and conservation of sea turtles) in Muro Alto, Cupe, Merepe, Porto de Galinhas, Maracaípe and Pontal de Maracaípe, totaling 12 km.

The Merepe beach, where the experiments were conducted (coordinates 08°27'15"S, 34°59'52"W), has a coastal length of 3.47 km. There is a great concentration of nesting in this area. The conditions are favorable for nesting because the area is free from reef barriers and the strip of sand is wide. In addition, the coastal vegetation and the terrain are flattered where compared to the other beaches. Along the Merepe beach there is a great number of hotels and resorts, generating a great amount of arti-

ficial lighting, but there are still some areas without interference of artificial illumination.

The data was collected under the authorization number 22741-1 issued by the ICMBio (Instituto Chico Mendes de Conservação da Biodiversidade). The experiments were conducted from March to May, 2012. To record the paths of hatchlings, ten experiments were performed with 15 hatchlings/tests at a time. Five experiments were conducted in areas without artificial illumination and five experiments were carried out in areas with the interference of artificial lighting, such as LED reflectors (methodology modified from Salmon and Witherington (1995).

For each experiment, hatchlings were randomly selected during the nesting period from 10 nests. The hatchlings were kept in thermal containers until we conducted the experiments and were shortly released near the sea. Circles with a 2 m radius were drawn on the sand. At the center, a small depression of 2–3 cm was made to simulate the natural conditions of emergence of the hatchlings. First, 15 hatchlings were placed in the center of the circles in order for us to observe how artificial light, or the absence thereof, influences their capacity to orient themselves towards the ocean.

After neonates crossed the edge of the circles, the tracks were photographed and drawn on a diagram to record the circular motion or the change of direction. For the diagram, a compass and a 360° protractor were used to estimate the orientation angles of each hatchling from the center of the circle to its edge in relation to the sea.

The direction of the hatchlings in relation to the ocean was considered deviated when it departed from it by more than 30° (Salmon and Witherington 1995).

Later, the data were subjected to circular statistical analyses using the Rayleigh V test (Zar 1999) and the software Bioestat version 5.0 to verify whether the orientation of the trajectories of neonates significantly differed from random. A uniform distribution of neonates around the circumference was considered the null hypothesis. The percentage of deviation of hatchlings was determined in illuminated and non-illuminated areas in relation to the lunar phase. The significant differences between the movement of the hatchlings from illuminated and non-illuminated nests were processed using the BioEstat 5.0 software, the ANOVA variance test and Tukey test conducted a posteriori. The Primer 6.0 software (Clarke and Gorley 2001) was used to evaluate the significance and similarities of groups formed using Multi-Dimensional Scaling (MDS).

RESULTS

The tracks of hatchlings from simulated nests placed in illuminated areas showed trajectory changes in 86.67% of the total (N = 65), with an angular range from 4 to 350° (Fig. 1). Neonates originating from simulated nests placed in non-illuminated areas showed deviations in only 33.33% (N = 25) of all tracks, with an angular change from 0 to 95° (Fig. 2).

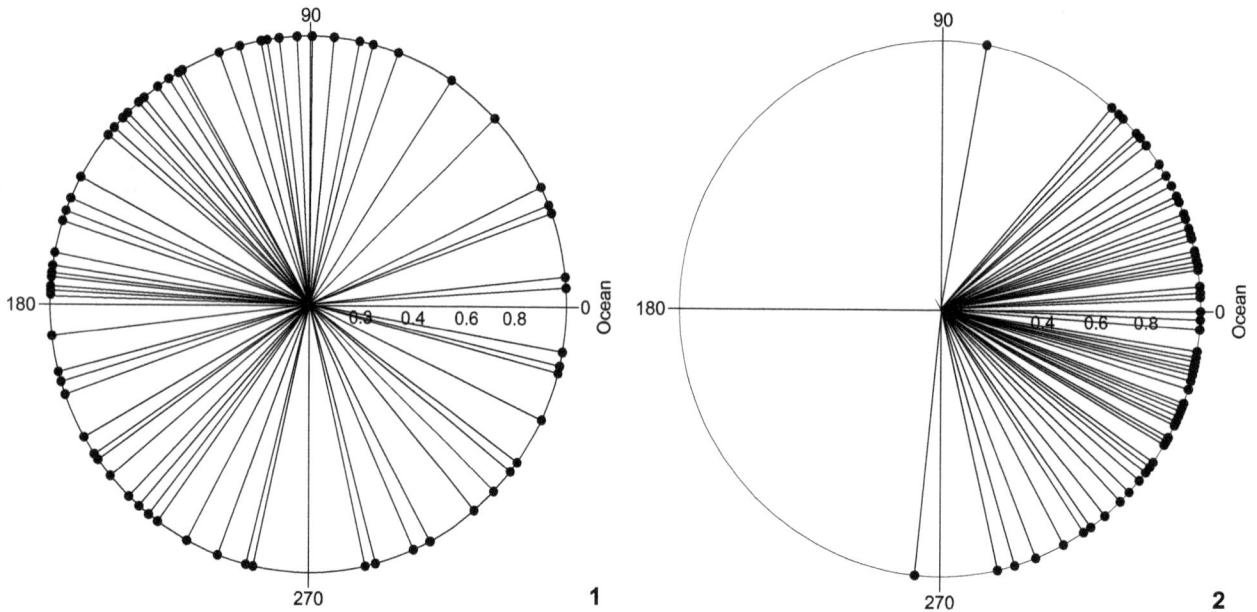

Figures 1–2. Trajectory deviations of hatchlings of the species *Eretmochelys imbricata* in (1) illuminated nests and (2) non-illuminated made in Merepe Beach, coast of Ipojuca, Pernambuco, Brazil.

The distribution of trajectories was considered random ($V = 19.4895$, p>0.05) for simulated nests located in artificially illuminated areas (Fig. 1). However, the experiments conducted in non-illuminated areas had an irregular distribution around the circumference, although the hatchlings' tracks were directed towards the sea ($Z = 63.4377$, $p < 0.01$) (Fig. 2). Movement patterns of hatchlings in illuminated areas were significantly different from non-illuminated areas ($F < 0.0001$, $p < 0.01$).

Through the Multi-Dimensional Scaling, it was possible to observe a separation of the two groups of hatchlings regarding the presence or absence of artificial light (Stress = 0.02) with a 75% similarity (Fig. 3).

We observed that during the crescent and first quarter moon for both illuminated and non-illuminated experiments the percentage of deviation was lower in comparison with the experiments conducted during the waning moon, which showed 77% of deviation (Fig. 4).

DISCUSSION

When the levels of artificial light are high, hatchlings may either ignore the natural light or be unable to perceive it (Lohmann and Lohmann 1996). This problem occurs in Florida's beaches, USA, where hatchlings originating from nests in illuminated areas had their trajectories changed in approximately 83% of the cases (McFarlane 1963, Salmon and Witherington 1995). This is similar to the results of the present study, in which

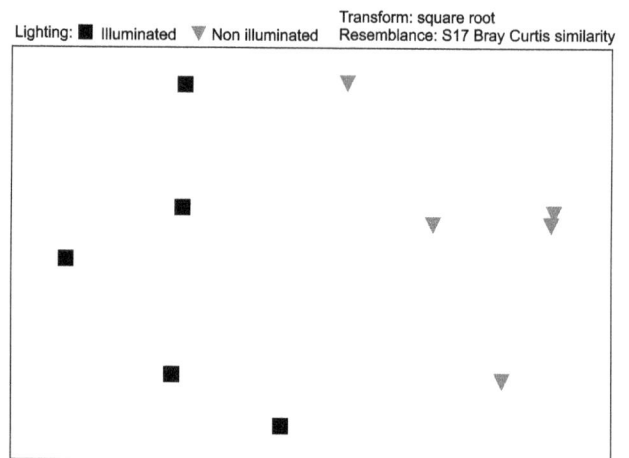

Figure 3. Multidimensional scaling (MDS) showing the existence of differences between the degree of orientation of hatchlings in relation to the ocean in illuminated and non-illuminated areas; formation of distinct groups (stress = 0.02) in samples collected in Merepe Beach, coast of Ipojuca, Pernambuco Brazil.

changes in the trajectories of hatchlings were observed in 86% of the cases (Fig. 1).

The largest angular variation reinforces the strong attraction that artificial lights exert on neonates. Similar results were recorded in experiments with *C. caretta* (Salmon and Witherington 1995), *C. mydas* (Mrosovsky and Carr 1967),

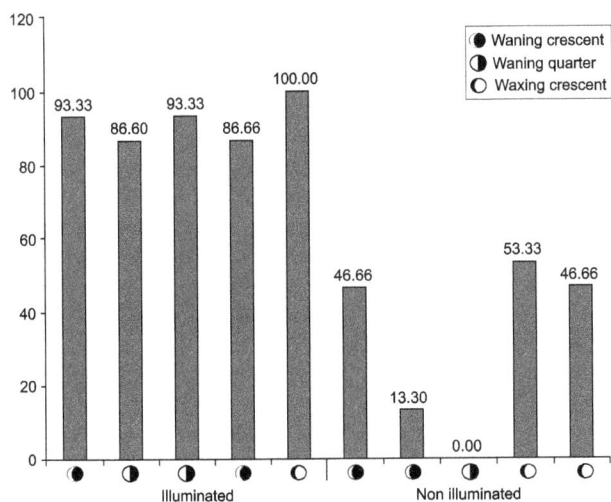

Figure 4. Percentage of emergence events diverted in function of the lunar phase and in illuminated and non-illuminated areas. Data based on ten experimental events in different lunar conducted in Merepe Beach, coast of Ipojuca, Pernambuco, Brazil.

Lepidochelys olivacea (Eschscholtz, 1829) (Mrosovsky and Carr 1967) and *Dermochelys coriacea* (Vandelli, 1761) (Bourgeois et al. 2009). In all of those studies, severe hatchling disorientation and increased disruption in their ability to find the sea were recorded. The disruption caused by artificial lights on neonate behavior includes crawling in the opposite direction of the ocean (Witherington and Martin 2003), or walking in circular paths, or abrupt changes in direction (Witherington and Bjorndal1991a, Salmon and Witherington 1995, Witherington and Martin 2003).

In addition, according to Wyneken (2000), disoriented hatchlings spend more energy to find the ocean and by doing that they become more vulnerable to predators. Nevertheless, the characteristics of the light influences the level of hatchling disorientation (Witherington and Bjorndal 1991b). In the absence of strong light stimulus, the hatchling will walk toward a dispersed light stimulus. The turtle *C. mydas*, for example, is attracted by light that has smaller wavelengths, while *C. caretta* not follows light with color yellow, like the ones with low sodium pressure (Mrosovsky and Shettleworth 1968, Witherington and Bjorndal 1991b).

Therefore, besides the fact that hatchlings are attracted to artificial illumination, we also know that the light's wavelength plays a role in the disorientation of sea turtles. Since is not known which wavelengths affect *E. imbricata*, more strongly, studies addressing this specific topic are needed to complement our findings.

According to Salmon and Witherington (1995), the lunar cycles are correlated with the trajectory of marine turtle hatchlings heading towards the sea. The highest number of path disruption occurs at night during periods close to the new moon (darker nights), while during the full moon, which is characterized by brighter nights, the hatchlings are oriented directly to the sea. However, when the levels of artificial light are high, hatchlings ignore these natural stimuli (Lohmann and Lohmann 1996).

Light pollution occurs mainly in highly developed locations with a higher population density. The coast of Ipojuca is one of the most developed beaches of Pernambuco and has a great number of hotels and resorts along its waterfront, which direct their lights toward the sea at night. Our findings reinforce the idea that the lights commonly used in this area is inadequate and harmful to *E. imbricata* hatchlings', contributing to their death by precipitating exhaustion, dehydration and increasing the risk of predation.

As a mitigation measure, the Ordinance #11 of 30 January 1995 was implemented in Brazil. It reads as follows:

"Article 1-Prohibit light sources that cause a light intensity higher than zero LUX in a strip of beach between the highest low-water mark up to 50 m (fifty meters) above the highest tide of the year (syzygy tide)."

Even though the ordinance covers the Fernando de Noronha District, and beach Boldro, Conceição, Caieira, Americano, Bode, Cacimba do Padre and Baía de Santo Antonio beaches in the state of Pernambuco, it does not cover Ipojuca's beaches.

In conclusion, the artificial lights in the studied region are negatively impacting the orientation of *E. imbricata* hatchlings' towards the sea and are affecting their survival in this region. We highlight the necessity for including Ipojuca's beaches within the protected area under Ordinance No. 11 and reinforce the need to control the level of light pollution in this turtle's nesting area.

ACKNOWLEDGEMENTS

We thank the Ecoassociados Non Governmental Organization for the logistical support in the development of the study and the Instituto Chico Mendes de Conservação da Biodiversidade for allowing the realization of the experiments under the license 22741-1.

REFERENCES

Bourgeois S, Gilot-Fromont E, Viallefont A, Boussamba F, Deem SL (2009) Influence of artificial lights, logs and erosion on leatherback sea turtle hatchling orientation at Pongara National Park, Gabon. Biological Conservation 142: 85–93. https://doi.org/10.1016/j.biocon.2008.09.028

Clarke KR, Gorley RN (2001) PRIMER Version 6 User Manual/Tutorial. Primer-Ltd, Plymouth.

Deem SL, Boussamba F, Nguema AZ, Sounguet G, Bourgeois S, Cianciolo J, Formia A (2007) Artificial lights as a significant cause of morbidity of leatherback sea turtles in Pongara National Park, Gabon. Marine Turtle News 116: 15–17.

Gallo BMG, Macedo S, Giffoni BB, Becker JH, Barata PCR (2006) Sea turtle conservation in Ubatuba, Southeastern Brazil, a feeding area with incidental capture in coastal fisheries. Chelonian Conservation and Biology 5: 93–101. https://doi.org/10.2744/1071-8443(2006)5[93:STCIUS]2.0.CO;2

Hamann M, Godfrey MH, Seminoff JA, Arthur K, Barata PCR, Bjorndal KA, Bolten AB, Broderick AC, Campbell LM, Carreras C, Casale P, Chaloupka M, Chan SKF, Coyne MS, Crowder LB, Diez CE, Dutton PH, Epperly SP, Fitz Simmons NN, Formia A, Girondot M, Hays GC, Cheng IJ, Kaska Y, Lewison R, Mortimer JA, Nichols WJ, Reina RD, Shanker K, Spotila JR, Tomás J, Wallace BP, Work TM, Zbinden J, Godley BJ (2010) Global research priorities for sea turtles: informing management and conservation in the 21st century. Endangered Species Research 11: 245–269. https://doi.org/10.3354/esr00279

IUCN (2016) Red List status assessment hawksbill turtle. Conservation International, and NatureServe. Available online at: http://www.iucnredlist.org/search [Accessed 13/01/20016]

Kraemer JE, Bennett SH (1981) Utilization of post-hatching yolk in the loggerhead sea turtles, Caretta caretta. Copeia 1981: 406–411. https://doi.org/10.2307/1444230

Limpus CJ (1971) Sea turtle ocean finding behaviour. Search 2: 385–387.

Lohmann KJ, Lohmann CMF (1996) Orientation and open-sea navigation in sea turtles. The Journal of Experimental Biology 199: 73–81.

Mann TM (1978) Impact of developed coastline on nesting and hatchling sea turtles in Southeastern Florida. Florida Marine Research Publications 33: 53–55.

Marcovaldi MA, Giffoni BB, Becker H, Fiedler FN (2009) Sea Turtle Interactions in Coastal Net Fisheries in Brazil. In: Proceedings of the Technical Workshop on Mitigating Sea Turtle By catch in Coastal Net Fisheries. IUCN, Regional Fishery Management Council, Hawaii.

McFarlane RW (1963) Disorientation of loggerhead hatchlings by artificial road lighting. Copeia 1963: 153. https://doi.org/10.2307/1441283

Mortimer JA (1979) Ascension Island: British jeopardize 45 years of conservation. Marine Turtle Newsletter 10: 7–8.

Mrosovsky N (1970) The influence of the sun's position and elevated cues on the orientation of hatchling sea turtles. Animal Behaviour 18: 648–651. https://doi.org/10.1016/0003-3472(70)90008-4

Mrosovsky N (1972) The Water-finding ability of Sea Turtles. Brain, Behavior and Evolution 5: 202–225. https://doi.org/10.1159/000123748

Mrosovsky N, Carr A (1967) Preference for light of short wavelengths in hatchling green sea turtles, Chelonia mydas, tested on their natural nesting beaches. Behaviour 28: 217–231. https://doi.org/10.1163/156853967X00019

Mrosovsky N, Shettleworth SJ (1968) Wavelength preferences and brightness cues in the water finding behaviour of sea turtles. Behaviour 32: 211–57. https://doi.org/10.1163/156853968X00216

Peters A, Verhoeven KJF (1994) Impact of artificial lighting on the seaward orientation of hatchling loggerhead turtles. Journal of Herpetology 28: 112–114. https://doi.org/10.2307/1564691

Philibosian R (1976) Disorientation of hawksbill turtle hatchlings, Eretmochelys imbricata, by stadium lights. Copeia 1976: 824. https://doi.org/10.2307/1443476

Salmon M (2003) Artificial night lighting and sea turtles. The Biologist 50: 163–168.

Salmon M, Witherington BE (1995) Artificial lighting and seafinding by loggerhead hatchlings: evidence for lunar modulation. Copeia 1995: 931–938. https://doi.org/10.2307/1447042

Salmon M, Wyneken J, Fritz E, Lucas M (1992) Seafinding by hatchling sea turtles: role of brightness silhouette and beach slope as orientation cues. Behaviour 122: 56–57. https://doi.org/10.1163/156853992X00309

Silman R, Vargas I, Troëng S (2002) Tortugas marinas. Guía Educativa. Corporación Caribeña para la Conservación, San Pedro, 38 pp. Available online at: http://www.lasecomujeres.org/files/SeaTurtleEducatorsGuide_esp.pdf [Accessed: 01/05/0217]

Tuxbury SM, Salmon M (2005) Competitive interactions between artificial lighting and natural cues during seafinding by hatchling marine turtles. Biological Conservation 121: 311–316. https://doi.org/10.1016/j.biocon.2004.04.022

Van Rhijn FA, Van Gorkom JC (1983) Optic orientation in hatchlings of the sea turtle Chelonia mydas III. Sea-finding behaviour: the role of photic and visual orientation in animals walking on the spot under laboratory conditions. Marine Behaviour and Physiology 9: 211–228. https://doi.org/10.1080/1023-6248309378594

Verheijen FJ (1960) The mechanisms of the trapping effect of artificial light sources upon animals. Archives Néerlandaises de Zoologie 13: 1–107. https://doi.org/10.1163/036551660X00017

Witherington BE (1997) The problem of photopollution for sea turtles and other nocturnal animals. In: Clemmons JR, Buchholz R (Eds) Behavioral Approaches to Conservation in the Wild. Cambridge University Press, Cambridge, 303–328.

Witherington BE (2000) Reducción de las amenazas al hábitat de anidación. In: Eckert KL, Bjorndal KA, Abreu-Grobois FA, Donnelly M (Eds). Técnicas de investigación y manejo para la conservación de las tortugas marinas. UICN/CSE Grupo especialista en Tortugas Marinas, Blanchard, 201–210.

Witherington BE, Bjorndal KA (1991a) Influences of Wavelength and Intensity on Hatchling Sea Turtle Phototaxis: implications for sea-finding behavior. Copeia 1991: 1060–1069.

Witherington BE, Bjorndal KA (1991b) Influences of artificial lighting on the seaward orientation of hatchling loggerhead turtles *Caretta caretta*. Biological Conservation 55: 139–149. https://doi.org/10.1016/0006-3207(91)90053-C

Witherington BE, Martin RE (2003) Understanding, assessing and resolving light-pollution problems on sea turtle nesting beaches. Florida Marine Research Institute Technical Report TR 2: 73.

Wyneken J (2000) The migratory behavior of hatchling sea turtles beyond the beach. In: Pilcher N, Ismail G (Eds) Sea Turtles of the Indo-Pacific. ASEAN Academic Press, London, 121–129.

Zar JH (1999) Biostatistical analysis. Prentice-Hall, New Jersey, 4th ed.

Author Contributions: TNS and CCMM designed the experiments; TNS and ACS conducted the experiments; TNS and CCMM analyzed the data; TNS, CCMM and ACS wrote the paper; TNS, CCMM and ACS conducted the paper review.

Competing Interests: The authors have declared that no competing interests exist.

3

Reproductive aspects of the Purple-throated Euphonia, *Euphonia chlorotica* (Aves: Fringillidae) in Southeastern Brazil, and first record of the species nesting inside a vespiary

Daniel F. Perrella[1], Paulo V. Davanço[2], Leonardo S. Oliveira[2], Livia M.S. Sousa[2], Mercival R. Francisco[2]

[1]*Programa de Pós-graduação em Ecologia e Recursos Naturais, Universidade Federal de São Carlos. Rodovia Washington Luís km 235, 13565-905 São Carlos, SP, Brazil.*
[2]*Departamento de Ciências Ambientais, Universidade Federal de São Carlos. Rodovia João Leme dos Santos km 110, 18052-780 Sorocaba, SP, Brazil.*
Corresponding author: Mercival R. Francisco (mercival@ufscar.br)

http://zoobank.org/35A480D1-8E00-478A-8225-B96E779289E6

ABSTRACT. Despite the fact that *E. chlorotica* (Linnaeus, 1766) is common and widely distributed in South America, the reproductive aspects of the species are poorly documented. Here we present data on 18 active nests found from August to February, between 2007 and 2012. Nests were globular with a lateral entrance, and measured 97.9 ± 14.4 mm in outside height, 110.6 ± 11.6 mm in outside diameter, and were 4.88 ± 2.09 m above ground. They were often supported from bellow and were composed mainly of tiny dry leafs and leaflets, fine petioles, and plumed seeds, all compacted with spider web silk. Eggs were laid on consecutive days or with one day interval, and clutch size varied from 1–3 eggs (2.1 ± 0.6, n = 9 nests). Only females incubated the eggs, but both sexes were involved in nest construction and nestling attendance at similar rates. Incubation and nestling periods were 14 and 21 days, respectively, and overall nest survival probability was 5%. A vespiary used for nesting was not occupied by wasps and nest material was deposited only to form the incubatory chamber. Although nesting near wasps or bees is a widespread strategy among birds in general, nesting inside the nests of social insects is a poorly documented behavior.

KEY WORDS. Birds, breeding biology, nesting behavior, Euphoniinae, wasps.

INTRODUCTION

Euphonias are small and conspicuous arboreal passerines, comprising 27 species that are restricted to the Neotropics (Hilty 2011). Sexual dimorphism is remarkable in most species, with males predominantly steely blue in the upperparts, and yellow in the underparts. Females, by contrast, are generally olive above and yellowish or grayish below (Sargent 1993, Ridgely and Tudor 1994, Sick 1997). These birds have been traditionally considered as Tanagers, but according to recent phylogenetic analyses they are finches of the family Fringillidae, where they compose the subfamily Euphoniinae together with the Chlorophonias (Zuccon et al. 2012). Distinctive features of the Euphonias and Chlorophonias are the highly specialized frugivorous diet, which is associated with the absence of gizzard in the Euphonias, and

the construction of globular nests with side entrance (Isler and Isler 1999, Hilty 2011, Zuccon et al. 2012).

Many behavioral and ecological aspects of the Euphonias, including reproductive biology, remain poorly documented. The nests of twenty of the currently recognized species are known, and egg characteristics have been described for 15 species. Information on nest construction and parental care is available, in varying levels of detail, for 14 species, whereas incubation and nestling periods are known for only five and four species, respectively (Nehrkorn 1910, Bertoni 1918, Bond 1943, Skutch 1945, 1985, Pinto 1953, Morton 1973, Ffrench 1980, Oniki and Willis 1983, 2003, Belton 1985, Isler and Isler 1999, and therein references, Pizo 2000, Greeney and Nunnery 2006, Solano-Ugalde et al. 2007, Janni et al. 2008, Kirwan 2009, Hilty 2011, and therein references, Marini et al. 2012).

Euphonia chlorotica (Linnaeus, 1766) is widely distributed in South America, occurring in most of Brazil, Guianas, and Paraguay, and in parts of Colombia, Venezuela, Peru, Bolivia, and Argentina. It inhabits forest borders, clearings, Cerrado, Caatinga (Ridgely and Tudor 1994, Sick 1997, Hilty 2011), and many types of secondary and anthropogenic habitats, such as orchards and urban vegetated areas. However, nesting information on this species is scattered and incomplete. Nests and/or egg descriptions, and data on clutch sizes, are given by Bertoni (1904), Snethlage (1928), De la Peña (1996), Lima (2006), Kirwan (2009), and Marini et al. (2012). Knowledge on parental activities is limited to the information that both males and females participate in nest construction (Oniki and Willis 1983, De la Peña 1996) and in nestling provisioning (Oniki and Willis 1983, Kirwan 2009), and incubation period is known from only one egg (Lima 2006).

Our specific goals in this paper were: 1) to provide supplemental information on nest, eggs, and nestling characteristics, clutch size, and incubation period; 2) to present for the first time information on nest measurements, nesting phenology, duration of the breeding season, nesting success, partitioning of parental activities, and the first nestling period, and 3) to provide the first report of a nest constructed inside a vespiary.

MATERIAL AND METHODS

Observations were conducted at the campus of Faculdade de Engenharia de Sorocaba (10.5 ha), Sorocaba, state of São Paulo, southeastern Brazil (23°28'S, 47°25'W), and at an adjacent smaller area (2 ha) of Cerrado sensu stricto, which was maintained within the urbanized area to protect a small and well-preserved stream. The campus presents extensive laws and gardens, with exotic trees, such as *Pinus* sp., *Eucalyptus* sp., *Mangifera* sp., and *Grevillea robusta* A. Cunn. ex R. Br., and also native trees typical of the Cerrado, with buildings and streets occupying about 30% of the area. The elevation is ~ 580 m asl., and the climate is classified as Cfa according to Koppen-Geiger (Kottek et al. 2006, Peel et al. 2007), with a humid, hot season from October-March (average rainfall 919 mm, and mean daily temperature ranging from 15.7 °C to 32.4 °C), and a dry, cold season from April-September (average rainfall 294 mm, and mean daily temperature ranging from 11.4°C to 30.6 °C).

Nests were routinely searched by walking along the whole area two to three times per week, from August to March, during three breeding seasons: 2007/2008, 2008/2009, and 2009/2010. Random searches were also performed in 2010/2011, and 2011/2012. Nests were found by following adults carrying material for nest construction, or delivering food to the nestlings (Martin and Geupel 1993), and monitored every 1–3 days since located. Nest type and egg shape were named following Winkler (2004), and they were measured with a caliper to the nearest 0.1 mm. Eggs were weighed to the nearest 0.1g with a spring scale. We analyzed nest material and took nest measurements only

after nests were no longer being used.

The incubation period was considered from the first day of incubation to the day before hatching, and nestling period from the day of hatching to the day before fledging (Winkler 2004, Oliveira et al. 2010). As neither eggs nor nestlings were marked, we assumed that the first eggs to be laid were the first to hatch, and that the first young to hatch were the first to fledge (Davanço et al. 2013). Incubation and nestling periods were estimated based only on nests for which laying and hatchling dates, respectively, were known. We performed 1hr observation sessions in a number of nests to estimate the frequency at which adults deposited material during nest construction, the proportion of time invested in the incubation of the eggs, and the frequency of nestling provisioning by adults. The proportional investment in these activities by males and females were compared using Mann-Whitney test, implemented in the Software BioEstat 5.3 (Ayres et al. 2007). Descriptive statistics were provided as mean ± standard deviation.

We considered nest predation when eggs or nestlings disappeared from a nest before fledging age, and abandonment when adults were no longer seen near a nest for at least three days. We considered a success when young being fed outside a nest were observed. The probability of nest survival for the whole nesting cycle was estimated following the method proposed by Mayfield (1961).

RESULTS

We found 18 active nests during five breeding seasons: six in 2007/2008, five in 2008/2009, three in 2009/2010, two in 2010/2011, and two in 2011/2012. The earliest nesting activity was observed on 22 August 2007 (a nest in construction stage), and the latest nestlings were observed on 7 February 2009. Most breeding activities in the first three of these seasons occurred from September to November (Fig. 1).

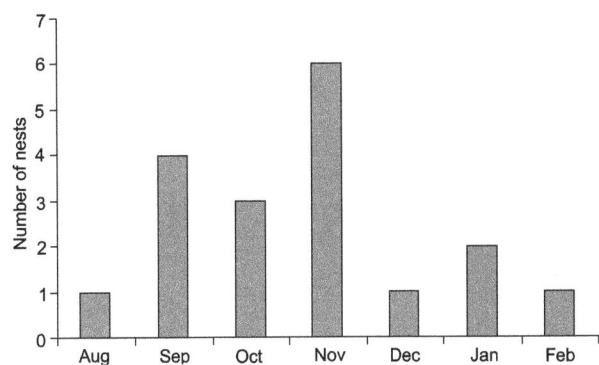

Figure 1. Monthly distribution of breeding activity of the Purple-throated Euphonia, from construction to nestling care, throughout three breeding seasons pooled together (2007/2008, 2008/2009, and 2009/2010).

Figures 2–5. Nest, eggs and nestlings of the Purple-throated Euphonia. (2) A typical nest, globular with a lateral entrance, made of dry vegetal material kept together with spider web silk. (3) Partial view of the head of a female sitting in the incubatory chamber of an atypical nest constructed inside a vespiary. (4) Eggs are typically pyriform, with brownish marks that may be more or less concentrated around the large pole. (5) Older nestlings beg for food at nest entrance, showing their red mouth lining and white oral flanges to their mother.

Nests were globular, but sometimes laterally flattened to adapt to the supporting branches. Nest wall was composed mainly of tiny dry leaves and leaflets, fine petioles, and plumed seeds, all highly compacted with a great amount of spider web silk. Pine needles and a few small stripes of dry grass were also found in some nests. Large dry leaves could be present, and all nests were dark brown externally (Fig. 2). The incubatory chamber was lined with very thin, light brown vegetal fibers, including peduncles of grass inflorescences, palm fibers, plumed seeds, and in one case, a few small feathers. The lateral entrance was well delimited by flexible fibers placed around it, and although the entrance was round in the beginning, it became elliptical in late nesting stages,

as nest roof tended to collapse during the nesting cycle. Nests were supported from below by a larger branch or fork, and often also laterally by a number of smaller branches or leaves. One nest, however, was hung from the tip of a descending branch of *Pinus* sp., to which it was attached only laterally. Another nest was constructed inside a large, tri-dimensional spider web, and it was totally supported by it. In this case, dry leaves of *G. robusta* fell down with the wind and remained attached to the spider web, serving to camouflage the nest. Measurements of six nests were 97.9 ± 14.37 mm (range = 77.5–114.7) in outside height, 110.61 ± 11.56 mm (range = 97.5–130.0) in outside diameter, 54.1 ± 4.55 mm (range = 48.9–57.4) in inside height, and 71.76 ± 7.34 mm (range = 64.0–78.6) in inside diameter (n = 6). The ellipsoid nest entrances measured 43.7 ± 3.04 (range = 39.1–46.5) x 28.45 ± 3.11 mm (range = 25.0–33.3) (n = 6), and the mean height of nests above ground was 4.88 ± 2.09 m (range = 2.25–9.0) (n = 9). Nesting trees were 2.75 to 17 m high (11.4 ± 5.4 m), and included *Pinus* sp. (n = 8), *G. robusta* (n = 3), *Yucca* sp. (n = 3), *Copaifera langsdorffii* Desf. (n = 1), *Caesalpinia peltophoroides* Benth. (n = 1), and *Gochnatia polymorpha* (Less.) Cabrera (n = 1). Three nests were deposited in the Ornithological Collection of the Museu de Zoologia da Universidade de São Paulo – MZUSP, under voucher numbers 2.283, 2.284, and 2.285.

Notably, in 2010 a female was found incubating two eggs in a nest constructed inside an abandoned nest of unidentified wasps. The vespiary was hung from a horizontal branch of *G. robusta* and it was very similar in size, shape (globular), and color to the nests of Purple-throated Euphonias (Fig. 3). Based on the size and round format of the entrance, and on the globular shape of the incubatory chamber, we believe that it was excavated by the birds and not by other animals. Nest material was deposited only to construct the incubatory chamber, and was similar to that we found in the other nests.

During nest construction, both males and females carried and deposited nest materials, and both were observed performing body movements to shape the incubatory chamber. In 16 hours of observation at nine nests, females brought nest material 1 to 21 times per hour, and males 0 to 14 times per hour, but the number of visits did not differ significantly between sexes (females 6.50 ± 5.09 times/hour, males 4.44 ± 4.15 times/hour, U = 94, p = 0.2). Notably, in 18% of the visits adults were observed carrying spider web stripes in their beaks that could be much longer than their own bodies (i.e. 15 cm), or they carried tufts of this material wrapped in the breast and belly. Clutch sizes were 1 (n = 1), 2 (n = 6) or 3 eggs (n = 2) (2.1 ± 0.6). Eggs were pyriform, with white background color, and dark and light brown blotches and spots, that could be round or elongated. These markings could be concentrated in the large end, or they could form a wreath near the large pole (Fig. 4). Egg measurements were (n = 5 eggs from two nests): length 16.24 ± 1.35 mm (range: 15–17.8), width 12.04 ± 0.09 mm (range: 11.9–12.1), and weight 1.16 ± 0.13g (range: 1–1.3).

In three nests, eggs were laid on consecutive days, and in one nest with one day interval. In these nests, incubation started the day the last egg was laid. In another nest in which only one egg was laid, incubation started two days later. Incubation period was 14 days (n = 2 eggs from one nest). In 16 hours of observation in seven nests, only females were recorded incubating the eggs. Average time spent incubating was 28.27 ± 17.34 min per hour (range = 0–57.57), females left the nests 1.2 ± 0.75 times per hour (range = 0–2), and incubation recesses lasted 11.55 ± 12.95 min (range = 0.87–60).

Hatching was synchronous (n = 2 nests) and nestling period was 21 days (n = 2 young from one nest). Hatchling skin was dark red with sparse gray down, and nestling presented bright red mouth lining and white swollen flanges (Fig. 5). In 13 hours of observation in five nests, nestlings were provisioned on average 4.84 ± 1.90 times per hour (range = 2–8), with equal participation of both sexes (females 2.69 ± 1.23 times/hour, males 2.15 ± 1.06 times/hour, U = 70, p = 0.10). Females brooded the young after provisioning until they were around 10 days old, and after that they were fed from the nest entrance, and their heads could be seen from outside while begging for food (Fig. 5). Of the 18 nests, three were abandoned in construction stage, and two were not monitored for success. Of the remaining 13 nests, two were abandoned during incubation, one was abandoned in unknown stage, two (including the nest constructed inside the vespiary) were predated in incubation stage, four were predated in nestling stage, and four fledged young. Mayfield average nest survival probability was 5% (three abandonments and six predations in 110 nest days).

DISCUSSION

The nesting season in our study area matched most of the records of active nests of the Purple-throated Euphonia from other localities, i.e. one nest recorded in late November in Paraguay (Bertoni 1904), one nest found in October in Santa Fé, Argentina (De La Peña 1996), one nest found in December in Bahia, northeastern Brazil (Lima 2006), and two nests found in September/October in Distrito Federal, Central Brazil (Marini et al. 2012). However, Kirwan (2009) observed a nest with nestlings on 7 August 2007 in the state of Mato Grosso, Brazil, meaning that breeding activities may have started in June, and in Manaus, northern Brazil, Oniki and Willis (1983) observed a nest that was active at least from 21 April (in construction) to 25 May 1974 (nestlings), indicating that the breeding season can be different in other South American regions.

Nest shape followed the general pattern found for other Euphoniinae, but nest materials used by the Purple-throated Euphonia have been only superficially described so far, impeding detailed comparisons with other regions. However, the use of materials of the supporting plants, also reported by other authors, may be an important source of variation in nest composition. Although we report the use of a spider web for the first time, we believe that this material may have been used also in the previously described nests, as it seemed to be an indispensable component to construct a globular structure using such small vegetal materials as those reported here. Although Bertoni (1904) has mentioned the presence of a false entrance on the top of a nest, and Kirwan (2009) has suspected, based on parental movements, that a nest observed in Mato Grosso could have a similar structure, we never found it in the nests we studied. The nest reported by De La Peña (1996) from Santa Fé, Argentina, was a deep cup, differing from the typical globular shape known for the entire subfamily, and seemed to be a rare exception.

The presence of nests among the leaves of a bromeliad (Oniki and Willis 1983), inside a spider web, as well as the use of other supporting plants, i.e. Urucum, *Bixa orellana* (Lima 2006), indicate that the Purple-throated Euphonia is a generalist with respect to nesting support, but the construction of a nest inside a vespiary was an unexpected finding. Other species of *Euphonia* have been recorded constructing their nests in abandoned structures made by other animals. For instance, Violaceous Ephonias *E. violacea* (Linnaeus, 1758), and Thick-billed Euphonias *E. laniirostris* (d'Orbigny and Lafresnaye, 1837) were observed, respectively, building their nests inside old nests of the Rusty-margined Flycatcher *Myiozetetes cayanensis* (Linnaeus, 1766) (Snethlage 1935, Oniki and Willis 2003), and of the Great Kiskadee *Pitangus sulphuratus* (Linnaeus, 1766) (Skutch 1969), and Skutch (1954) reported a pair of the Yellow-crowned Euphonia *E. luteicapilla* (Cabanis, 1861) constructing a nest "in a pocket between the layers of brood-cells in an old wasps' nest". In all of these cases the abandoned structures served as protection or support, and Euphonias have constructed the entire structure of their own nests. Our case is different because the vespiary itself served as the nest wall, and the only building material carried by the birds was that used for lining of the incubatory chamber. We do not believe this was an anti-predatory strategy, because the vespiary was inactive, and there is no evidence from other nests of an association between Purple-Throated Euphonias and wasps or bees. Although the association of nesting birds with venomous insects is a relatively common strategy (see Hansell 2002 for a review), nesting inside the structures constructed by these insects seems to be very uncommon. To our knowledge, the only other reported case for Neotropical birds was that of a Violaceous Trogon *Trogon violaceus* (Gmelin 1788) which also nested inside a vespiary (Skutch 1976).

The clutch sizes of Purple-throated Euphonias were similar to those reported by other authors, being three eggs in the nest found by De La Peña (1996), three in the nest found by Lima (2006), and two in each of the two nests found by Marini et al. (2012). Among the Euphonias, six of the species with known nests have exceptionally large clutch sizes of four to five eggs (Barnard 1954, Sargent 1993). Among birds, in general, closed nesters tend to present larger clutch sizes, but 4–5 eggs is too large compared to most of the Neotropical closed-nester species (see Sargent 1993). Clutch sizes reported for the Purple-throated Euphonia, as well as for some other species of the genus, as the Chestnut-bellied Euphonia *E. pectoralis* (Latham, 1801) (Isler and Isler 1999, Pizo 2000), and the Golden-rumped Euphonia *E. cyanocephala* (Vieillot, 1818) (Hilty 2011), fall within the regular range of 2–3 eggs expected for Neotropical passerines, indicating that exceptionally large clutch sizes is not a characteristic disseminated among the whole genus. Notably, some of the species that lay large clutches are from equatorial regions as the Yellow-throated Euphonia *E. hirundinacea* (Bonaparte, 1838) (mostly 5 eggs) (Sargent 1993) and the Jamaican Euphonia *E. jamaica* (Linnaeus, 1766) (3–4 eggs) (Bond 1961, March 1863), whereas smaller clutch sizes have been reported for some species or populations from tropical/subtropical locations i.e. the Orange-bellied Euphonia *E. xanthogaster* (Sundevall, 1834) (3

eggs) (Solano-Ugalde et al. 2007), and the Violaceous Euphonia (2 eggs in Southeastern Brazil) (Pinto 1953). This counteracts the general theory of larger clutch sizes in higher latitudes, observed in all of the continents (Jetz et al. 2008), indicating that other ecological and evolutionary aspects may be involved in this variation, which deserves investigation.

Although our data on incubation and nestling periods are based on a single nest, together with the incubation period of one egg provided by Lima (2006), these are the only information available so far. Assuming that incubation has started in the day the last egg was laid (as often observed here), and using our method of estimation, the incubation period presented by Lima (2006) would be 13 days. Thus, nesting cycle of the Purple-throated Euphonia was similar to that observed for the Yellow-throated Euphonia (14–16 days of incubation, and 18–20 days of nestling period: Skutch 1945, 1954, Sargent 1993), for the White-vented Euphonia *E. minuta* (Cabanis, 1849) (15–17 days, and 18–20 days, respectively: Skutch 1972, 1976), and for the Yellow-crowned Euphonia (13–14 days, and 22–24 days, respectively: Skutch 1954).

Many bird species can benefit from reproducing in anthropic habitats, where nest survival can be higher due to the increased protection provided by man-made structures, or due to the absence of certain nest predators (Møller 2010). However, nest survival in our study population was very low when compared to other passerines studied in urban parks or university campuses in southeastern Brazil, e.g. the Lined Seedeater *Sporophila lineola* (Linnaeus, 1758) (40%: Oliveira et al. 2010), the Yellowish Pipit *Anthus lutescens* (Pucheran, 1855) (87.0%: Freitas and Francisco 2012a), the Grassland Yellow-Finch *Sicalis luteola* (Sparrman, 1789) (47.0%: Freitas and Francisco 2012b), the Pale-breasted Thrush *Turdus leucomelas* (Vieillot, 1818) (54%: Davanço et al. 2013), and the Red-crested Finch *Coryphospingus cucullatus* (Statius Müller, 1776) (28.2%: Zima and Francisco 2016). Although some Euphonias seem to be dependent on specific habitats, the Purple-throated Euphonia has adapted to disturbed areas, but our data provide evidence that at least some of the areas inhabited by this species may act as reproductive traps (for a review see Battin 2004). Despite nests and eggs having been described for most of the species of *Euphonia*, this genus is still poorly known in terms of other reproductive aspects. Due to their widespread distribution and apparent geographic variations in reproductive parameters, such as clutch size and breeding phenology, gathering data on multiple populations and species of Euphonia may permit to test important theories about Neotropical birds life history evolution (Davanço et al. 2013), and the information presented here is a contribution to fill these knowledge gaps.

ACKNOWLEDGMENTS

We are grateful to the Faculdade de Engenharia de Sorocaba (FACENS) for authorizing field work at the campus.

PV Davanço received a fellowship from Universidade Federal de São Carlos (PIADRD programe), and LS Oliveira was supported by Conselho Nacional de Desenvolvimento Científico e Tecnológico (PIBIC/CNPq). We also thank D.C. Silva for assistance in the preparation of the images. This study was part of a major project on bird reproductive biology that was approved by IBAMA/CEMAVE (#3023/1 and #3023/2).

REFERENCES

Ayres M, Ayres Jr M, Ayres DL, Santos AS (2007) BioEstat 5.3: aplicações estatísticas nas áreas das ciências biológicas e médicas. Brasília, Sociedade Civil Mamirauá, MCT, CNPq.

Barnard GC (1954) Notes on the nesting of the Thick-billed Euphonia in the Panama Canal zone. The Condor 56: 98–101. https://doi.org/10.2307/1364666

Battin J (2004) When good animals love bad habitats: ecological traps and the conservation of animal populations. Conservation Biology 18: 1482–1491. https://doi.org/10.1111/j.1523-1739.2004.00417.x

Belton W (1985) Birds of Rio Grande do Sul, Brasil – Part 2 – Formicariidae through Corvidae. Bulletin of the American Museum of Natural History 180: 1–242.

Bertoni AW (1904) Contribución para el conocimiento de las aves del Paraguay. Anales Científicos Paraguayos 3: 1–10.

Bertoni AW (1918) Sobre nidificacion de los euphonidos (Ornit.). Anales Cientificos Paraguayos 2: 242–244.

Bond J (1943) Nidification of the passerine birds of Hispaniola. The Wilson Bulletin 55: 115–125.

Bond J (1961) Birds of West Indies. Collins Clear-Type Press, London.

Davanço PV, Oliveira LS, Sousa LMS, Francisco MR (2013) Breeding life-history traits of the Pale-breasted Thrush (Turdus leucomelas) in southeastern Brazil. Ornitologia Neotropical 24: 401–411.

De la Peña MR (1996) Descripción de nidos nuevos o poco conocidos de la avifauna Argentina. El Hornero 14: 085–086.

Ffrench RP (1980) A Guide to the birds of Trinidad and Tobago. Harrowood Books, Newtown.

Freitas MS, Francisco MR (2012a) Reproductive life history traits of the Yellowish Pipit (Anthus lutescens). The Wilson Journal of Ornithology 124: 119–126. https://doi.org/10.1676/11-038.1

Freitas MS, Francisco MR (2012b) Nesting behavior of the Yellow-Finch (Sicalis luteola) in southeastern Brazil. Ornitologia Neotropical 23: 341–348.

Greeney HF, Nunnery T (2006) Notes on the breeding of northwest Ecuadorian birds. Bulletin of the British Ornithologists' Club 126: 38–45.

Hansell MR (2002) Bird nests and construction behaviour (3rd ed.). Cambridge University Press.

Hilty S (2011) Family Thraupidae (Tanagers). In: Del Hoyo J, Elliot A, Christie DA (Eds) Handbook of the Birds of the World: Tanagers to New World Blackbirds. Lynx Edicions, Barcelona, 46–329.

Isler ML, Isler PR (1999) The Tanagers: natural history, distribution and identification. Smithsonian Institution, Washington, DC.

Janni O, Boano G, Pavia M, Gertosio G (2008) Notes on the breeding of birds in Yanachaga-Chemillén National Park, Peru. Cotinga 30: 42–46.

Jetz W, Sekercioglu CH, Böhning-Gaese K (2008) The worldwide variation in avian clutch size across species and space. Plos Biology 6: 2650–2657. https://doi.org/10.1371/journal.pbio.0060303

Kirwan GM (2009) Notes on the breeding ecology and seasonality of some Brazilian birds. Revista Brasileira de Ornitologia 17: 121–136.

Kottek M, Grieser J, Beck C, Rudolf B, Rubel F (2006) World map of the Köppen-Geiger climate classification updated. Meteorologische Zeitschrift 15: 259–263. https://doi.org/10.1127/0941-2948/2006/0130

Lima PC (2006) Euphonia chlorotica chlorotica (Linnaeus, 1766), um ninho diferente: adaptação ou evolução? Atualidades Ornitológicas 129: 8–9.

March WT (1863) Notes on the Birds of Jamaica (continued). Proceedings of the Academy of Natural Sciences of Philadelphia 15: 283–304.

Marini MA, Borges FJA, Lopes LE, Sousa NOM, Gressler DT, Santos LR, Paiva LV, Duca C, Manica LT, Rodrigues SS, França LF, Costa PM, França LC, Heming NH, Silveira MB, Pereira ZP, Lobo Y, Medeiros RCS, Roper JJ (2012) Breeding biology of birds in the Cerrado of Central Brazil. Ornitologia Neotropical 23: 385–405.

Martin TE, Geupel GR (1993) Nest-monitoring plots: methods for locating nests and monitoring success. Journal of Field Ornithology 64: 507–519.

Mayfield H (1961) Nesting success calculated from exposure. The Wilson Bulletin 73: 255–261.

Møller AP (2010) The fitness benefit of association with humans: elevated success of birds breeding indoors. Behavioral Ecology 21: 913–918. https://doi.org/10.1093/beheco/arq079

Morton ES (1973) On the evolutionary advantages and disadvantages of fruit eating in tropical birds. The American Naturalist 107: 8–22. https://doi.org/10.1086/282813

Nehrkorn A (1910) Katalog der eiersammlung, nebst Beschreibungeh der aussereuropäischen Eier. Verlag Von R. Friedländer and Sohn, Berlin.

Oliveira LS, Sousa LMS, Davanço PV, Francisco MR (2010) Breeding behaviour of the Lined Seedeater (Sporophila lineola) in southeastern Brazil. Ornitologia Neotropical 21: 251–261.

Oniki Y, Willis EO (1983) Breeding records of birds from Manaus, Brazil. Part 5. Icteridae to Fringillidae. Revista Brasileira de Biologia 43: 55–64.

Oniki Y, Willis EO (2003) Re-uso de ninhos por aves neotropicais. Atualidades Ornitológicas 116: 4–7.

Peel MC, Finlayson BL, McMahon TA (2007) Updated world map of the Köppen-Geiger climate classification. Hydrology and Earth System Sciences 11: 163–1644. https://doi.org/10.5194/hess-11-1633-2007

Pinto O (1953) Sobre a coleção Carlos Estevão de peles, ninhos e ovos das aves de Belém (Pará). Papéis Avulsos de Zoologia 11: 111–222.

Pizo MA (2000) Attack on Chestnut-Bellied Euphonia nestlings by Army Ants. The Wilson Bulletin 112: 422–424. https://doi.org/10.1676/0043-5643(2000)112[0422:AOCBEN]2.0.CO;2

Ridgely RS, Tudor G (1994) The birds of South America: Volume II: The Suboscine Passerines. University of Texas Press, Austin.

Sargent S (1993) Nesting biology of the Yellow-throated Euphonia: large clutch size in a neotropical frugivore. The Wilson Bulletin 105: 285–300.

Sick H (1997) Ornitologia brasileira. Nova Fronteira, Rio de Janeiro.

Skutch AF (1945) Incubation and nesting periods of Central American birds. The Auk 62: 8–37. https://doi.org/10.2307/4079958

Skutch AF (1954) Life histories of Central American birds (families Fringillidae, Thraupidae, Icteridae, Parulidae, and Coerebidae). Pacific Coast Avifauna 31: 1–448.

Skutch AF (1969) Life histories of Central American birds (families Cotingidae, Pipridae, Formicariidae, Furnariidae, Dendrocolaptidae, and Picidae). Pacific Coast Avifauna 35: 1–580.

Skutch AF (1972) Studies of tropical American birds. Nuttall Ornithological Club 10: 1–228.

Skutch AF (1976) Parent birds and their young. University of Texas Press, Austin and London.

Skutch AF (1985) Clutch size, nesting success, and predation on nests of neotropical birds, reviewed. Ornithological Monographs 36: 575–594. https://doi.org/10.2307/40168306

Snethlage H (1928) Meine reise durch Nordostbrasilien. III. Bausteine zur biologie der angetroffenen arten. Journal für Ornithologie 76: 668–738. https://doi.org/10.1007/BF01923575

Snethlage H (1935) Beitrage zur brutbiologie brasilianischer. Journal für Ornithologie 83: 1–24. https://doi.org/10.1007/BF01908740

Solano-Ugalde A, Arcos-Torres A, Greeney HF (2007) Additional breeding records for selected avian species in northwest Ecuador. Boletín SAO 17: 17–25.

Winkler DW (2004) Nests, eggs, and young: the breeding biology of birds. In: Podulka S, Rohrbaugh Jr RW, Bonney R (Eds) Handbook of bird biology (2nd ed.). Cornell Lab of Ornithology, Ithaca, 8.1–8.152.

Zima PVQ, Francisco MR (2016) Reproductive behavior of the Red-crested Finch Coryphospingus cucullatus (Aves: Thraupidae) in southeastern Brazil. Zoologia 33: e20160071. https://doi.org/10.1590/S1984-4689zool-20160071

Zuccon D, Prys-Jones R, Rasmussen PC, Ericson PGP (2012) The phylogenetic relationships and generic limits of finches (Fringillidae). Molecular Phylogenetics and Evolution 62: 581–596. https://doi.org/10.1016/j.ympev.2011.10.002

Author Contributions: PVD, LSO, LMSS and MRF designed and conducted the experiments; DFP and MRF analyzed the data and wrote the paper.

Competing Interests: The authors have declared that no competing interests exist.

4

Oxygen consumption remains stable while ammonia excretion is reduced upon short time exposure to high salinity in *Macrobrachium acanthurus* (Caridae: Palaemonidae), a recent freshwater colonizer

Carolina A. Freire[1], Leonardo de P. Rios[1], Eloísa P. Giareta[1], Giovanna C. Castellano[1]

[1]*Departmento de Fisiologia, Setor de Ciências Biológicas, Universidade Federal do Paraná. Centro Politécnico, Jardim das Américas, 81531-980 Curitiba, PR, Brazil.*
Corresponding author: Carolina Arruda Freire (osmolab98@gmail.com; cafreire@ufpr.br)

http://zoobank.org/E11A9A9C-55C4-445A-BDC4-CC1FAF8889E9

ABSTRACT. Palaemonid shrimps occur in the tropical and temperate regions of South America and the Indo-Pacific, in brackish/freshwater habitats, and marine coastal areas. They form a clade that recently (i.e., ~30 mya) invaded freshwater, and one included genus, *Macrobrachium* Bate, 1868, is especially successful in limnic habitats. Adult *Macrobrachium acanthurus* (Wiegmann, 1836) dwell in coastal freshwaters, have diadromous habit, and need brackish water to develop. Thus, they are widely recognized as euryhaline. Here we test how this species responds to a short-term exposure to increased salinity. We hypothesized that abrupt exposure to high salinity would result in reduced gill ventilation/perfusion and decreased oxygen consumption. Shrimps were subjected to control (0 psu) and experimental salinities (10, 20, 30 psu), for four and eight hours (n = 8 in each group). The water in the experimental containers was saturated with oxygen before the beginning of the experiment; aeration was interrupted before placing the shrimp in the experimental container. Dissolved oxygen (DO), ammonia concentration, and pH were measured from the aquaria water, at the start and end of each experiment. After exposure, the shrimp's hemolymph was sampled for lactate and osmolality assays. Muscle tissue was sampled for hydration content (Muscle Water Content, MWC). Oxygen consumption was not reduced and hemolymph lactate did not increase with increased salinity. The pH of the water decreased with time, under all conditions. Ammonia excretion decreased with increased salinity. Hemolymph osmolality and MWC remained stable at 10 and 20 psu, but osmolality increased (~50%) and MWC decreased (~4%) at 30 psu. The expected reduction in oxygen consumption was not observed. This shrimp is able to tolerate significant changes in water salt concentrations for a few hours by keeping its metabolism in aerobic mode, and putatively shutting down branchial salt uptake to avoid massive salt load, thus remaining strongly hyposmotic. Aerobic metabolism may be involved in the maintenance of cell volume, concomitant with reduced protein/aminoacid catabolism upon increase in salinity. More studies should be conducted to broaden our knowledge on palaemonid hyporegulation.

KEY WORDS. Ammonia, lactate, osmoregulation, palaemonidae.

INTRODUCTION

Palaemonid shrimps have a wide global distribution. They occur in a great variety of aquatic environments, from seawater up to full freshwater (Augusto et al. 2009, Anger 2013). The family Palaemonidae has a marine origin and passed through various independent events of freshwater invasion (Murphy and Austin 2005, Ashelby et al. 2012, McNamara and Faria 2012, Anger 2013). These events have occurred quite recently, ~30 mya (Tertiary), and are still happening (Murphy and Austin 2005, Augusto et al. 2007a, b, 2009, Pileggi and Mantelatto 2010, Collins et al. 2011, McNamara and Faria 2012). The family comprises about 116 genera. Among these, *Macrobrachium* Bate, 1868 was the most successful colonizer of freshwater and estuarine waters, showing a wide geographical distribution (Anger 2013). One species in particular, commonly referred to as "giant freshwater prawn," *Macrobrachium rosenbergii* (De Man, 1879) has large economic importance globally (Bond-Buckup and Buckup 1989, Valenti et al. 1989).

The fact that freshwater Palaemonid shrimps have recently transitioned from saline waters to more dilute waters renders them quite tolerant to increased salinity (Freire et al. 2008a, 2013). Freshwater Palaemonid shrimp adults - especially the more coastal and diadromous species – those whose larvae depend on brackish waters for their proper development (Charmantier 1998, Anger 2003, McNamara and Faria 2012), are particularly euryhaline (McNamara 1987).

Estuarine, but especially freshwater crustaceans, are good hyper-osmoregulators, that is, they keep steep osmotic and ionic gradients with respect to the surrounding water, aided by the low permeability of their cuticle (Péqueux 1995, Freire et al. 2003, 2008a, b). Palaemonid shrimps also follow this pattern (Freire et al. 2003, Murphy and Austin 2005, Augusto et al. 2009, Boudour-Boucheker et al. 2013). It is during ecdysis that cell volume may be challenged in these shrimps, as their internal medium fluctuates (compared to the intermoult period), diluting strongly in fresh waters. However, crustaceans in general, palaemonid shrimps in particular, can face this challenge by regulating cell/tissue hydration quite efficiently (see Freire et al. 2013). These hyper-regulation mechanisms, one of the most prominent features of freshwater crustaceans (second to the cuticle) have been frequently studied in Palaemonidae (e.g., Freire et al. 2008b, McNamara and Faria 2012). Hyporegulation mechanisms, by contrast, have been much less investigated and remain elusive (Freire et al. 2008b, McNamara and Faria 2012).

Salinity challenges can also result in changes in metabolic responses. For instance, a decrease in the respiratory rate of the marine shrimps *Marsupenaeus (Penaeus) japonicus* (Bate, 1888) (Setiarto et al. 2004), and *Litopenaeus (Penaeus) setiferus* (Linnaeus, 1767) (Rosas et al. 1999), when exposed to a decrease in salinity, was observed. Some palaemonid shrimps exposed to salinity increases, for instance *Macrobrachium heterochirus* (Wiegmann, 1836) and *Macrobrachium potiuna* (Müller, 1880), experience a decrease in metabolic rates, while the diadromous *Macrobrachium acanthurus* (Wiegmann, 1836) and *Macrobrachium olfersii* (Wiegmann, 1836) experience a peak in their metabolism-salinity curves ("dome-shaped curve") close to their isosmotic point, ~21 psu (Moreira et al. 1983). In another diadromous shrimp, *Macrobrachium amazonicum* (Heller, 1862), oxygen consumption was lower in freshwater than in 18 psu in zoea II, and was higher in freshwater than in 12 and 18 psu in zoea V (Mazzarelli et al. 2015). The freshwater shrimp *Macrobrachium tuxtlaense* Villalobos and Alvarez, 1999, when exposed to increased salinities up to 30 psu, has shown an increase in oxygen consumption in 5 and 10 psu, and a decrease in this parameter in the other salinities, with respect to the control salinity, fresh water (Ordiano et al. 2005). With this variability in the metabolic response of shrimps in the background, the aim of this study was to test whether a short term exposure to increased salinity would result

in a "shut down" of oxygen uptake, a putative "escape response", potentially activating anaerobic metabolism and lactate production in this diadromous palaemonid.

MATERIAL AND METHODS

Specimens of *M. acanthurus* were bought from local fishermen from Rio dos Barrancos (25°36'32.0"S, 48°24'02.5"W), municipality of Pontal do Paraná, Paraná, Brazil, who sell them as live bait. Shrimps were transported to the laboratory for approximately two hours, in plastic gallons with constant aeration. The animals were acclimated for about five days in 35 liters aquaria with fresh water (double filtered tap water, charcoal and cellulose filters), in temperature of 20±1°C, constant aeration, and natural photoperiod (~12 h light: 12 h dark). Some ions were assayed in our tap water (mean±standard deviation, in mM, n = 6 for all): chloride 0.23 ± 0.29; magnesium: 0.16 ± 0.06; sodium 4.67 ± 1.94; potassium 0.57 ± 0.20, and osmolality of 26.2 ± 4.3 mOsm/kg H_2O. Shrimps were fed fragments of fish fillet on alternate days.

Shrimps (5.2 ± 0.7 cm, n = 64) were individually subjected to salinities 0 (control), 10, 20, or 30 psu, for 4 or 8 hours (n = 8 for each coupled condition of salinity x time), in 250 ml containers, water temperature of 21.1 ± 0.05 °C. Saline waters were obtained through proportional mixture of filtered tap water with natural sea water. The experiments were conducted without aeration, in order to allow the determination of oxygen consumption by the shrimp, but the initial water was saturated with oxygen, through overnight aeration, before the start of the experiments (initial oxygen concentration of 7.49 ± 1.19 mg/l for 0 psu, 7.36 ± 0.60 mg/l for 10 psu, 6.73 ± 0.15 mg/l for 20 psu, and 6.55 ± 0.15 mg/l for 30 psu, n = 16 for each salinity). The following water parameters were analyzed at the initial (before placing the shrimp in the container) and final (after removing the shrimp from the container) times of exposure: dissolved oxygen, pH, and ammonia. Differences between the initial and final concentrations of dissolved oxygen and ammonia represented, respectively, oxygen consumption and ammonia excretion by the shrimp (N-NH₃). There was essentially no ammonia in the water at the beginning of the experiments (0.008 ± 0.003 mg/l of N-NH₃, n = 70 samples). Experiments were also conducted in containers with water but without animals, as blanks for water parameters (n = 6 for each experimental condition, yielding a total of 72 containers).

After the stipulated times of exposure, the animals were cryoanesthetized (covered with ground ice) for about 1 minute, until fully immobile. Then, hemolymph samples were collected through cardiac puncture, with a micropipette inserted under the exosqueleton, for determinations of lactate and osmolality. Finally, the exosqueleton was removed, and a fragment of abdominal muscle was collected for determination of water content.

Other individuals were subjected to the same protocol of salinity increase (n = 3 for each condition of salinity x time), to evaluate whether shrimps were ventilating their gills equally in high salinity media, as they do in their habitat, fresh water. The hypothesis was that their gills would get stained from the dye added to the water, after some minutes of exposure, from gill ventilation. At the end of the experimental exposures to high salinity, five drops of 1% methylene blue were added to each of the 250 ml containers. Shrimps were maintained in these conditions for 5 minutes, after which they were removed from the containers and had both sides of their cephalotorax photographed, with focus on their gills. The intensity of the blue staining of their gills was qualitatively evaluated. The same procedure with the dye was conducted for control shrimps in fresh water (see Suppl. material 1).

The levels of dissolved oxygen were detected in the water through an oxymeter (YSI model 55, USA). Water pH was determined using a bench pHmeter (inoLAB pH Level 1WTW, Germany). The concentration of ammonia was assayed through colorimetric commercial kits (Alfakit, Brazil), and absorbance was read at 630 nm (Spectrophotometer Ultrospec 2100 PRO Amersham Pharmacia biotech, Sweden).

Hemolymph lactate was assayed through colorimetric commercial kits (Labtest, Brazil), with absorbance read at 550 nm. Hemolymph osmolality was determined using a vapour pressure osmometer in undiluted samples (Vapro 5520, Wescor, USA). For the determination of muscle water content, tissue fragments were weighed (wet weight, analytical balance Bioprecisa FA2104N, Brazil, precision of 0.1 mg), dried in an oven at 60 °C for 24 hours, then weighed again (dry weight). The difference between wet and dry weights, as a percentage, represents the muscle water content, or its hydration.

Two-way ANOVAs (factors were salinity and time) with *post hoc* tests of Holm-Sidak were conducted for each of the following parameters: oxygen consumption, lactate, osmolality, and muscle water content. Initial and final pH values did not pass the normality and equal variance tests. These data were transformed to meet the requirements of the parametric two way ANOVA. Ammonia values could not be normalized, and for this reason they were treated differently. Two non-parametric (Kruskal-Wallis) "one-way-ANOVAs" were conducted, one for 4 hours, one for 8 hours. The respective values of each salinity, 4 vs 8 hours, were compared using t-tests. The initial versus final values of dissolved O_2, and pH in the water of containers were compared through paired t-tests for each experimental condition. Pearson correlations were performed for factors salinity, oxygen consumption, lactate, osmolality, muscle water content, excreted ammonia, final pH, and total length. The adopted significance level was 0.05.

RESULTS

Experimental blanks for water parameters

The initial and final water parameters (O_2, pH, and NH_3) in the blanks, vials without any shrimp – for all salinities and times of exposure – are shown in Table 1.

Table 1. Initial and final concentrations of oxygen and ammonia, and values of pH in the water of "blank" containers, without any shrimp, in salinities 0 (control), 10, 20, and 30 psu, for 4 and 8 hours of exposure (n = 6 for each group).

	Initial	After 4 hours	After 8 hours
O_2 (mg/l)			
0 psu	6.88 ± 0.41	6.89 ± 0.24	6.74 ± 0.16
10 psu	6.98 ± 0.21	6.84 ± 0.12	6.69 ± 0.05
20 psu	6.63 ± 0.19	6.48 ± 0.10	6.27 ± 0.04
30 psu	6.29 ± 0.09	6.08 ± 0.03	5.85 ± 0.02
pH			
0 psu	6.90 ± 0.16	7.05 ± 0.15	6.88 ± 0.09
10 psu	7.61 ± 0.05	7.47 ± 0.04	7.42 ± 0.02
20 psu	8.03 ± 0.06	7.81 ± 0.05	7.71 ± 0.05
30 psu	8.24 ± 0.03	8.05 ± 0.05	7.93 ± 0.05
NH_3 (mg/l)			
0 psu	0.03 ± 0.01	0.03 ± 0.01	0.01 ± 0.01
10 psu	0.01 ± 0.01	0.01 ± 0.01	0.01± 0.01
20 psu	0.00 ± 0.00	0.00 ± 0.00	0.00 ± 0.00
30 psu	0.00 ± 0.01	0.01 ± 0.01	0.01 ± 0.01

Water oxygen consumption and lactate concentration in the hemolymph

The two-way ANOVA revealed that time (F = 14.1, p < 0.001) and salinity (F = 2.9, p = 0.043), but not their interaction (F = 0.66, p = 0.58) affected water oxygen consumption in *M. acanthurus*. The initial oxygen concentration in the water was always higher than the final concentration, for all salinities and times of exposure, indicating oxygen consumption by the shrimp (Fig. 1). However, consumption, as quantified by the difference between initial and final oxygen levels in the water of the container, was stable and did not vary among the experimental treatments and controls. The two-way ANOVA revealed that time (F = 4.3, p = 0.042), but not salinity (F = 0.068, p = 0.977), or their interaction (F = 0.15, p = 0.93) affected hemolymph lactate in *M. acanthurus*. Production of lactate by the shrimp, measured from its hemolymph, was essentially invariable in the several experimental conditions (Fig. 2).

pH and excreted ammonia

The two-way ANOVA on initial pH values revealed no effect of time (F = 0.79, p = 0.38), but an effect of salinity (F = 132, p < 0.001), and no interaction between time and salinity (F = 0.82, p = 0.49). The final pH of the water , according to the two-way ANOVA, was affected by time (F = 19.5, p < 0.001), and salinity (F = 56.3, p < 0.001), but not by their interaction (F = 1.100, p = 0.36). The pH of the water was always higher at the start (initial) of the experiment than after 4 and 8 hours (final, Fig. 3). The Kruskal-Wallis tests revealed that salinity had an effect on water ammonia concentrations after 4 hours (p < 0.001), and after 8 hours (p < 0.001). The amount of excreted

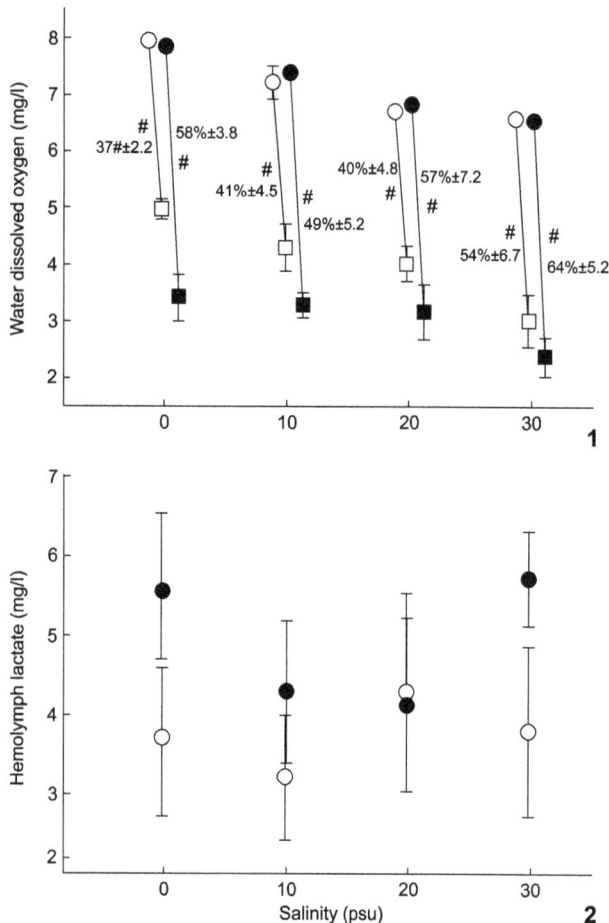

Figures 1–2. Initial (circles) and final (squares) levels of dissolved oxygen in the water (1) and lactate concentration in the hemolymph (2) of *M. acanthurus* exposed to salinities 0 (control), 10, 20, and 30 psu for 4 (white symbols) and 8 hours (black symbols). Values near to the lines represent oxygen consumption (mean ± std dev) as a percentage of the initial oxygen concentration (considered 100%, as a reference value). (#) Initial and final levels of dissolved oxygen are different. There were no significant differences between salinities or times (4 and 8 hours) for oxygen consumption and lactate concentration.

ammonia ($N-NH_3$) decreased with increasing salinity, with the highest value in 0 and 10 psu, and the lowest values in 20 and 30 psu. The effect of time on the amount of excreted ammonia was noted in the controls in 0 only, with higher ammonia levels measured in the water after 8 hours than after 4 hours (Fig. 3).

Osmolality of the hemolymph and muscle water content

The two-way ANOVA revealed that time ($F = 12.9$, $p < 0.001$) and salinity ($F = 79.8$, $p < 0.001$), and their interaction ($F = 7.3$, $p < 0.001$) had an effect on the osmolality of the hemo-

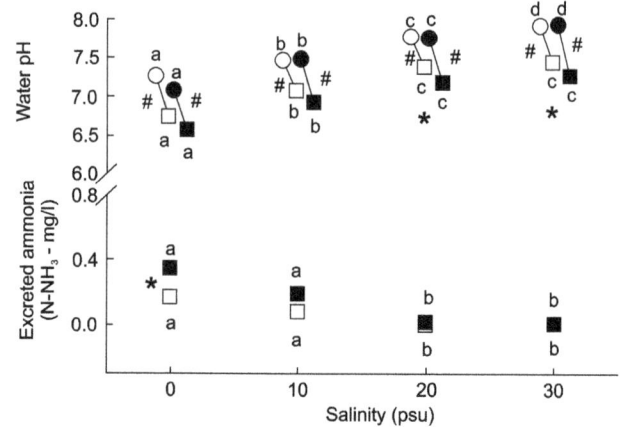

Figure 3. Initial (circles) and final (squares) pH and excreted ammonia ($N-NH_3$) in water of *M. acanthurus* exposed to salinities 0 (control), 10, 20, and 30 psu for 4 (white symbols) and 8 hours (black symbols). Different letters mean differences between salinities within each time of exposure. * = value for 4 hours is different from value for 8 hours within a same salinity, # = initial and final values of water pH are different.

lymph of *M. acanthurus*. It increased by 20 psu, and further by 30 psu with respect to the control (0 psu), after 4 and 8 hours. An effect of time was noted at 30 psu: osmolality was higher after 8 hours than after 4 hours (Fig. 4). Coherently, muscle water content showed the opposite trend, that is, it decreased as salinity increased. The two way ANOVA revealed that salinity ($F = 25.3$, $p < 0.001$) and the interaction between salinity and time ($F = 2.95$, $p = 0.041$) affected the hydration of the muscle, but time alone did not ($F = 2.3$, $p = 0.14$). In the highest salinity, 30 psu, muscle water content after 8 hours was lower than after 4 hours, matching the raised osmolality of the hemolymph (Fig. 5).

Pearson correlation

Salinity had a positive correlation with oxygen consumption (Weak Correlation coefficient 0.254, P value 0.0430), with water pH (Strong Correlation coefficient 0.808, P value 7.54×10^{-16}), and osmolality (Strong Correlation coefficient 0.813, P value 3.53×10^{-16}). Conversely, salinity had a negative correlation with excreted ammonia (Strong Correlation coefficient -0.775, P value 6.02×10^{-14}), and muscle water content (Strong Correlation coefficient -0.72, P value 1.82×10^{-11}).

DISCUSSION

Water dissolved oxygen (DO) was consumed by *M. acanthurus* during the experiments, and was consistently detected by our assay method, which is evidenced by the reduction in water DO. No decrease in water DO was detected under the same experimental conditions (volume, temperature, previous

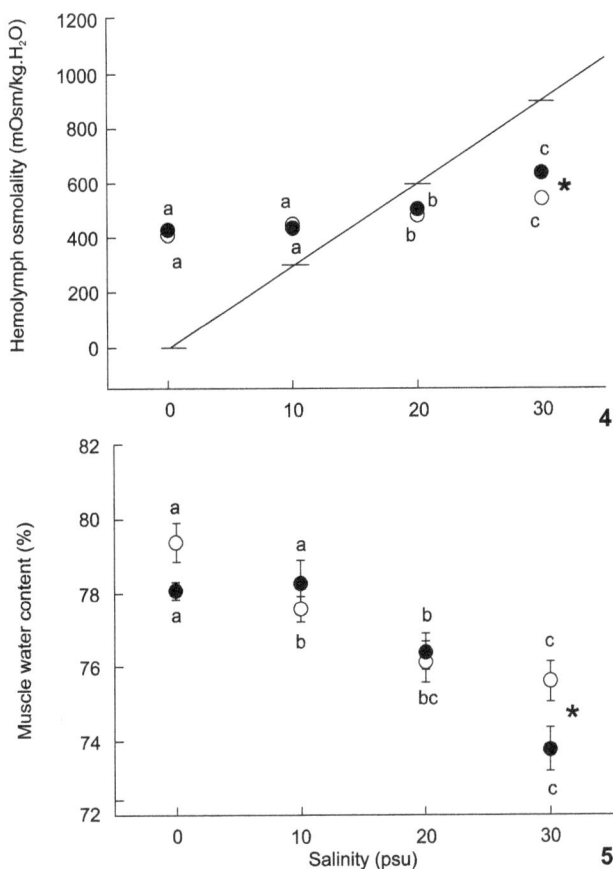

Figures 4–5. Hemolymph osmolality (4) and muscle water content (5) of *M. acanthurus* exposed to salinities 0 (control), 10, 20, and 30 psu for 4 (white circles) and 8 hours (black circles). Dashed line represents water expected values, from the relationship 1 psu = 30 mOsm/kg H_2O, short horizontal lines indicate value of calculated water osmolality for the tested salinities. Different letters mean differences between salinities within each time of exposure. * = value for 4 hours is different from value for 8 hours within a same salinity.

DO saturation protocol, DO electrode), but without a shrimp in the vial ("blanks"), supporting the conclusion that oxygen was indeed consumed by the shrimp. Unexpectedly, salinity did not affect oxygen consumption by *M. acanthurus* after 4 or 8 hours of exposure; the correlation between these two variables was significant, but weak.

The relationship between salinity and oxygen consumption in shrimps is rather complex and variable. In the marine palaemonid *Palaemon serratus* (Pennant, 1777) no change in oxygen consumption between salinities 34 and 15 psu (Salvato et al. 2001) was observed. In contrast, in the marine shrimp *L. (P.) setiferus* the effect of salinity on the rate of oxygen consumption was influenced by the developmental stage of the individual: post-larvae PL10-PL15 showed the highest oxygen

consumption at 10 psu, while post-larvae PL15-PL21 showed the highest oxygen consumption at 40 psu (salinity ranged from 5 to 40 psu – Rosas et al. 1999).

Importantly, in the study cited immediately above (Rosas et al. 1999), oxygen consumption in penaeid shrimps was measured in these cited salinity levels after 120-264 hours of exposure to them. In addition, penaeid marine shrimps are hyper-hypo-regulators (Péqueux 1995, Freire et al. 2008b), as opposed to palaemonid shrimps, which are essentially hyper-regulators, as already mentioned. In the results of an experiment with *M. amazonicum* at different ontogenetic stages (zoea I, II, V, and IX) exposed to different salinities (0.5, 6, 12 or 18 psu), there was a greater consumption of oxygen at 0.5 psu in the zoea stage V, probably due to the great amounts of energy required for the active transport of salts through the epithelia (Mazzarelli et al. 2015). A similar pattern was observed in *M. tuxtlaense*, an strictly freshwater prawn exposed to a salinity gradient (0, 5, 10, 15, 20, 25, and 30 psu): higher rates of oxygen consumption in the shrimp *M. tuxtlaense* were observed at low salinities (0, 5 and 10 psu), to account for hyper-regulation of the osmolality of the hemolymph (Ordiano 2005). In the results of Moreira et al. (1983) working with *M. acanthurus*, the oxygen consumption curve after 24 hours of exposure was dome-shaped, peaking at 21 psu. These results show that the metabolic response to variations in salinity is indeed variable in these decapod crustaceans, is dependent on the osmoregulatory history and strategy of the species, and is also time-dependent.

Consistent with the results on oxygen consumption rates, the concentration of hemolymph lactate in *M. acanthurus* remained constant, they did not change when salinity increased. The hypothesis here was that increased salinity would lead to a reduction in gill perfusion and, consequently, reduced oxygen consumption. Reduced oxygen consumption would result in anaerobic metabolism and lead to increased levels of hemolymph lactate (Booth et al. 1982). In many decapod crustaceans, lactate is the main product of metabolism when there is a hypoxic condition (Bridges and Brand 1980, Taylor and Spicer 1989, Maciel et al. 2008). These shrimps continued to take up oxygen from the water, and did not enter into functional hypoxia. Had they entered into intense anaerobic metabolism, increased hemolymph lactate would have been detected. However, metabolic carbon dioxide was produced and released through the gills, causing water acidification (Henry and Wheatly 1992), which was detected in our results.

Ammonia release decreased with increased salinity (strong and significant negative Pearson correlation). One possible factor that could at least partially account for this inverse relationship is the fact that NH_3 can be excreted as NH_4^+, especially in acidic water, and in animals with acidosis, replacing K^+ in the Na^+/K^+-ATPase (e.g., Claiborne et al. 1982, Wall 1995, Furriel et al. 2004). If Na^+/K^+-ATPase activity is reduced with increased salinity in this shrimp (see Maraschi et al., 2015), then NH_4^+ transport from the hemolymph also putatively decreases, leading to less ammonia in the water, as observed here. Alternatively, ammonia may also be transported and eliminated through Rh-proteins (see Weirauch et al. 2009).

This freshwater shrimp strongly hyper-regulates in fresh-water (gradient of +400 mOsm/kg H$_2$O), its natural habitat in the adult phase, and continues to show hyper-osmotic hemolymph after 4-8 hours in 10 psu (+100 mOsm/kg H$_2$O). However, after 4-8 hours in 20 or 30 psu, although there is some increase in hemolymph osmolality (strong positive correlation between salinity and hemolymph osmolality), it becomes hyposmotic to the water at -150 and -300 mOsm/kg H$_2$O, respectively.

When there is a significant salt load, for instance 10, 20, 30 psu, what happens to the salt uptake system of the gills of freshwater shrimps, which normally steeply absorb salt from freshwater? The first hypothesis that can explain the relative osmotic stability of the hemolymph is that gill ventilation/perfusion would drastically decrease, especially when the exposure is short (up to a few hours). When this happens, consumption of oxygen from the water also decreases. Such decrease in oxygen consumption, however, was not observed in our data. In fact, under all experimental conditions tested here, when a vital dye (methylene blue) was pipetted next to the shrimp, and the branchial chamber and gills were observed under a stereomicroscope, the gills were stained blue (data not shown, see Suppl. material 1). There was great variability in the resulting blue color of the gills of the shrimps. The idea behind this experiment using the blue stain was that, if the shrimp reduced its gill ventilation upon increased water salinity, oxygen consumption from the water would be reduced and hence its gills would remain clear, whitish, when tested. Apparently, however, compatible with the oxygen consumption data, *M. acanthurus* apparently perfuses its gills even under severe salinity increase/stress.

Among estuarine palaemonids, apparent hyporegulation was verified in *Palaemon pandaliformis* (Stimpson, 1871) at 20-30 psu (Freire et al. 2003, Foster et al. 2010) and in *Macrobrachium equidens* (Dana, 1852) at 20-40 psu (Denne 1968). Among fresh-water palaemonids, hyporegulation response was observed in the following species of *Macrobrachium*: *M. acanthurus* at 22-26 and 30 psu after 168 hours (Signoret and Brailovsky 2004), at 30 psu after 0.5, 1, 2, 3,6, 16, and 24 hours (Foster et al. 2010), at 25 psu after 24 hours (Maraschi et al. 2015), *M. rosenbergii* at 18-35 psu for 6 hours-15 days (Cheng et al. 2003), *M. brasiliense* at 20 psu, and *M. olfersi* and *M. potiuna* at 20-30 psu for 1-10 days of exposure (Freire et al. 2003). This apparent hyporegula-tion consists of a response of supression of the hyperegulation, strongly employed in freshwater, in these shrimps, and an still elusive salt secretion, possibly. The response may also be called a hypo-conformation, as in Moreira et al. (1983) for *M.acanthurus* and other species of the genus. This strong tolerance to increases in salinity is in fact expected from the components of a clade that has invaded the freshwater relatively recently (Murphy and Austin 2005, Augusto et al. 2007a, b, 2009, Pileggi and Mantelatto 2010, Collins et al. 2011, McNamara and Faria 2012).

When salinity rises beyond the organisms' homeostatic range of osmoregulation, the osmolality of the hemolymph in-creases with respect to values in lower salinities but still remains below the osmolality of the water. And this, in turn, beyond a certain limit, leads to an inability to control tissue hydration and volume. The water content in the muscle of *M. acanthurus* was inversely proportional to salinity (strong negative Pearson correlation), and was maintained within a narrow range of variation, with a decrease of ~4-5% in 30 psu with respect to the control (0 psu), after 4-8 hours of exposure. Conversely, at 30 psu, hemolymph osmolality increased by ~50%. This means that, even when the hemolymph experienced a great increase in osmotic concentration, the hydration of the muscle varied very little, indicating that this tissue has high capacity to regulate water concentrations. A similar result was observed in *M. acan-thurus* at 30 psu (Foster et al. 2010). The great ability of tissue to maitain its water concentration was also documented for *M. acanthurus* and the other palaemonids *M. potiuna* and *P. pandal-iformis*, through an "*in vitro*" experiment in which tissues were exposed to hypo- and hyperosmotic saline solutions that corre-sponded to a 50% change with respect to the isosmotic control. In this study, the hydration of the shrimp tissues varied in only about 10% (Freire et al. 2013). This high capacity to maintain tissue water levels is in part responsibe for the euryhalinity of *M. acanthurus*, which, throughout its life cycle, switches between freshwater and brackish water (Freire et al. 2008a, 2013)

The maintenance of tissue hydration happens through the regulation of the flux of inorganic ions and concentration of aminoacids or other nitrogenous compounds in the tissues or body fluids (e.g., Pierce 1982, Gilles 1987). Under hyperos-motic challenges, osmolyte concentrations increase in tissues, as already shown for the crustaceans *M. amazonicum*, *M. olfersii*, *Dilocarcinus pagei* (Augusto et al. 2007a, b). As a consequence, aminoacid catabolism should decrease, which is compatible with our results. Thus, the decrease in ammonia excretion observed in our study under hypersaline challenges probably means that there is a mechanism to retain aminoacid or nitrogenous compounds, which allows the maintenance of tissue hydration in high salinities. Compatible with this idea, excretion of am-monia in our results was inversely proportional to salinity (as pointed by the Pearson correlation). The role of aminoacids in the maintenance of tissue hydration has been documented in recent freshwater invaders (Augusto et al. 2007a,b). In summary, when there is an increase in salinity, metabolic energy is routed to controlling extracellular and intracellular homeostasis, by shutting down branchial salt uptake and reducing protein/aminoacid catabolism. The palaemonid shrimp studied here, a recent freshwater invader, does not display "avoidance" or "escape" response when faced with severe salt challenges, even considering its strong and effective apparatus to perform salt uptake from freshwater. An avoidance behaviour would result in reduced oxygen consumption from the water. Rather, although certainly reducing salt uptake, this shrimp maintains its hemolymph hyposmotic with respect to the ambient water at 20-30 psu for 4-8 hours, during which it uses more oxygen to control water tissue levels . Additional studies are needed to elucidate the limits, degrees, and mechanisms of hypo-regulation or hypo-conformation in palaemonid shrimps.

ACKNOWLEDGEMENTS

The authors acknowledge the financial support from the Brazilian Federal Agencies CAPES (Masters fellowship to EPG - 40001016008P4), and CNPq (Masters fellowship to LPR - 40001016072P4, PhD fellowship to GCC - 141213/2013-2, and Research Fellowship/Grant to CAF - 306630/2011-7). Authors hold a permit from the Environmental Ministry to collect specimens of *M. acanthurus* from the wild (IBAMA/SISBIO 20030-4).

REFERENCES

Anger K (2003) Salinity as a key parameter in the larval biology of decapod crustaceans. Invertebrate Reproduction and Development 43(1): 29–45. https://doi.org/10.1080/07924259.2003.9652520

Anger K (2013) Neotropical *Macrobrachium* (Caridea: Palaemonidae): on the biology, origin, and radiation of freshwater-invading shrimp. Journal of Crustacean Biology 33: 151–183. https://doi.org/10.1163/1937240X-00002124

Ashelby CW, Page TJ, De Grave S, Hughes JM, Johnson ML (2012) Regional scale speciation reveals multiple invasions of freshwater in Palaemoninae (Decapoda). Zoologica Scripta 41: 293–306. https://doi.org/10.1111/j.1463-6409.2012.00535.x

Augusto A, Greene LJ, Laure HJ, Mcnamara JC (2007a) Adaptive shifts in osmoregulatory strategy and the invasion of freshwater by brachyuran crabs: evidence from *Dilocarcinus pagei* (Trichodactylidae). Journal of Experimental Zoology A 307: 688–698. https://doi.org/10.1002/jez.a.422

Augusto A, Greene LJ, Laure HJ, McNamara JC (2007b) The ontogeny of isosmotic intracellular regulation in the diadromous, freshwater palaemonid shrimps, *Macrobrachium amazonicum* and *M. olfersi* (Decapoda). Journal of Crustacean Biology 27: 626–634. https://doi.org/10.1651/S-2796.1

Augusto A, Pinheiro AS, Greene LJ, Laure HJ, McNamara JC (2009) Evolutionary transition to freshwater by ancestral marine palaemonids: evidence from osmoregulation in a tide pool shrimp. Aquatic Biology 7: 113–122. https://doi.org/10.3354/ab00183

Bond-Buckup G, Buckup L (1989) Os Palaemonidae de águas continentais do Brasil meridional (Crustacea, Decapoda). Revista Brasileira de Biologia 49: 883–896.

Boudour-Boucheker N, Boulo V, Lorin-Nebel C, Elguero C, Grousset E, Anger K, Charmantier G (2013) Adaptation to freshwater in the palaemonid shrimp *Macrobrachium amazonicum*: comparative ontogeny of osmoregulatory organs. Cell and Tissue Research 353: 87–98. https://doi.org/10.1007/s00441-013-1622-x

Booth, CE, McMahon BR and Pinder AW (1982) Oxygen uptake and the potentiating effects of increased hemolymph lactate on oxygen transport during exercise in the blue crab, *Callinectes sapidus*. Journal of Comparative Physiology 148: 111–121. https://doi.org/10.1007/BF00688894

Bridges CR, Brand AR (1980) The effect of hypoxia on oxygen consumption and blood lactate levels of some marine Crustacea. Comparative Biochemistry and Physiology A 65: 399–409. https://doi.org/10.1007/BF00688894

Charmantier G (1998) Ontogeny of osmoregulation in crustaceans: a review. Invertebrate Reproduction and Development 33: 177–190. https://doi.org/10.1080/07924259.1998.9652630

Cheng W, Liu CH, Cheng CH, Chen JC (2003) Osmolality and ion balance in giant river prawn *Macrobrachium rosenbergii* subjected to changes in salinity: role of sex. Aquaculture Research 34: 555–560. https://doi.org/10.1046/j.1365-2109.2003.00853.x

Claiborne JB, Evans DH, Goldstein L (1982) Fish branchial Na/NH_4 exchange is via basolateral Na, K-activated ATPase. Journal of Experimental Biology 96: 431–434.

Collins PA, Giri F, Williner V (2011) Biogeography of the freshwater decapods in the La Plata basin, South America. Journal of Crustacean Biology 31: 179–191. https://doi.org/10.1651/10-3306.1

Denne LB (1968) Some aspects of osmotic and ionic regulation in the prawns *Macrobrachium australiense* (Holthuis) and *M. equidens* (Dana). Comparative Biochemistry Physiology 26: 17–30. https://doi.org/10.1016/0010-406X(68)90309-5

Foster C, Amado EM, Souza MM, Freire CA (2010) Do osmoregulators have lower capacity of muscle water regulation than osmoconformers? A study on decapod crustaceans. Journal of Experimental Zoology A 313: 80–94. https://doi.org/10.1002/jez.575

Freire CA, Cavassin F, Rodrigues EN, Torres AH, McNamara JC (2003) Adaptive patterns of osmotic and ionic regulation, and the invasion of freshwater by the palaemonid shrimps. Comparative Biochemistry Physiology A 136: 771–778. https://doi.org/10.1016/j.cbpb.2003.08.007

Freire CA, Amado EM, Souza LR, Veiga MPT, Vitule JRS, Souza MM, Prodocimo V (2008a) Muscle water control in crustaceans and fishes as a function of habitat, osmoregulatory capacity, and degree of eurihalinity. Comparative Biochemistry Physiology A 149: 435–446. https://doi.org/10.1016/j.cbpa.2008.02.003

Freire CA, Onken H, McNamara JC (2008b) A structure-function analysis of ion transport in crustacean gills and excretory organs. Comparative Biochemistry Physiology A 151: 272–304. https://doi.org/10.1016/j.cbpa.2007.05.008

Freire CA, Souza-Bastos LR, Amado EM, Prodocimo V, Souza MM (2013) Regulation of muscle hydration upon hypo-or hyper-osmotic shocks: differences related to invasion of the freshwater habitat by decapod crustaceans. Journal of Experimental Zoology A 319: 297–309. https://doi.org/10.1002/jez.1793

Furriel RPM, Masui DC, Mcnamara JC, Leone FA (2004) Modulation of gill Na+, K+-ATPase activity by ammonium ions: Putative coupling of nitrogen excretion and ion uptake in the freshwater shrimp *Macrobrachium olfersii*. Journal of Experimental Zoology A 301: 63–74. https://doi.org/10.1002/jez.a.20008

Gilles R (1987) Volume regulation in cells of euryhaline invertebrates. Current Topics in Membrane and Transport 30: 205–247. https://doi.org/10.1016/S0070-2161(08)60372-X

Henry RP, Wheatly MG (1992) Interaction of respiration, ion regulation, and acid-base balance in the everyday life of aquatic crustaceans. American Zoologist 32: 407–416. https://doi.org/10.1093/icb/32.3.407

Maciel JES, Souza F, Valle S, Kucharski LC, da Silva RSM (2008) Lactate metabolism in the muscle of the crab *Chasmagnathus granulatus* during hypoxia and post-hypoxia recovery. Comparative Biochesmistry Phisiology A 151: 61–65. https://doi.org/10.1016/j.cbpa.2008.05.178

Maraschi AC, Freire CA, Prodocimo V (2015) Immunocytochemical localization of V-H⁺-ATPase, Na⁺/K⁺-ATPase, and carbonic anhydrase in gill lamellae of adult freshwater euryhaline shrimp *Macrobrachium acanthurus* (Decapoda, Palaemonidae). Journal of Experimental Zoology A 323: 414–421. https://doi.org/10.1002/jez.1934

Mazzarelli CCM, Santos MR, Amorim RV, Augusto A (2015) Effect of salinity on the metabolism and osmoregulation of selected ontogenetic stages of an amazon population of *Macrobrachium amazonicum* shrimp (Decapoda, Palaemonidae). Brazilian Journal of Biology 75: 372–379. https://doi.org/10.1590/1519-6984.14413

McNamara JC (1987) The time course of osmotic regulation in the freshwater shrimp *Macrobrachium olfersii* (Wiegmann)(Decapoda, Palaemonidae). Journal of Experimental Marine Biology and Ecology 107: 245–251. https://doi.org/10.1016/0022-0981(87)90041-4

McNamara JC, Faria SC (2012) Evolution of osmoregulatory patterns and gill ion transport mechanisms in the decapod Crustacea: a review. Journal of Comparative Physiology B 182: 997–1014. https://doi.org/10.1007/s00360-012-0665-8

Moreira GS, McNamara JC, Shumway SE, Moreira PS (1983) Osmoregulation and respiratory metabolism in brazilian *Macrobrachium* (Decapoda, Palaemonidae). Comparative Biochemistry Physiology A 74: 57–62. https://doi.org/10.1016/0300-9629(83)90711-9

Murphy NP, Austin CM (2005) Phylogenetic relationships of the globally distributed freshwater prawn genus *Macrobrachium* (Crustacea: Decapoda: Palaemonidae): biogeography, taxonomy and the convergent evolution of abbreviated larval development. Zoologica Scripta 34: 187–197. https://doi.org/10.1111/j.1463-6409.2005.00185.x

Ordiano A, Alvarez F, Alcaraz G (2005) Osmoregulation and oxygen consumption of the hololimnetic prawn, *Macrobrachium tuxtlaense* at varying salinities (Decapoda, Palaemonidae). Crustaceana 78: 1013–1022. https://doi.org/10.1163/156854005775197316

Péqueux A (1995). Osmotic regulation in crustaceans. Journal of Crustacean Biology 15: 1–60. https://doi.org/10.2307/1549010

Pileggi LG, Mantelatto FL (2010) Molecular phylogeny of the freshwater prawn genus *Macrobrachium* (Decapoda, Palaemonidae), with emphasis on the relationships among selected American species. Invertebrate Systematics 24: 194–208. https://doi.org/10.1071/IS09043

Pierce SK (1982) Invertebrate cell volume control mechanisms: a coordinated use of intracellular amino acids and inorganic ions as osmotic solute. Biological Bulletin 163: 405–419. https://doi.org/10.2307/1541452

Rosas C, Ocampo L, Gaxiola G, Sánchez A, Soto LA (1999) Effect of salinity on survival, growth, and oxygen consumption of postlarvae (PL10-PL21) of *Litopenaeus setiferus*. Journal of Crustacean Biology: 244–251. https://doi.org/10.2307/1549230

Salvato B, Cuomo V, Di Muro P, Beltramini M (2001) Effects of environmental parameters on the oxygen consumption of four marine invertebrates: a comparative factorial study. Marine Biology 138: 659–668. https://doi.org/10.1007/s002270000501

Setiarto A, Augusto SC, Takashima F, Watanabe S, Yokota M (2004) Short-term responses of adult kuruma shrimp *Marsupenaeus japonicus* (Bate) to environmental salinity: osmotic regulation, oxygen consumption and ammonia excretion. Aquaculture Research 35: 669–677. https://doi.org/10.1111/j.1365-2109.2004.01064.x

Signoret GP, Brailovsky DS (2004) Adaptive osmotic responses of *Macrobrachium acanthurus* (Wiegmann) and *Macrobrachium carcinus* (Linnaeus)(Decapoda, Palaemonidae) from the southern Gulf of Mexico. Crustaceana 77: 455–465. https://doi.org/10.1163/1568540041643364

Taylor AC, Spicer JI (1989) Interspecific comparison of the respiratory response to declining oxygen tension and the oxygen transporting properties of the blood of some palaemonid prawns (Crustacea: Palaemonidae). Marine and Freshwater Behavior and Physiology 14: 81–91. https://doi.org/10.1080/10236248909378695

Valenti WC, Mello J de TC de, Lobão VL (1989) Fecundidade de *Macrobrachium acanthurus* (Wiegmann, 1836) do Rio de Iguape (Crustacea, Decapoda, Palaemonnidae). Revista Brasileira de Zoologia 6: 9–15. https://doi.org/10.1590/S0101-81751989000100002

Wall SM (1995) Ammonium transport and the role of the Na, K-ATPase. Electrolyte Metabolism 22: 311–317.

Weirauch D, Wilkie MP, Walsh PJ (2009) Ammonia and urea transporters in gills of fish and aquatic crustaceans. Journal of Experimental Biology 212: 1716–1730. https://doi.org/10.1242/jeb.024851

Author Contributions: CAF, EPG, LPR and GCC designed the experiments, EPG, LPR and GCC conducted the experiments and assays, analysed the data, prepared the figures, and wrote a first preliminary draft of the manuscript. CAF had the original idea, proposed the explanation for the data, and rewrote the text entirely. **Competing Interests:** The authors have declared that no competing interests exist.

5

Astyanax taurorum a new species from dos Touros River, Pelotas River drainage, an upland Southern Brazilian river (Characiformes: Characidae)

Carlos Alberto S. de Lucena[1], Amanda Bungi Zaluski[1], Zilda Margarete Seixas de Lucena[1]

[1]Museu de Ciências e Tecnologia, Pontifícia Universidade Católica do Rio Grande do Sul. Avenida Ipiranga 6681, Caixa Postal 1491, 90619-900 Porto Alegre, RS, Brazil.
Corresponding author: Carlos Alberto S. de Lucena (lucena@pucrs.br)

http://zoobank.org/2C4F6889-11BA-4C8A-9E1C-4C6CF36D50C6

ABSTRACT. A new species of Astyanax belonging to the Astyanax scabripinnis complex is described from dos Touros River, tributary of the Pelotas River, Uruguay River basin. Astyanax taurorum **sp. nov.** is distinguished from other species of the Astyanax scabripinnis species complex by having two humeral spots, the first vertically elongated; teeth of inner row of premaxilla with three to five cusps; 2–3 (modes 2 or 3) maxillary teeth; 20–23 (mode 22) branched anal-fin rays; 13–15 (mode 14) gill rakers on lower branch of the first branchial arch; 20–23 (mode 21) total gill rakers in first branchial arch; 33–36 (mode 35) perforated lateral line scales. Astyanax taurorum **sp. nov.** is similar to Astyanax paris; nevertheless, it can be readily distinguished from it by having a smaller head depth (73.6-83.1% vs. 86.4–95.6%) and smaller interorbital width (24.1–28.0% vs. 30.8–32.8%). In addition, it differs from A. paris by the presence a posttemporal hook-shaped posterodorsal margin.

KEY WORDS. Taxonomy, Rio Grande do Sul, Uruguay River, distribution.

INTRODUCTION

The fishes of the genus Astyanax Baird & Girard, 1854 inhabit Neotropical drainages from the Colorado River in Texas and New Mexico in the United States to Northern Patagonia, Argentina (Menni 2004, López et al. 2008, Ornelas-Garcia et al. 2008). As suggested by phylogenetic analyses based on morphological (Mirande 2010) and molecular evidence (Javonillo et al. 2010, Oliveira et al. 2011), Astyanax is not monophyletic. The genus comprises 150 valid species (Eschmeyer et al. 2016), and it is still defined as in Eigenmann (1921, 1927) (for the characters, see Marinho and Ohara 2013). The Astyanax scabripinnis complex is a non-monophyletic group with 29 species (Ingenito and Duboc 2014: tab. 1). It is characterized, according to Bertaco and Lucena (2006), by possessing the deepest and most robust body area close to the middle length of the pectoral fins, a robust head, snout short and abrupt, body depth smaller than 41% of SL, reduced number of branched anal-fin rays (13-23, rarely 22 or 23 rays), presence of one or two humeral spots, and a dark mid-lateral body stripe extending to the tip of the middle caudal-fin rays.

Currently, there are 19 recognized species of Astyanax from the Uruguay River, Laguna dos Patos system to the Tramandaí River drainage: A. aramburui Protogino, Miquelarena & López, 2006; A. bagual Bertaco & Vigo, 2015; A. brachypterygium Bertaco & Malabarba, 2001; A. cremnobates Bertaco & Malabarba, 2001; A. douradilho Bertaco, 2014; A. dissensus Lucena & Thofehrn, 2013; A. eigenmanniorum (Cope, 1894); A. henseli Melo & Buckup, 2006; A. lacustris (Luetken, 1875); A. laticeps (Cope, 1894); A. obscurus (Hensel, 1870); A. ojiara Azpelicueta & Garcia, 2000; A. paris Azpelicueta, Almirón & Casciotta, 2002; A. pirabitira Lucena & Bertcaco, 2013; A. procerus Lucena, Castro & Bertaco, 2013; A. saguazu Casciotta, Almirón & Azpelicueta, 2003; A. stenohalinus Messner, 1962), Astyanax sp. aff. fasciatus, sensu Melo and Buckup (2006); and A. xiru, Lucena, Castro & Bertaco, 2013. While studying the genus Astyanax from the Pelotas River drainage, a new species of the Astyanax scabripinnis complex was found and it is described herein.

MATERIAL AND METHODS

The examined material belongs to the following institutions: Museu de Ciências e Tecnologia, Pontifícia Universidade Católica do Rio Grande do Sul, Porto Alegre (MCP); Museu de Zoologia, Universidade de São Paulo, São Paulo (MZUSP); Muséum d'histoire naturelle, Genève (MHNG); Universidade Federal do Rio Grande do Sul, Porto Alegre (UFRGS).

Counts and measurements follow Fink and Weitzman (1974) and Bertaco and Lucena (2006) with the addition of the head depth, measured at the vertical through posterior margin of the orbit. Measurements were preferentially taken on the left side of specimens using callipers (0.1 mm approximation). Counts of vertebrae, supraneurals, teeth of dentary, unbranched dorsal, and anal-fin rays taken from cleared and stained (c&s) specimens prepared according to the protocol of Taylor and van Dyke (1985). Vertebral counts included the four vertebrae of the Weberian apparatus, and the terminal centrum counted as a single element. In the description, the frequency of each count is given in parentheses after the respective value. In the material examined, the total number of specimens in the lot follows each catalogue number, and in parentheses is the number of specimens measured and counted with their respective standard length range. HL stands for head length throughout.

The Laguna dos Patos system and the Tramandaí River drainage, follow definitions of Malabarba (1989) and Malabarba and Isaia (1992), respectively. Data for *A. paris* are from Azpelicueta et al. (2002), except when said otherwise.

TAXONOMY

Astyanax taurorum sp. nov.

http://zoobank.org/F15C02B5-AF1B-4052-A67A-741560E1468F
Fig. 1, Table 1

Types series. Brazil, Rio Grande do Sul, Bom Jesus. Holotype: Tributary of dos Touros River ca. 4 km northeastern of the road BR-285, Pelotas River drainage, 1,056 m a.s.l., 28°41′06″S 50°12′51″W, 12 Feb 2016, J. Pezzi da Silva and E. Pereira leg., MCP 49468, 80.7 mm SL. Paratypes: Tributary of dos Touros River, on the road Silveira-Rondinha, ca. 28°39′08″S 50°18′25″W, 14 Jan 1989, C. Lucena, P. Azevedo and E. Pereira leg., MCP 14370, 20 (17, 22.2-82.6 mm SL, 3 c&s, 29.6–62.6 mm SL). Same locality of MCP 14370, MZUSP 120697, 1, 29.9 mm SL. Dos Touros River, dowstream dam, Pelotas River drainage, road Rondinha – Silveira, 998 m a.s.l., 28°38′44″S, 50°17′06″W, 12 Feb 2016, J.P. Silva and E. Pereira leg., MCP 49467, 1, 75.7 mm SL.

Diagnosis. *Astyanax taurorum* sp. nov. belongs to the *A. scabripinnis* species complex and is distinguished from the species of that complex by having two humeral spots (vs. one in *A. courensis* Bertaco, Carvalho & Jerep, 2010, *A intermedius* Eigenmann, 1908, *A. jenynsii* (Steindachner, 1877), *A. jordanensis* Vera Alcaraz, Pavanelli & Bertaco, 2009, *A. laticeps* (Cope, 1894), *A. microschemos* Bertaco & Lucena, 2006, *A. serratus* Garavello & Sampaio, 2010, *A. totae* Haluch & Abilhoa, 2005, *A. rivularis* and *A. varzeae* Abilhoa & Duboc, 2007); 20–23 total gill rakers in first branchial arc (vs. 16 in *A. jacobinae* Zanata and Carmelier, 2008, 16–17 in *A. gymnogenys* Eigenmann, 1911, 18 in *A. burgerai* Zanata & Carmelier, 2009, 17–18 in *A. troya* Azpelicueta, Casciotta & Almirón, 2002, 18–19 in *A. epiagos* Zanata & Carmelier, 2008,

Table 1. Morphometric data of *Astyanax taurorum* sp. nov. The range includes the holotype; n = number of specimens; SD = standard deviation.

	Holotype	n	Range	Mean	SD
Standard length (mm)	80.7	14	54.7–82.6	69.6	–
Percents of standard length					
Depth at dorsal-fin origin	34.9	14	33.8–37.6	35.9	0.99
Predorsal length	52.8	14	51.6–54.9	53.5	0.99
Prepectoral length	29.0	14	27.8–30.0	29.1	0.61
Preanal length	66.4	14	65.0–68.7	66.4	1.04
Prepelvic length	50.5	14	49.1–52.2	51.0	0.88
Dorsal-fin length	25.9	14	22.3–26.6	24.6	1.24
Pectoral-fin length	22.7	14	18.6–22.7	20.6	1.22
Pelvic-fin length	15.6	14	14.4–16.8	15.2	0.62
Anal-fin base length	26.9	14	23.6–28.0	26.3	1.08
Anal-fin lobe length	17.0	13	13.2–18.3	16.3	1.56
Caudal peduncle length	14.0	14	12.7–15.7	14.2	0.88
Caudal peduncle depth	11.4	14	11.0–12.7	11.5	0.42
Head length	29.1	14	29.0–30.5	29.9	0.75
Percents of head length					
Head depth	83.1	14	75.1–86.7	81.6	2.98
Snout length	25.1	14	23.8–28.8	27.1	1.58
Interorbital width	26.7	14	24.7–28.0	26.4	0.99
Horizontal orbit diameter	30.3	14	29.0–32.4	30.7	1.22
Upper jaw length	38.4	14	36.7–41.5	39.2	1.60

17–19 in *A. ojiara* Azpelicueta & Garcia, 2000, *A. cremnobates* Bertaco & Malabarba, 2001 and *A. leonidas* Azpelicueta, Casciotta & Almirón, 2002); 13–15 gill rakers on the lower branch of the first branchial arch (vs. 9–10 in *A. turmalinensis* Triques, Voino & Caiafa, 2003); 33–36 perforated lateral line scales (vs. 39–41 in *A. gymnogenys* and *A. eremus* Ingenito & Duboc, 2014, 40–43 in *A. guaricana* Oliveira, Abilhoa & Pavanelli, 2013); 20–23, usually 21 or 22, branched anal-fin rays (vs. 18–19 in *A. burgerai*, 13–16 in *A. goyanencis* Miranda Ribeiro, 1944, 16–20, usually 16 or 17 in *A. serratus* Garavello & Sampaio, 2010, 14–18 in *A. microschemos*, 15–18 in *A. totae*, 12–16 in *A. brachypterychium*, 14–18 in *A. cremnobates*, and 13–17 in *A. epiagos* and *A. jordanensis*); inner row of premaxilla with teeth bearing three to five cusps (vs. heptacuspid in *A. ita* Almirón, Azpelicueta & Casciotta, 2002 and *A. pirabitira* Lucena, Bertaco & Berbigier, 2013); 2–3 maxillary teeth (vs. 1 in *A. obscurus* Hensel, 1870, *A. ojiara*, *A. troya* Azpelicueta, Casciotta & Almirón, 2002, *A. guaricana*, *A. courensis* and *A. ita* Almirón, Azpelicueta & Casciotta, 2002; 0–1 in *A. pirapuan* Tagliacollo, Britzke, Silva & Benine, 2011); length of anal-fin base 23.6–28.0% (mean = 26.3%) of SL (vs. 19.8–24.3% (mean = 21.5%) of SL in *A. eremus* and 30.2% in *A. scabripinnis* Jenyns, 1842 in the holotype); body depth 33.8–37.6% of SL (vs. 26.9–29.7% in *A. microschemos*, and 27.3–31.3% in *A. eremus*); head length 29.0–30.5% of SL (vs. 22.9–25.1% in *A. gymnogenys*, 21.9–27.1% in *A. courensis*, 26.6–28.2% in *A. turmalinensis*, and 23.9–26.6% of SL in *A. guaricana*); eye diameter 29.0–32.4% of

Figure 1. *Astyanax taurorum* sp. nov., MCP 49468, 80,7 mm SL, holotype, tributary of dos Touros River, Rio Grande do Sul, Brazil.

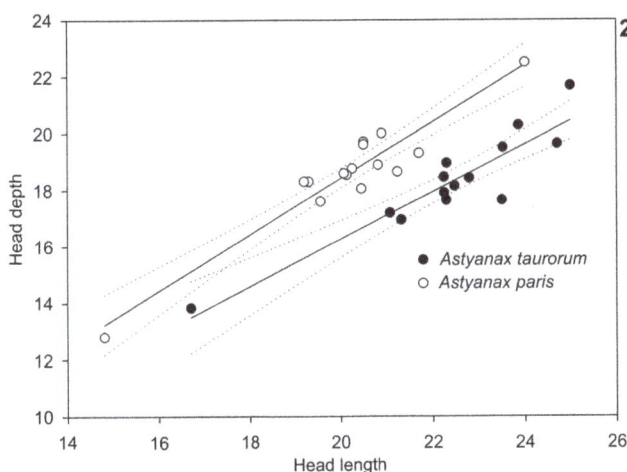

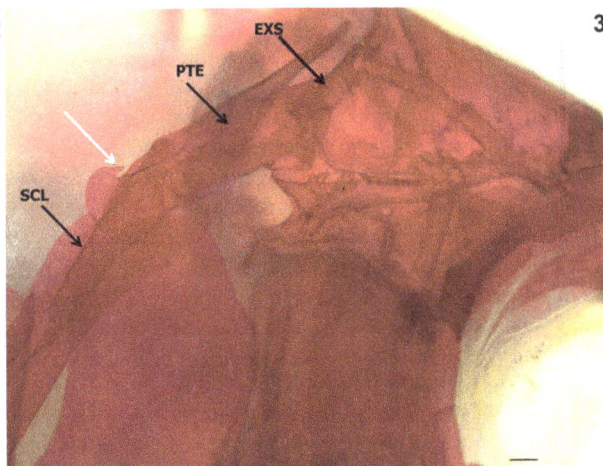

Figures 2–3. (2) Head depth as function of head length for *Astyanax paris* (y = -1.517 +(0.996X), R = 0.952, and *Astyanax taurorum* sp. nov. (y = -5·158+(1.037X), R = 0.932. Dotted lines, confidence interval of 95%; (3) *Astyanax taurorum* sp. nov. Posterodorsal region of head (lateral view, right side); EXS = Extrascapular, PTE = posttemporal, SCL = supracleithrum. White arrow indicates the hook on posterodorsal region of posttemporal bone (see text). Scale bar = 0.2 mm.

HL (vs. 24.4–26.1% in *A. gymnogenys*, and 36.8–40.3% in *A. jacobinae*); snout length 23.8–28.8% of HL (vs. 16.0–20.4% in *A. paranae*); and interorbital length 24.7–28.0% of HL (vs. 35.2–37.8% in *A. gymnogenys*, 29.6–37.3% in *A. jacobinae*, 30.4–34.5% in *A. goyanensis*, 37.5–47.1% in *A. intermedius*, 29.8–37.7% in *A. varzea*, 32.7–40.9% in *A. guaricana*, 30.6–35.7% in *A. jordanensis*, 40.7% in *A. scabripinnis* holotype), and 31.7–39.2% in *A. pirapuan*). Within the *Astyanax scabripinnis* complex, *Astyanax taurorum* sp. nov. is most similar to *A. paris* Azpelicueta, Almirón & Casciotta, 2002 – species known from the type locality, Arroio Fortaleza, tributary of upper Uruguay River, Argentina – with which most counts and morphometric percentages overlap. Nevertheless, *Astyanax taurorum* sp. nov. differs from *A. paris* by the presence

of hooks on branched anal-fin rays (vs. secondary sexual dimorphism absent in *A. paris*), interorbital width 24.7–28.0% HL (vs. 28.4–32.8% HL) and head depth 73.6–86.7% HL (vs. 86.4–96.6% HL) (Fig. 2) (Table 2), and by having posterodorsal margin of the posttemporal hook-shaped (Fig. 3) (see Discussion). *Astyanax taurorum* sp. nov. is distinguished from the other species in the genus by the following combination of characters: presence of two conspicuous humeral spots, the first one vertically elongated with the upper portion enlarged, but narrowing ventrally; dark midlateral horizontal stripe; conspicuous caudal spot extending posteriorly to the middle of caudal-fin rays; 20–23 branched anal-fin rays; 20–23 total gill rakers in first branchial arc; 33–36 perforated lateral line scales; 5–7 scale rows between lateral line

Table 2. Morphometric data of *Astyanax paris*.* = Values from Azpelicueta et al. (2002). **The range includes the holotype. Museo de La Plata (MLP), Muséum d'histoire naturelle (MNHG), Museu de Ciências e Tecnologia PUCRS (MCP). n = number of specimens, m = mean.

Measurements	Holotype MLP 9584	Paratypes						Topotypes MCP 34461	
		MLP 9586			MNHG 2623.65				
		n	Range**	m	n	Range**	m	n	Range	m
Standard length (mm)	75.6*	7	51.3–86.1*	72.4	3	70.3–73.1	71.6	4	66.9–73.5	70.4
Percentages of head lenght										
Head depth	95.6	7	86.4–96.6	93.1	3	90.7–92.6	91.8	4	88.2–92.6	89.6
Interorbital width	32.8*	15	28.4–32.8*	30.8*				4	30.8–32.1	31.2

and pelvic-fin origin; outer row of premaxilla with tricuspid teeth; teeth in inner row of premaxilla with three to five cusps; 2–3 tricuspid teeth in the maxilla, head length 29.0–30.5% of SL; body depth 33.8–37.6% of SL; interorbital width 24.7–28.0% of HL; eye diameter 29.0–32.4% of HL, and length of anal-fin base 23.6–28.0% of SL.

Description. Morphometric data summarized in Table 1. Body compressed and moderately elongate, greatest body depth at vertical through near middle length of pectoral fin. Dorsal profile of head convex from tip of snout to vertical through nostrils, straight from that point to vertical through posterior border of orbital, slanted until tip of supraoccipital spine. Snout relatively slender. Dorsal profile of body convex from tip of su-praoccipital bone to dorsal-fin origin; straight from that point to end of caudal peduncle. Ventral body profile convex from mandibular symphysis to pelvic-fin origin, nearly straight from that point to anal-fin origin, and slanted along anal-fin base. Dorsal and ventral profiles of caudal peduncle nearly straight.

Mouth terminal or slightly subterminal, slit below horizontal passing through middle of eye. Posterior tip of maxilla extending between vertical through anterior margin of orbit and the vertical through middle of orbit. Two tooth rows in premaxilla; outer row with 3*(2), 4(13), or 5(5) tricuspid teeth; inner row with five teeth, usually bearing four cusps on first tooth, five cusps on second to fourth tooth, three cusps on fifth tooth. Maxilla with 2(8) or 3(8) tricuspid teeth. Dentary with four large pentacuspid teeth, followed by seven small tricuspid teeth and one conical tooth (two c&s). Median cusp in all cuspidate teeth longer than remaining cusps; cusp tips slightly curved inwardly on dentary (Fig. 4).

Dorsal-fin rays ii, 9 (23); first unbranched ray short, one-half length of second ray. Distal margin of dorsal fin slightly convex. Dorsal-fin origin slightly behind middle of SL. Adipose-fin origin at vertical through base of fifth or sixth last anal-fin rays. Anal-fin rays iii-iv 20(2), 21(6), 22(9), or 23(4). Anal-fin origin posterior to vertical through base of last dorsal-fin ray. Pectoral-fin rays i, 11(1), 12(7), 13(11), or 14(1). Tip of pectoral-fin tip ending one scale before or, occasionally, reaching pelvic-fin insertion. Pelvic-fin rays i, 7(23), tip of fin not reaching anal-fin origin. Axillary scale present.

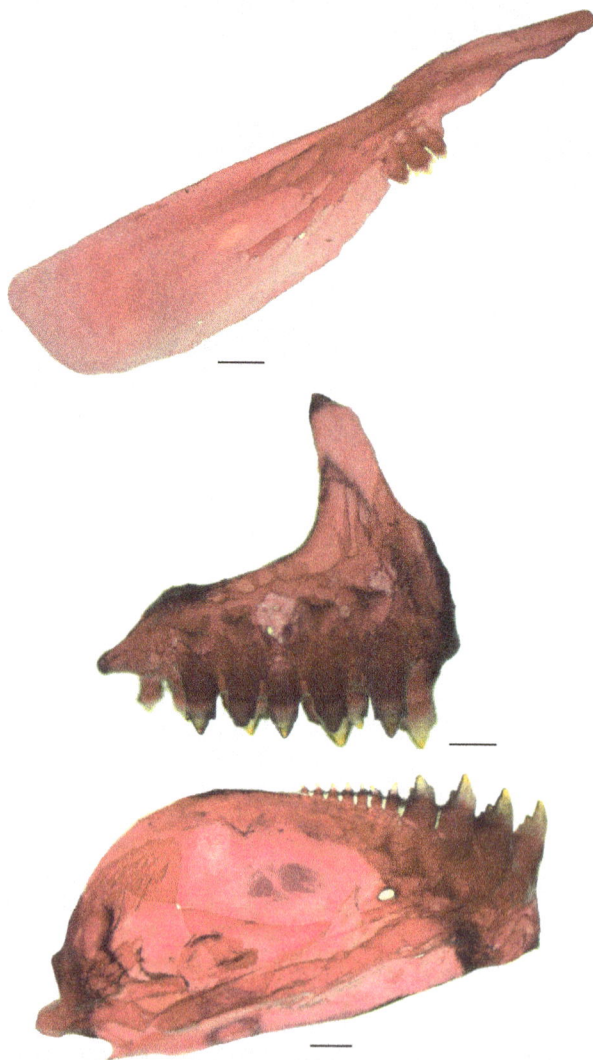

Figure 4. Maxilla, premaxilla and dentary of *Astyanax taurorum* sp. nov., MCP 14370, paratype, 62.6 mm SL, lateral view of right side. Scale bar = 0.5 mm.

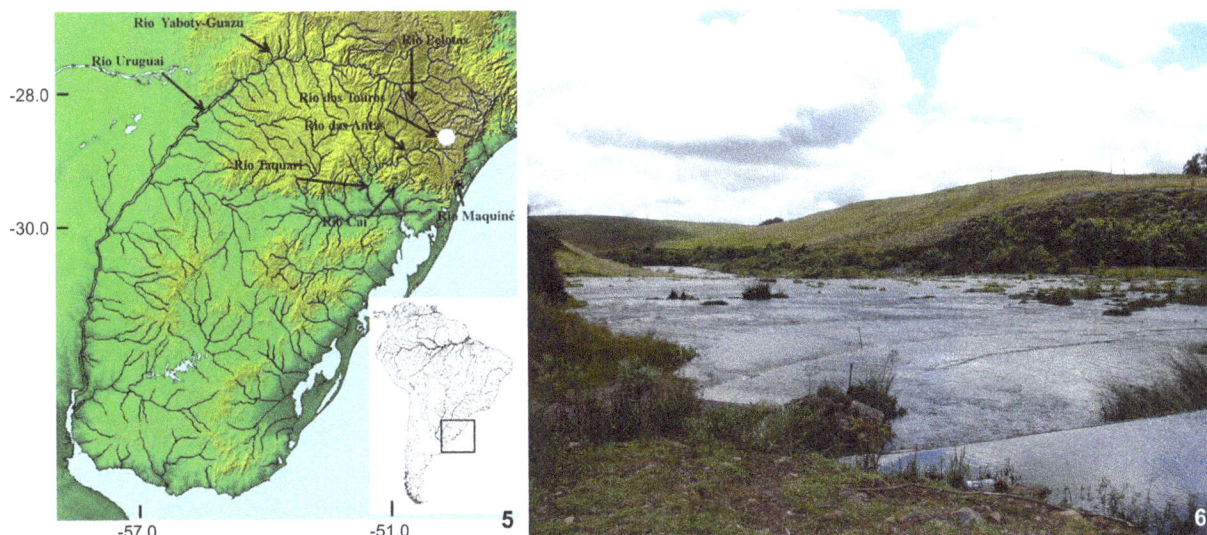

Figures 5–6. (5) Distribution of *Astyanax taurorum* sp. nov., white circle = type-locality. The symbol represents more than one locality; (6) stream tributary of dos Touros River, type locality of *Astyanax taurorum* sp. nov.

Caudal-fin forked, lobes similar in size.

Lateral line complete with 33(2), 34(2), 35(8), or 36(5) perforated scales. Scale rows between dorsal-fin origin and lateral line 6(5) or 7(12); scale rows between lateral line and pelvic-fin origin 5(5), 6(13), or 7(1); scale rows between lateral line and anal-fin origin 5(1), 6(15), or 7(1); scale rows around caudal peduncle 14 (8), 15(5), or 16(2).

Precaudal vertebrae 13 (3); caudal vertebrae 18(2) or 19(1); total vertebrae 31(2) or 32(2). Supraneurals 5(3). Gill rakers on upper branch 6(1), 7(11) or 8(11) and on lower branch 13(7), 14(14), or 15(2) in first branchial arch; total gill rakers in first branchial arch 20(3), 21(13), 22(6), or 23(1).

Color in alcohol. Dorsal and dorsolateral portions of head and body dark brown. Scales on lateral of body with dark brown chromatophores sometimes concentrated on anterior border. Two conspicuous humeral spots. Anterior humeral spot vertically elongate with upper portion wider, located on second to third or fourth scale vertical series, extending three horizontal scale series above lateral line; lower portion narrow, extending on the lateral line and one or two horizontal scale series below it. Posterior humeral spot large, absent in small specimens (22.8–29.4 mm SL), reaching but not surpassing lateral line ventrally, extending on two or three horizontal scale series and three vertical scale series. Humeral spots separated by a clear area occupying two or three vertical scale series. Dark midlateral strip inconspicuous anteriorly, but conspicuous posteriorly from about vertical through middle of dorsal-fin base to caudal peduncle; absent in small specimens (22.8–29.4 mm SL). Caudal peduncle spot triangular, extending over median caudal-fin rays. Scattered dark chromatophores on dorsal, anal, and caudal fins. Pectoral and pelvic-fins hyaline or covered by sparse dark chromatophores.

Color in life. Overall body olive green, silvery below lateral line. Humeral spots and caudal peduncle spot conspicuous. Dorsal, anal, pelvic and caudal fins reddish. Pectoral-fin yellowish. Dark brown blotches located on anterior portion of scales.

Sexual dimorphism. Hooks on anal-fin rays of four specimens with 75.4 to 80.6 mm SL (MCP 14370). Hooks short, conical or slightly retrorse found on the first or third to eighth branched rays, along the posterolateral margin of the posterior branch. One pair on each segment. One specimen with a single hook on the third branched anal-fin ray and another specimen with a very small hook on the first branched ray and small protuberances on other rays.

Distribution and habitat. *Astyanax taurorum* sp. nov. is known from the dos Touros River drainage, tributary of Pelotas River, which in turn is a tributary of Uruguay River (Fig. 5). The Pelotas River drainage is located in the region named "Campos de Altitude do Planalto das Araucárias (= Araucaria Plateau in Bertaco et al. 2016)" or "Campos de Cima da Serra", which has a high level of endemism of fishes (Malabarba et al. 2009, Bertaco et al. 2016: 430) and other groups of animals (for example: sponges, Ribeiro et. al. 2009; crustaceans, Bond-Buckup et al. 2009). The dos Touros River tributary, type locality of *Astyanax taurorum* sp. nov., has a low to medium flow, transparent waters with stones and rocks on the bottom and moderate emergent marginal vegetation (Fig. 6). Four characid species were caught along with *Astyanax taurorum* sp. nov.: *Bryconamericus patriciae* Silva, 2004, *B. iheringi* Boulenger, 1887, *Cheirodon interruptus* Jenyns, 1842, and *Oligosarcus brevioris* Menezes, 1987.

Etymology. The specific name *taurorum*, is derived from the Latin masculine noun *taurus* (second declension, meaning bull) inflected in the plural and genitive case. Therefore *taurorum* means "of the bulls" in reference to "rio dos Touros" (= Portuguese, which means "river of the bulls") the type locality.

Conservation status. *Astyanax taurorum* sp. nov. is likely rare and occurs in low densities. All type specimens were collected in the dos Touros River drainage, during two field trips in 1989 and 2016. Over the last four decades (from 1980 to 2016), six field trips to the dos Touros River system have been conducted by the MCP team, two of which with the sole purpose of collecting specimens of *A. taurorum*. Unfortunately, no specimens were collected in 2015, and only two were found in 2016. The Museu de Ciências Naturais (FZB, Porto Alegre) and Universidade Federal do Rio Grande do Sul also conducted field surveys in that region, but no specimens of *A. taurorum* sp. nov. were obtained. Despite the reduced number of specimens collected and the apparently restricted geographical distribution of the new species, we did not assign *Astyanax taurorum* sp. nov. to any threat category because we the lack biology data for it. Instead, we considered *A. taurorum* sp. nov. as data deficient (DD) (IUCN 2014).

Additional material. Types: Argentina: *Astyanax paris*, MNHG 2623.065 paratypes, 3 (70.3–73.1 mm SL), arroio Fortaleza. MCP 34461 topotypes, 5 (4, 66.9–73.5 mm SL, 1 c&s 68.2 mm SL). *Astyanax troya*, MCP 28438 paratypes. All c&s specimens: Types: Brazil: *Astyanax cremnobates*, paratypes MCP 11650 (2); *Astyanax dissensus*, paratypes MCP 17361 and MCP 47518 (1); *Astyanax douradilho*, paratypes MCP 25700 (3);*Astyanax elachylepis*, paratype MCP 16054 (1); *Astyanax eremus*, paratypes MCP 46942 (1); *Astyanax jordanensis*, paratype MCP 41915(1); *Astyanax microschemos*, paratype MCP 34366 (1); MCP 19783 (5); *Astyanax pelecus*, paratype MCP 17919 (1); *Astyanax pirabitira*, paratypes MCP 14390 (6); *Astyanax utiariti*, paratypes MCP 40041 (3); *Astyanax procerus*, paratype MCP 25513 (1); *Astyanax xiru*, paratype MCP 21730 (1). Non-types: *Astyanax lacustris*, MCP 20339 (1); *Astyanax eigenmanniorum* MCP 25122 (1); *Astyanax henseli* MCP 48121 (3); *Astyanax* aff *fasciatus* MCP 21627 (1); *Astyanax laticeps*, MCP 25690 (1), MCP 17614 (1), MCP 27619 (2); *Astyanax obscurus* MCP 26125 (2)); *Astyanax saguazu* MCP 16808 (1), and MCP 4000 (3); *Astyanax* sp. UFRGS 14052 (6), UFRGS 14051 (6) and UFRGS 14055 (6); *Hyphessobrycon anisitsi* MCP 21633 (2); *Markiana nigripinnis* MCP 17086 (1).

DISCUSSION

The *Astyanax scabripinnis* species complex is not a monophyletic group; this clustering, however artificial, facilitates discussions and comparisons on the diversity of the genus according to Bertaco and Lucena (2006). These authors presented a series of morphological characters that delimit this complex, which are mostly found in *A. taurorum* sp. nov. except for the abrupt snout (slender in *A. taurorum* sp. nov.).

Mirande (2010) presented the most encompassing, morphology-based phylogeny of Characidae. *Astyanax taurorum* sp. nov. has all synapomorphies that define node 201 in that analysis (sister group of Tetragonopterinae clade): dorsal expansion

in the rhinosphenoid absent (character 48, state 0), and tubule with anterior branch parallel to anterior margin of maxilla, reaching a third of its length (character 98, state 1). Node 201 has two branches, the *Hyphessobrycon luetkenii* clade and node 200. The new species shares all synapomorphies of node 200: fourth infraorbital approximately square, or more developed longitudinally than dorsoventrally (67, state 0); coronomeckelian situated dorsal to Meckelian cartilage (character 110, state 1); 24 or less branched anal-fin rays (288, state 0). Node 200 has two branches named *Astyanax paris* and node 199 clades. For now, we note that *A. taurorum* shares, with the former clade, "the abrupt decrease in size of dentary teeth (character 148, state 1)" and not the synapomorphy of clade 199 "ventral margin of horizontal process of anguloarticular perpendicular to laterosensory canal of dentary from medial view". We have examined some species of the clade at node 199, mostly included in the *Astyanax* clade (node 267) (e.g., *Astyanax lacustris* Lutken, 1875 and *A. eigenmanniorum* Cope, 1894, *Markiana nigripinnis* Perugia, 1891, and *Hyphessobrycon anisitsi* Eigenmann, 1907), and other representatives of *Astyanax* not analyzed by Mirande (see Additional material). None of them have a posterodorsal margin with posttemporal hook-shape, as it is the case with *A. taurorum*, though some species have an enlarged posterodorsal margin, which is not hook-shaped (see Mirande 2010: fig. 48).

Astyanax is represented in the Laguna dos Patos, Tramandaí River, and Uruguay River drainages, by 20 species, eight of which have restricted distribution within these drainages (Fig. 5): *A. bagual* Bertaco & Vigo, 2015 (middle Taquari-Antas River), *A. brachypterigium* (upper Pelotas River and upper das Antas River), *A. cremnobates* (upper das Antas River, upper Caí River and upper Maquiné River), *A. douradilho* (middle and upper portions of Maquiné River), *A. obscurus* (upper das Antas and upper Caí Rivers), *A. ojiara* (upper Yaboty-Guazu River), *A. paris* (upper Yaboty-Guazu River), and *A. taurorum* sp. nov. (upper dos Touros River). With the exception of *A. bagual* and *A. douradilho*, these species are included in the *A. scabripinnis* complex (Ingenito and Duboc 2014) and are mainly found in headwater streams. *Astyanax laticeps*, in contrast, is the only species of the *Astyanax scabripinnis* complex that is widely distributed, occurring in the three main drainages of Rio Grande do Sul, besides the southern and southeastern coastal rivers of Brazil (Lucena and Bertaco 2010). Recent descriptions of new species in different genera (e.g., characiforms *Bryconamericus patriciae*, *Hollandichthys taramandahy* Bertaco & Malabarba, 2013, siluriforms *Trichomycterus tropeiro* Ferrer & Malabarba, 2011, or perciforms *Australoheros taura* Ottoni & Cheffe, 2009 and *Crenicichla lucenai* Mattos, Schindler, Ottoni & Cheffe, 2014), with restricted distributions on the headwaters mentioned above, as well as the occurrence of 43 undescribed fish species in these drainages (Bertaco et al. 2016: 428), demonstrate the importance of conservation of this type of environment, as remarked by other authors (e.g., Ferrer and Malabarba 2011, Ferrer et al. 2015).

ACKNOWLEDGEMENTS

We are grateful to the following colleagues: Vinicius Bertaco (Museu de Ciências Naturais-FZB) for offering suggestions to an earlier version of this manuscript; Carlos Oliveira (Universidade Estadual de Maringá) for providing information about the genus *Astyanax*; Alessio Datovo (MZUSP), Luiz R. Malabarba (UFRGS), and Sonia Fisch-Muller (MHNG), who provided material for this study; and José Pezzi da Silva, Rafael Angrazani and Edson Pereira for their efforts in the field. Paulo Lucinda, Universidade Federal do Tocantins, for his help with the specific name. We thank CNPq/PUCRS for grants received to ABZ.

REFERENCES

Azpelicueta M, Almiron A, Casciotta JR (2002) *Astyanax paris*: A new species from the Río Uruguay Basin of Argentina (Characiformes, Characidae). Copeia 2002: 1052–1056. https://doi.org/10.1643/0045-8511

Bertaco VA, Lucena CAS (2006) Two new species of *Astyanax* (Ostariophysi: Characiformes: Characidae) from eastern Brazil with a synopsis of the *Astyanax scabripinnis* species complex. Neotropical Ichthyology 4: 53–60. https://doi.org/10.1590/S1679-62252006000100004

Bertaco VA, Lucena CAS (2010) Redescription of *Astyanax obscurus* (Hensel, 1870) and *A. laticeps* (Cope, 1894) (Teleostei: Characidae): two valid freshwater species originally described from rivers of Southern Brazil. Neotropical Ichthyology 8: 7–20. https://doi.org/10.1590/S1679-62252010000100002

Bertaco VA, Ferrer J, Carvalho FR, Malabarba LR (2016) Inventory of the freshwater fishes from a densely collected area in South America – a case study of the current knowledge of Neotropical fish diversity. Zootaxa 4138: 401–440. https://doi.org/10.11646/zootaxa.4138.3.1

Bond-Buckup G, Buckup L, Araujo PB, Zimmer A, Quadros A, Sokolowicz C, Castiglioni D, Barcelos D, Gonçalves R (2009) Crustáceos. In: Boldrin I (Org.) Biodiversidade dos campos do Planalto das Araucárias. Ministério do Meio Ambiente, Secretaria de Biodiversidade e Florestas, Brasília, 110–129.

Eigenmann CH (1921) The American Characidae. Part 3. Memoirs of the Museum of Comparative Zoology 43: 209–310.

Eigenmann CH (1927) The American Characidae. Part 4. Memoirs of the Museum of Comparative Zoology 43: 311–428.

Eschmeyer W (2016) Catalog of Fishes. California Academy of Sciences. Avaliable online at: http://researcharchive.calacademy.org/research/ichthyology/catalog/fishcatmain.asp [Accessed: 23/08/2016]

Ferrer J, Malabarba LR (2011) A new *Trichomycterus* lacking pelvic fins and pelvic girdle with a very restricted range in Southern Brazil (Siluriformes: Trichomycteridae). Zootaxa 2912: 59–67.

Ferrer J, Donin L, Malabarba LR (2015) A new species of *Ituglanis* Costa & Bockmann, 1993 (Siluriformes: Trichomycteridae) endemic to the Tramandaí-Mampituba ecoregion, southern

Brazil. Zootaxa 4020: 375–389. https://doi.org/10.11646/zootaxa.4020.2.8

Fink WL, Weitzman SH (1974) The So-called Cheirodontin Fishes of Central America with Descriptions of Two New species (Pisces: Characidae). Smithsonian Institution. Smithsonian Contributions to Zoology 172: 1–46. https://doi.org/10.5479/si.00810282.172

Ingenito LFS, Duboc LF (2014) A new species of *Astyanax* (Ostariophysi: Characiformes: Characidae) from the upper rio Iguaçu basin, southern Brazil. Neotropical Ichthyology 12: 281–290. https://doi.org/10.1590/1982-0224-20130117

IUCN (2014) Guidelines for Using the IUCN Red List Categories and Criteria. Standards and Petitions Subcommittee, version 11, Avaliable online at: http://www.iucnredlist.org/documents/RedListGuidelines.pdf [Accessed: 30/11/2015]

Javonillo R, Malabarba LR, Weitzman SH, Burns JR (2010) Relationships among major lineages of characid fishes (Teleostei: Ostariophysi: Characiformes), based on molecular sequence data. Molecular Phylogenetics and Evolution 54: 498–511. https://doi.org/10.1016/j.ympev.2009.08.026

López H, Menni R, Donato M, Miquelarena A (2008) Biogeographical revision of Argentina (Andean and Neotropical Regions): an analysis using freshwater fishes. Journal of Biogeography 35: 1564–1579. https://doi.org/10.1111/j.1365-2699.2008.01904.x

Malabarba LR (1989) Histórico sistemático e lista comentada das espécies de peixes de água doce do Sistema da Laguna dos Patos, Rio Grande do Sul, Brasil. Comunicações Museu Ciências PUCRS, Série Zoologia, 2: 107–179.

Malabarba LR, Isaia EA (1992) The fresh water fish fauna of the Rio Tramandaí drainage, Rio Grande do Sul, Brazil with a discussion of its historical origin. Comunicações Museu Ciências PUCRS, Série Zoologia, 5: 197–223.

Malabarba LR, Fialho CB, Anza JA, Santos JF, Mendes GN (2009) Peixes. In: Boldrin I (Ed.) Biodiversidade dos campos do Planalto das Araucárias. Ministério do Meio Ambiente, Secretaria de Biodiversidade e Florestas, Brasília, 133–155.

Marinho MF, Ohara WM (2013) Redesciption of *Astyanax guaporensis* Eigenmann, 1911 (Characiformes: Characidae), a small characid from the rio Madeira basin. Zootaxa 3652: 475–484. https://doi.org/10.11646/zootaxa.3652.4.5

Menni RC (2004) Peces y Ambientes em La Argentina continental. Monografias del Museo Argentino de Ciencias Naturales, Buenos Aires, 316 pp.

Mirande JM (2010) Phylogeny of the family Characidae (Teleostei: Characiformes): from characters to taxonomy. Neotropical Ichthyology 8: 385–568. https://doi.org/10.1590/S1679-62252010000300001

Oliveira C, Avelino GS, Abe K, Mariguela T, Benine R, Orti G, Vari R, Castro RMC (2011) Phylogenetic relationships within the speciose family Characidae (Teleostei: Ostariophysi: Characiformes) based on multilocus analysis and extensive ingroup sampling. BMC Evolutionary Biology, 11, 1–25. https://doi.org/10.1186/1471-2148-11-275

Ornelas-Garcia CP, Domínguez-Domínguez O, Doadrio I (2008) Evolutionary history of the fish genus *Astyanax* Baird & Girard (1854) (Actinopterygii, Characidae) in Mesoamerica reveals multiple morphological homoplasies. BMC Evolutionary Biology, 8, 1–17. https://doi.org/10.1186/1471-2148-8-340

Ribeiro CV, Barbosa R, Machado V, Cunha G (2009) Esponjas. In: Boldrin I (Org.) Biodiversidade dos campos do Planalto das Araucárias. Ministério do Meio Ambiente, Secretaria de Biodiversidade e Florestas, Brasília, 99–108.

Taylor WR, van Dyke GC (1985) Revised procedures for staining and clearing small fishes and other vertebrates for bone and cartilage study. Cybium 9: 107–119.

Author Contributions: CASL and ZMSL analyzed the data and wrote the paper. ABZ examined specimens and take the counts and measurements.

Competing Interests: The authors have declared that no competing interests exist.

6

Comparative analysis of the integument of different tree frog species from *Ololygon* and *Scinax* genera (Anura: Hylidae)

Henrique Alencar Meira da Silva[1], Thiago Silva-Soares[2], Lycia de Brito-Gitirana[1]

[1]Laboratório de Histologia Integrativa, Programa de Pesquisa em Glicobiologia, Instituto de Ciências Biomédicas, Universidade Federal do Rio de Janeiro. Avenida Carlos Chagas Filho 373, Bloco B1-019, 21941-902 Rio de Janeiro, RJ, Brazil.
[2]Laboratório de Zoologia, Museu de Biologia Prof. Mello Leitão, Instituto Nacional da Mata Atlântica. Avenida José Ruschi, 29650-000 Santa Teresa, ES, Brazil.
Corresponding author: Lycia de Brito-Gitirana (lyciabg@histo.ufrj.br)

http://zoobank.org/94B1A213-D259-4366-8EB4-86E2A8E05443

ABSTRACT. The integuments of ten treefrog species of two genera from Scinaxnae – *O. angrensis* (Lutz, 1973), *O. flavoguttata* (Lutz & Lutz, 1939), *O. humilis* (Lutz & Lutz, 1954), *O. perpusilla* (Lutz & Lutz, 1939), *O. v-signata* (Lutz, 1968), *Scinax hayii* (Barbour, 1909), *S. similis* (Cochran, 1952), *O. trapicheroi* (Lutz & Lutz, 1954) and *S. x-signatus* (Spix, 1824) – were investigated using conventional and histochemical techniques of light microscopy, and polarized light microscopy. All integuments showed the basic structure of the anuran integument. Moreover, the secretory portions of exocrine glands, such as serous merocrine and apocrine glands, were found to be restricted to the spongious dermis. Lipid content occurred together with the heterogeneous secretory material of the glands with an apocrine secretion mechanism. In addition, clusters of these apocrine glands were present in the ventrolateral integument of some species. Melanophores were also visualized in all examined hylids. However, the occurrence of iridophores, detected through polarized light microscopy, varied according to the species. The Eberth-Katschenko layer occurred in the dorsal integument from both genera, but it was only present in the ventral integument of *O. albicans*, *O. angrensis*, *O. flavoguttata*, *O. perpusilla* and *O. v-signata*. Although the integument of all treefrogs showed the same basic structure, some characteristics were genus-specific; however, these features alone may not be used to distinguish both genera.

KEY WORDS. Brazilian Atlantic forest, histochemistry, hylids, treefrog.

INTRODUCTION

Many challenges confront biologists studying amphibians, since human activities have been prejudicial to natural biota. Hylidae is a large anuran family, and one of the most abundant and prominent groups of frogs in the Neotropics. Overall, 345 species of Hylidae are known to exist in Brazil (SBH 2016).

The integument of anurans performs several functions, such as protection against diverse environmental circumstances (Elkan 1968, Fox 1986a, b, Greven et al. 1995, de Brito-Gitirana and Azevedo 2005, Azevedo et al. 2006, de Brito-Gitirana and Azevedo 2005), mechanical protection (Fox 1986a, b, Azevedo et al. 2006), chemical defense (Delfino et al. 1995, Daly 1995, Jekel et al. 2015), sensory perception (Mearow and Diamond 1988, Koyama et al. 2001), ionic transport and water absorp-

tion (Sullivan et al. 2000, Azevedo et al. 2006), and respiration (Duellman and Trueb 1994).

In adult anurans, the integument consists of two firmly attached layers: the epidermis and the dermis, which is located just beneath the epidermis. The epidermis, formed by a stratified squamous epithelium, overlies the dermis of connective tissue, which is subdivided into two layers: the spongious dermis and the compact dermis. The spongious dermis is formed by loose connective tissue with pigment cells, such as melanophores (melanin producing cells) and iridophores (with reflective or iridescent structures) located just beneath the basal lamina (Bagnara et al. 1968). In addition to the pigment cells, blood vessels, alveolar mucous, serous, mixed and granular glands also occur. The compact dermis is composed of a series of alternating layers of collagenous fiber bundles arranged in a crisscross manner (Fox

1986a, b, Azevedo et al. 2005, de Brito-Gitirana and Azevedo 2005). Nevertheless, integument structure varies according to the body region (de Brito-Gitirana and Azevedo 2005, Felsemburgh et al. 2007). In addition, different types of exocrine glands occur in the anuran integument, withtheir secretory portions restricted to the spongious dermis (Duellmann and Trueb 1994, Brizzi et al. 2002).

In general, anurans show a wide array of colors related to specialized cells named chromatophores. Color and reflectivity are important mechanisms that allow anurans to change their integument color, enabling the maintenance of body temperature and avoiding detection by predators (Duellman and Trueb 1994). Three types of chromatophores are important for amphibian coloration: melanophores (with a black or brownish pigment named melanin), xantophores (with yellow colored pigments), and iridophores (with reflective or iridescent structures). These pigment cells are located in the spongious dermis just beneath the basal lamina, and their arrangement was called the Dermal Chromatophore Unit (DCU) (Bagnara et al. 1968). However, the exact mechanism of DCU organization is still unknown.

Some anurans, like arboreal frogs, exhibit diverse morphology in their typical integument outline, showing cutaneous adaptations to avoid evaporative water loss (Blaylock et al. 1976, Warburg et al. 2000, Barbeau and Lillywhite 2005). Such adaptations include exocrine glands secreting lipids, which are spread over the body surface, aided by an elaborate series of behavioral movements to form a useful barrier against water loss. In addition, the histological aspects of the anuran integument can contribute to the differentiation of anuran species from *Proceratophrys* Miranda-Ribeiro, 1920 (Felsemburgh et al. 2007).

The abundant fauna of amphibians of Brazil (SBH 2016) contrast with the present loss and degradation of natural habitats and the scarce knowledge on the morphological, taxonomical, biological, and geographical distribution of most species. The aim of the present work was to investigate the morphological features of the integuments of some hylids in order to characterize this tissue and determine whetherthere are species-specific structures, contributing to the knowledge of the integument biology.

MATERIAL AND METHODS

Nine species of male adult tree frogs belonging to Scinaxinae from both *Ololygon* Fitzinger, 1843 and *Scinax* Wagler, 1830 genera were collected in Serra dos Órgãos National Park (PARNASO, 22°29'31"S, 42°59'11.48"W), in the municipality of Teresópolis, during two years (from March 2007 to November 2008). *Ololygon angrensis* (Lutz, 1973) was collected in Rio de Janeiro state at the Lídice district (22°46'53.82"S, 44°13'55.96"W), municipality of Rio Claro (from March 2009 to March 2010) (Table 1). Tree frogs were collected under license permit number 15396-1 and 219/2011, issued by the Instituto Chico Mendes de Conservação da Biodiversidade (ICMBio).

Samplings were carried out at night from 7 to 10 pm and all individuals collected (Figs 1–10) were carefully placed in

Table 1. List of individuals used in this study.

	Number of individuals	Common name (Frost 2016)
Ololygon albicans (Bokermann, 1967)	10	Teresopolis Snouted Treefrog
Ololygon angrensis (Lutz, 1973)	5	Serra da Bocaina Snouted Treefrog
Ololygon flavoguttata (Lutz & Lutz, 1939)	5	Yellowbelly Snouted Treefrog
Ololygon humilis (Lutz & Lutz, 1954)	3	Rio Babi Snouted Treefrog
Ololygon perpusilla (Lutz & Lutz, 1939)	3	Bandeirantes Snouted Treefrog
Ololygon trapicheroi (Lutz & Lutz, 1954)	5	Three-lined Snouted Treefrog
Ololygon v-signata (Lutz, 1968)	2	Forest Snouted Treefrog
Scinax similis (Cochran, 1952)	3	Cochran's Snouted Treefrog
Scinax hayii (Barbour, 1909)	3	Hay's Snouted Treefrog
Scinax x-signatus (Spix, 1824)	3	Venezuela Snouted Treefrog

small humid plastic bags, transported from the collectionsite to the field laboratory with a walk of no longer than one hour. They were kept in captivity until the next morning, when the fixation procedures were carried out. Individuals were euthanized with 0.5% xylocaine according to regulations of the Federal Council of Veterinary Medicine (Law 5.517/68 article 16, "f"). The animals were fixed in 10% formaldehyde and subsequently maintained in 70% (v/v) alcohol. At the end of field expedition, the treefrogs were taken to the laboratory, to undergo histological analysis.

For light microscopic (LM) analysis, 3-5 mm thick sections from the ventral, ventrolateraland dorsal regions of the integument were processed according to standard histological techniques for paraffin embedding before sectioning, i.e., the sections were quickly washed in water, dehydrated (70%, 90%, twice in 100% ethanol; 30 minutes each), clarified twice in xylene (30 minutes each), infiltrated and embedded in paraffin. The 5-μm thick serial histological sections were stained with hematoxylin-eosin (HE) (Lillie and Fulmer 1976), which is the standard stain for histological examination of animal tissues, staining thenuclei and cytoplasm in blue with the extracellular matrix in pale pink. Mallory's trichrome (Lillie and Fulmer 1976) was used since it is a good stain for distinguishing cellular from extracellular elements and is especially suitable for studying connective tissue, staining the collagenous fibers in blue, red blood cells in orange, and nuclei in red. Staining with 1% Alcian blue (AB) 8GX at pH 2.5 (Mowry 1963, Kiernan 1990) was utilized to detect sulfated and carboxylated glycoconjugates (in light blue). In addition, HE-stained slices were observed under a light microscopeusing polarized light in order to detect iridophores (an iridescent cell), since these cell types possess reflective structures (Bagnara et al. 1968) that can be easily recognized under polarized light. Sections were analyzed using a Leica DM750 microscope, and the images were captured using a Leica DFC452 digital camera.

RESULTS

In general, the integuments of all species showed the basic morphological structure of the anuran integument, essentially

Figures 1–10. Photograph of the species: (1) *O. albicans*; (2) *O. angrensis*; (3) *O. flavoguttata*; (4) *S. hayii*; (5) *O. humilis*; (6) *O. perpusilla*; (6) *S. similis*; (8) *O. trapicheroi*; (9) *O. v-signata*; (10) *S. x-signatus*.

being formed by an epidermal and dermal layer (Figs 11–55).

The epidermis was relatively thin and consisted of a partially keratinized stratified squamous epithelium supported by a dermis. Although the keratinocyte was the predominant cell type, native flask cells were also visualized. In addition, epidermal cells were organized into a basal layer, an intermediate layer and an outermost layer. The epidermal cells of the outermost layer were partially keratinized, since their nuclear profiles could be easily visualized.

Considering the number of epidermal cell layers of the integument, subtle differences betweenspecies and body regions were observed (Table 2), i.e., the ticker epidermis was observed in *Ololygon humilis* (Lutz & Lutz, 1954) and *Scinax x-signatus* (Spix, 1824).

The dermis was subdivided into the spongious dermis, composed of loose connective tissue, and the compact dermis, which rested on the hypodermis. Moreover, the compact dermis was formed by collagenous fibers organized in a series of alternating layers, compactly arranged in a crisscross manner. Between the spongious dermis and the compact dermis, irregular basophilic deposits occurred scattered through this boundary region, corresponding to the Eberth-Katschenko (EK) layer.

Table 2. Number of epithelial cell alyers of *Oloygon* spp. and *Scinax* spp.

	Number of epithelial cell layers	
	Dorsal	Ventral
O. albicans	4	4–5
O. angrensis	4	4–5
O. flavoguttata	4	4–5
O. humilis	4	8
O. perpusilla	3–4	4–5
O. trapicheroi	5	5
O. v-signata	4	4
S. similis	4	4
S. hayii	4	4–5
S. x-signatus	5	7

Another typical feature of these treefrogs was the occurrence of different exocrine glands, whose secretory portions were housed in the spongious dermis (Figs 11–55). From the secretory portion, a single unbranched excretory duct passed over the epidermis and opened on the integument surface. These ducts were lined by two layers of cuboid cells.

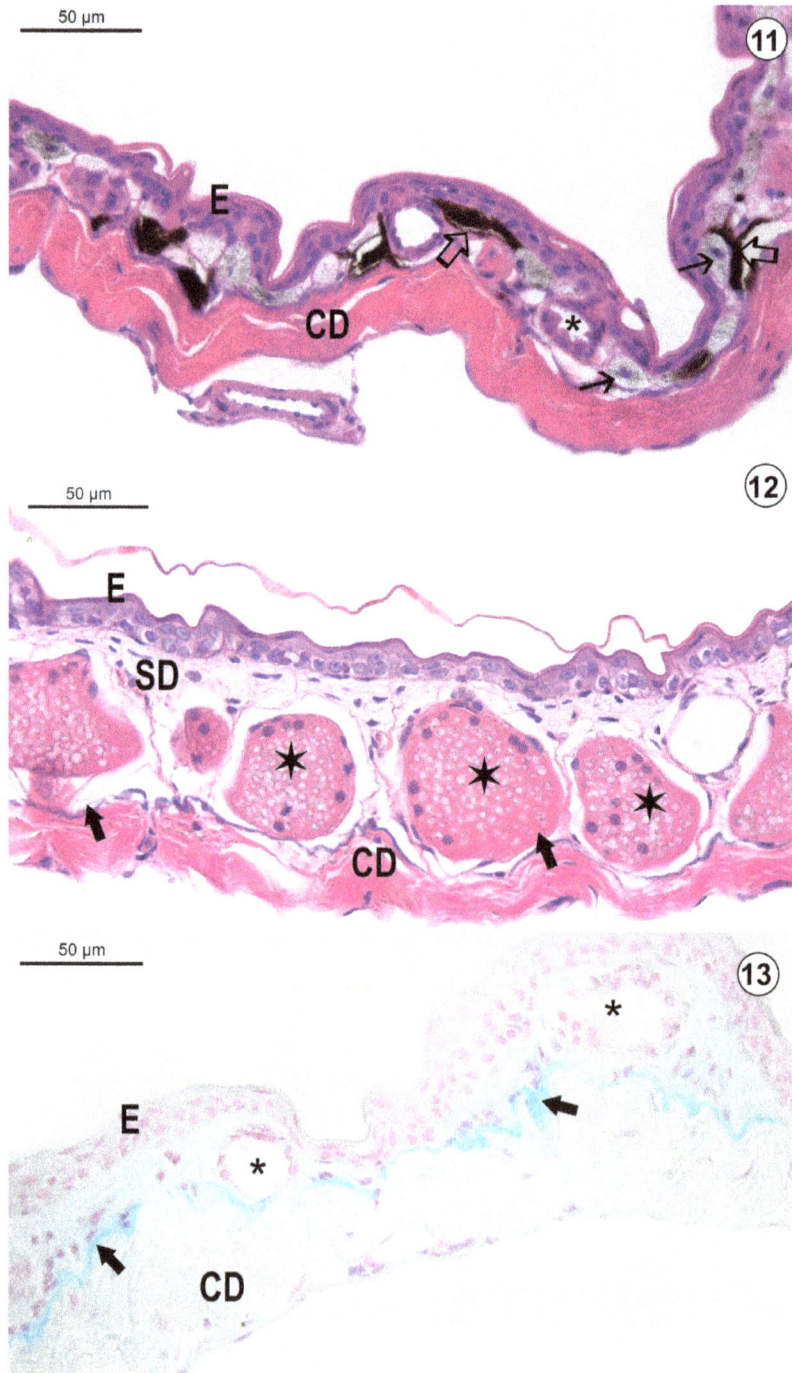

Figures 11–13. Light micrograph of the integument of *O. albicans*: (11) Dorsal region (HE-staining); (12) Ventrolateral region (HE-staining); (13) Ventral region (AB-method). In all integument regions, the epidermis (E) rests on the dermis, which is subdivided into the spongious dermis (SD) and the compact dermis (CD). Iridophores (→) occur in the dorsal region; however, they are absent in both ventrolateral and ventral regions. Melanophores (⇨) in the spongious dermis. Both serous (✱) and apocrine glands (*) occur in the spongious dermis. No glandular cell reacts to AB-method, suggesting that secretory units is made up of serous cells. The EK-layer (➡) exhibits its typical basophilic staining. Note clusters of apocrine glands (*) with heterogeneous content in the ventrolateral integument. The EK-layer (➡) exhibits typical alcianophilic reaction of its glycoconjugate content. Large blood vessels occur in the hypodermis.

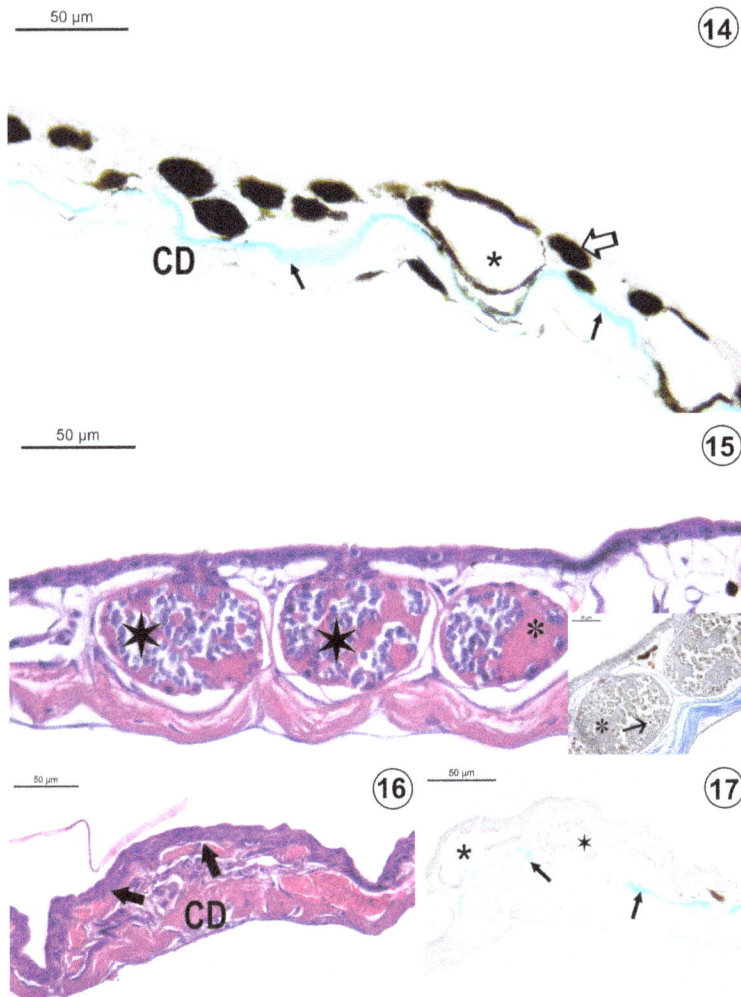

Figures 14–17. Light micrograph of the integument of *O. angrensis*: (14) Dorsal region (AB-staining); (15) Ventrolateral region (HE-staining), inset (Mallory´s trichrome staining); (16) Ventral region (HE-staining); (17) Ventral region (AB-method). The melanophores (⇨) are numerous and located just beneath of the epidermis as well as around de glandular secretory units. They occur also in the hypodermis, but absent in the ventral integument. Iridophores (→) occur in the spongious dermis of the dorsal region just beneath the epidermis. No iridophore is visualized in the ventral region of the integument. Clusters of apocrine glands (∗) with heterogeneous intake predominate at ventrolateral integument; inset of Fig. 15: Observe the granular content with dense stained core (➔) intermingled with cytoplasm material (*). The EK-layer (➔) is continuous in the ventral region, but discontinuous in the ventral integument. Serous glands (★) occur in all body regions. Note blood vessels (➡) in the spongious dermis. CD = compact dermis.

Depending on the dye affinity to their secretion, the merocrine glands of the serous and mixed types were visualized. Serous glands are formed by only serous-secreting cells, which have spherical nuclei and an acidophilic cytoplasm. Mixed secretory units are made up of mucous cells (nuclei generally flattened and displaced to the basal portion of the secretory cells and basophilic cytoplasm) and serous cells. It is noteworthy that mucous glands essentially formed by mucous cells did not occur in the integuments of all tree frogs (Table 3 and Figs 11–55).

While serous glands were visualized in all species, mixed glands occurred in *Ololygon albicans* (Bokermann, 1967), *O. angrensis*, *O. flavoguttata* (Lutz & Lutz, 1939), *O. humilis*, *O. trapicheroi* (Lutz & Lutz, 1954), *O. v-signata* (Lutz, 1968), *Scinax similis* (Cochran, 1952) and *S. x-signatus*, but they did not occur in *S. hayii* (Barbour, 1909) and *O. perpusilla* (Lutz & Lutz, 1939) (Table 3).

Considering the apocrine glands, their secretory portion consisted of syncytial units, varying according to their content and dye affinity. Their secretory products were made up of small

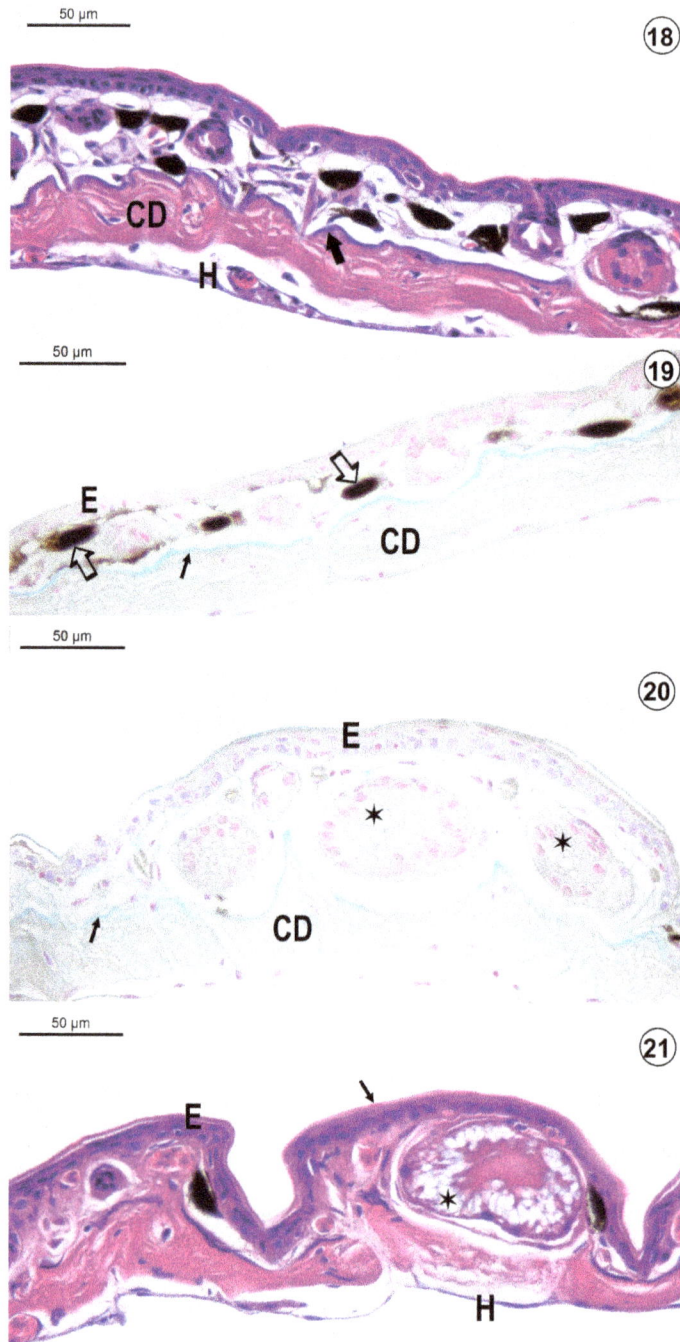

Figures 18–21. Light micrograph of the integument of *O. flavoguttata*: (18) Dorsal region (HE-staining); (19) Dorsal region (AB-method); (20) Ventrolateral region (HE-staining) (21) Ventral region (HE-staining). The epidermis (E) is partially keratinized, and the outermost cell layer exhibits the nuclear profiles (→). Melanophores (⇨) occur in the spongious dermis of both dorsal and ventrolateral regions; they are more frequent in the dorsal integument. Small clusters of apocrine glands with heterogeneous content occur in the ventrolateral integument. In the ventral region, the lipid content (∗), mixed with basophilic material, is easily visualized in the apocrine glands (Fig. 21). In the dorsal region, the EK-layer (➔) is continuous, but discontinuous in the ventrolateral integument, being less alcianophilic in the ventral region when compared to other integument regions. CD = compact dermis; H = hypodermis.

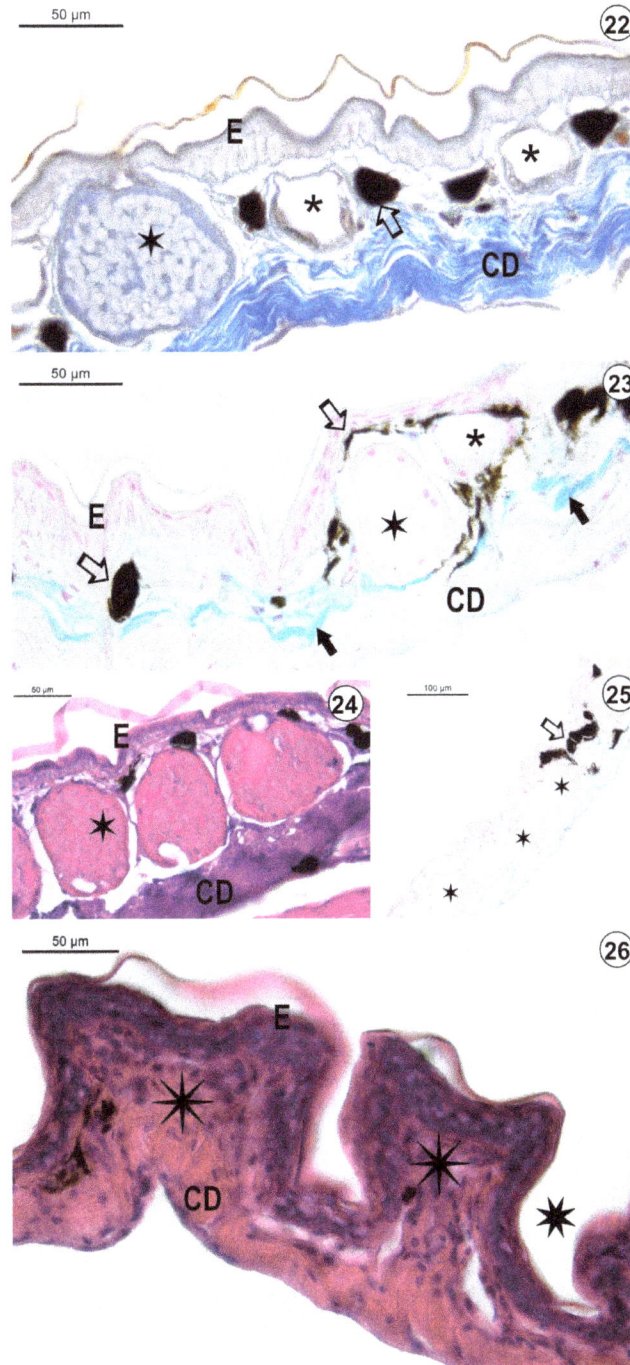

Figures 22–26. Light micrograph of the integument of *S. hayii*: (22) Dorsal region (Mallory´s trichrome staining); (23) Dorsal region (AB-method); (24) Ventrolateral region (HE-staining) (25) Ventral region (AB-method); (26) Ventral region (HE-staining). In the dorsal integument, exocrine glands are more frequent, mainly the serous glands (✱). Melanophores (⇨) occur in the spongious dermis, even around the secretory portion of glands. Note clusters of apocrine glands with heterogeneous content (✳)in the spongious dermis of the ventrolateral integument. The EK-layer (➡) is continuous in the dorsal integument, but absent in the ventral region. Slight cutaneous elevations (✳) in the ventral integument are formed by the epidermis and the dermis, mainly the spongious dermis. They are separated by groves (✽). CD = compact dermis.

Table 3. Occurence of exocrine merocrine glands in the integument of *Oloygon* spp. and *Scinax* spp.

| | Merocrine glands | | | | | | | | |
| | Mucous gland | | | Serous gland | | | Mixed glands | | |
	Dorsal	Ventrolateral	Ventral	Dorsal	Ventrolateral	Ventral	Dorsal	Ventrolateral	Ventral
O. albicans	–	–	–	+	+	+	–	–	+
O. angrensis	–	–	–	+	+	+	–	+	–
O. flavoguttata	–	–	–	+	+	+	–	+	–
O. humilis	–	–	–	+	+	+	+	+	+
O. perpusilla	–	–	–	+	+	–	–	–	–
O. trapicheroi	–	–	–	+	+	+	+	+	+
O. v-signata	–	–	–	+	–	–	+	+	+
S. similis	–	–	–	+	+	+	+	+	+
S. hayii	–	–	–	+	+	+	–	–	–
S. x-signatus	–	–	–	+	–	–	+	+	+

(+) Present, (–) absent.

Table 4. Occurence of apocrine glands in the integument of *Oloygon* spp. and *Scinax* spp.

| | Heterogeneous content | | | | | |
| | Basophilic | | | Lipid | | |
	Dorsal	Ventrolateral	Ventral	Dorsal	Ventrolateral	Ventral
O. albicans	+	+/×	+	+	+	+
O. angrensis	+	–	–	+	+	–
O. flavoguttata	+	+/×	+	–	+	+
O. humilis	+	+	+	+	+	+
O. perpusilla	+	+	–	+	+	–
O. trapicheroi	+	+	+	+	–	–
O. v-signata	+	+	+	+	+	+
S. similis	–	–	–	+	+	–
S. hayii	+	+/×	–	+	+	+
S. x-signatus	–	+	–	+	+	–

(+) Present, (–) absent, (×) occurrence in clusters.

acidophilic granules, seemingly due to their protein content. Nevertheless, some apocrine glands revealed that this secretion is a mixture of basophilic and lipid contents, intermingled with cytoplasmic material; their rounded nuclei were displaced to the cell basal domain. In some glands, the secretory product was constituted of acidophilic cytoplasmic material mixed with heterogeneous material, which revealed slight basophilic reaction usually associated with lipid material (Table 4).

In *O. angrensis*, the secretory product of the apocrine glands exhibited basophilic granules with an acidophil core after staining with Mallory's trichrome (Figs 14–17). Furthermore, these secretory granules were also observed in *S. similis* (Figs 36–40) and contained lipids associated with basophilic material. We suggest that the basophilic staining was due to carbohydrates of glycolipids, since they showed no alcianophilic reaction typical of acid polysaccharides.

Table 5. Occurence of pigment cells in the integument of *Oloygon* spp. and *Scinax* spp.

| | Melanophores | | Iridophores | |
	Dorsal	Ventral	Dorsal	Ventral
O. albicans	–/+	–/+	+	+
O. angrensis	+	–/+	–/+	–
O. flavoguttata	+	–/+	–	–
O. humilis	+	–/+	+	–
O. perpusilla	+	–	–	–
O. trapicheroi	–/+	–	+	–
O. v-signata	+	–	–	–
S. similis	+	–/+	+	+
S. hayii	+	–/+	–	–
S. x-signatus	+	–	–	–

(+) Widespread, (–/+) occasional, (–) not visualized.

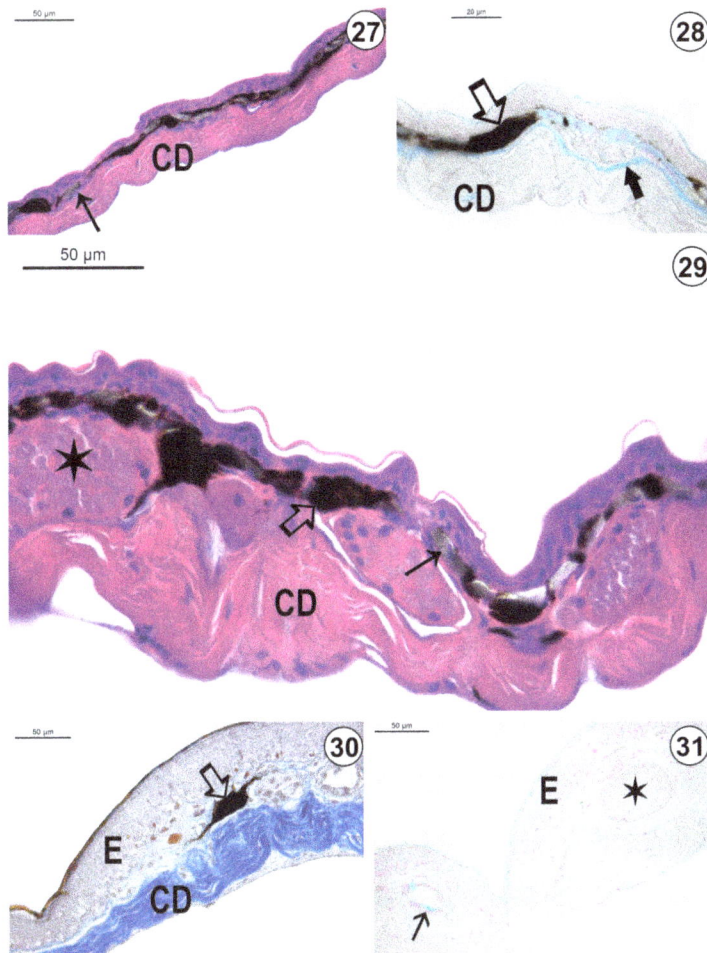

Figures 27–31. Light micrograph of the integument of *O. humilis*: (27) Dorsal region (HE-staining); (28) Dorsal region (AB-method); (29) Ventrolateral region (HE-staining) (30) Ventral region (Mallory´s trichrome staining); (31) Ventral region (AB-method). In the dorsal region, the spongious dermis is poorly developed. Melanophores (⇨) are visualized in all integument regions; however, iridophores (→) are visualized only in both dorsal and ventrolateral integument. Both pigment cells are located just beneath the epidermis. Alcianophilic reaction is observed in cytoplasm of iridophores as well as in the EK-layer (➡) of the dorsal integument. The EK-layer is absent in the ventral integument. Apocrine glands with heterogeneous content (✱) occur in both ventrolateral and ventral integument. In *S. humilis*, mixed glands (➔) are visualized in the ventral region, being formed by serous and mucous cells. Mucous cells exhibit alcianophilic reaction. E = epidermis; CD = compact dermis.

In *O. albicans, O. angrensis, O. flavoguttata* and *S. hayii*, apocrine glands with heterogeneous content were more frequent in the ventrolateral integument, occurring as small clusters (Figs 11–26).

Pigment cells, such as melanophores and iridophores, were identified in the integument, occurring in the spongious dermis, just beneath the basal lamina. While melanophores were identified through the typical brownish color of their melanin granules under light microscopy, iridophores were visualized by polarized light microscopy through their reflective or iridescent pigments (Figs 11–55). Melanophores also occurred in the hypodermis of *O. angrensis, O. v-signata* and *S. similis* (Figs 4–11, 36–40, 52–55),

while iridophores were only detected in the hypodermis of *S. similis*. The occurrence of both pigment cells varied according to the species and integument region (Table 5).

The Eberth-Katschenko (EK) layer occurred between the spongious and compact dermis and was recognized through its typical basophilic and alcianophilic reaction after using the HE- and AB-methods, respectively (Table 6; Figs 11–55). The EK-layer was visualized in the dorsal region of the integument of all *Oloygon* species, but absent in all species of *Scinax*.

Cutaneous elevations occurred in the ventral integument of *O. angrensis, O. flavoguttata, O. perpusilla, S. similis, O.*

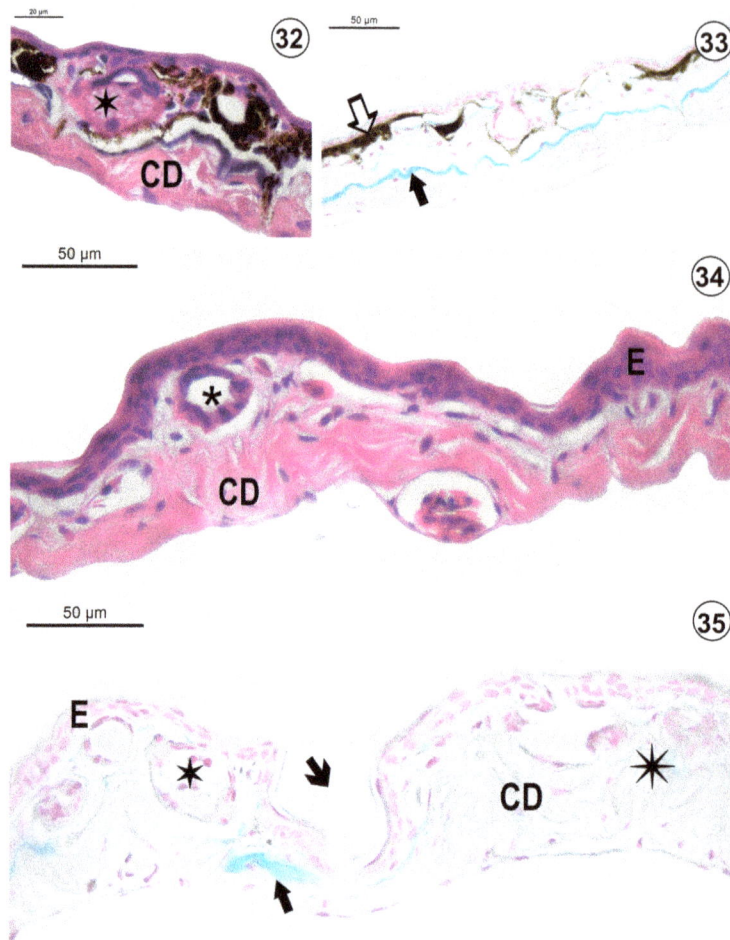

Figures 32–35. Light micrograph of the integument of *O. perpusilla*: (32) Dorsal region (HE-staining); (33) Dorsal region (AB-method); (34) Ventrolateral region (HE-staining) (35) Ventral region (AB-method). The spongious dermis houses both apocrine glands with heterogeneous content (✳) and serous glands (★). Melanophores (⇨) occur in the dorsal integument, but they are not identified in either the ventrolateral or ventral region. Iridophores did not occur in all integument regions. The EK-layer (➡) is a well defined continuous layer occurs as irregular deposits between the spongious and compact dermis of the ventral region. Cutaneous elevations (✳) are separated by grooves (➔) in the ventral region. CD = compact dermis.

trapicheroi and *S. x-signatus* (Figs 14–21, 32-40, 52–55). These slight elevations were formed by the epidermis, followed by the dermis, being separated by grooves.

DISCUSSION

Although some studies of the anuran integument are available, such as those of bufonids (de Brito-Gitirana and Azevedo 2005, de Brito-Gitirana et al. 2007, Almeida et al. 2007, Felsemburgh et al. 2009), ranids (Azevedo et al. 2006, Pelli et al. 2010), and leptodactylids (Goniakowska-Witalinska and Kubiczek 1998, Warburg et al. 2000, Nosiet al. 2002, Barbeau and Lillywhite 2005, Felsemburgh et al. 2007, Gonçalves and de

Brito-Gitirana 2008, Rigolo et al. 2008), information about the integument structure of hylids is still insufficient.

In this study, the integument showed the basic structure as already described for other anurans, i.e., the epidermis rests on a dermis, which is divided into a spongious and a compact dermis. The majority of anurans display this structural pattern of the integument (Elkan 1968, Goniakowska-Witalinska and Kubiczek 1998, Warburg et al. 2000, de Brito-Gitirana and Azevedo 2005, Delfino et al. 2006, Felsemburgh et al. 2007, Gonçalves and de Brito-Gitirana 2008, Felsemburgh et al. 2009).

In general, the epidermis varied from 4-5 cell layers. However, in *O. humilis* and *S. x-signatus* the ventral epidermis was thicker, and it may be related to species habitat. Nevertheless,

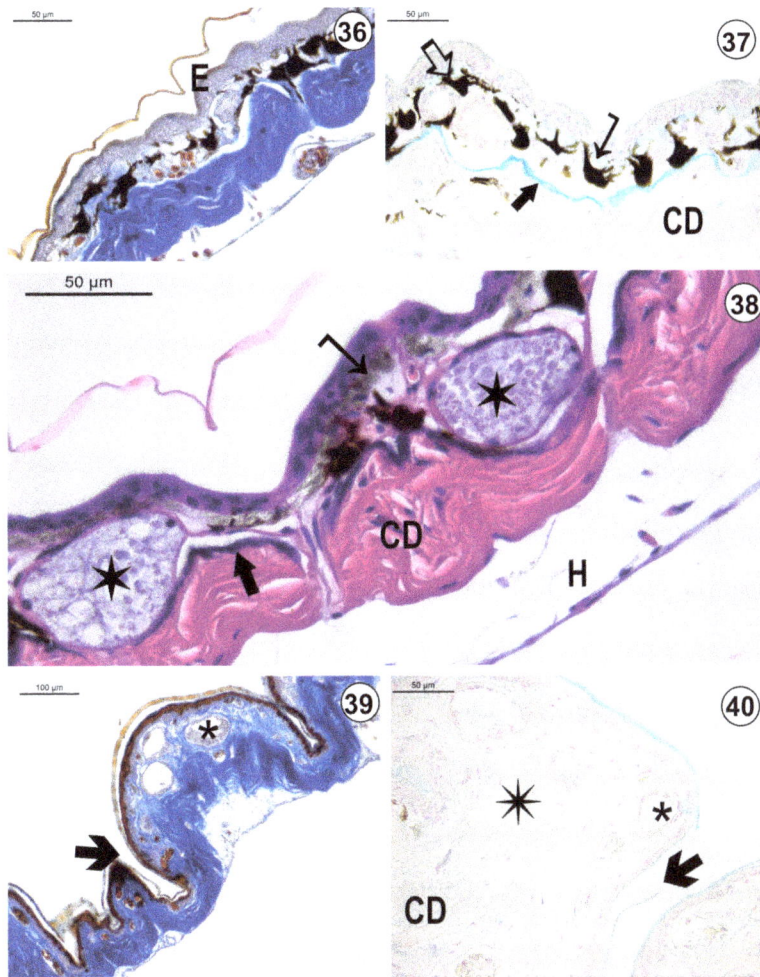

Figures 36–40. Light micrograph of the integument of *S. similis*: (36) Dorsal region (Mallory´s trichrome staining); (37) Dorsal region (AB-method); (38) Ventrolateral region (HE-staining); (39) Ventral region (Mallory´s trichrome staining); (40) Ventral region (AB-method). Melanophores (⇨) and iridophores (→) organized as chromatophore units occur in both dorsal and ventrolateral integument. Iridophores exhibit alcianophilic reaction. Note serous glands (★) and apocrine glands with granular content (✳) in the spongious dermis. In the ventral region, cutaneous elevations (✲) are separated by prominent grooves (➔). The EK-layer (➡) occur in the dorsal integument but not in the ventral integument. Moreover, the epidermis (E) of the ventral region is more developed than other integument regions. E = epidermis; CD = compact dermis; H = hypodermis.

we did not find detailed behavioral data in the literature to support this explanation. In mammals, in some body areas, friction and other forces dictate the thickness of the lining epithelium, since the number of epithelial cell layers is related to epithelial resistance (Ham 1977, de Brito-Gitirana 2015).

Anuran glands have received significant attention. They have been described as being of different types, like mucous, serous, lipid (or wax), and mixed (seromucous) glands. However, some authors have named the granular glands as poison or serous glands (Mills and Prum 1984, Duellmann and Trueb 1994, Brizzi et al. 2002).

In this study, well-established histological criteria to categorize the cutaneous gland of mammals (Ham 1977, Kierzenbaum 2004) were used in order to adopt a coherent histological classification, especially in reference to anuran glands, since they exhibit variable morphology. Thus, on the basis of how their secretory products are released to the external environment, the exocrine glands can be classified as holocrine, apocrine or merocrine. Holocrine glands release both secretions and entire cells, while apocrine glands release the secretory product and cytoplasmic matrix of the apical portion of the cell. In merocrine cells, no cytoplasm is lost and the gland uses exocytosis

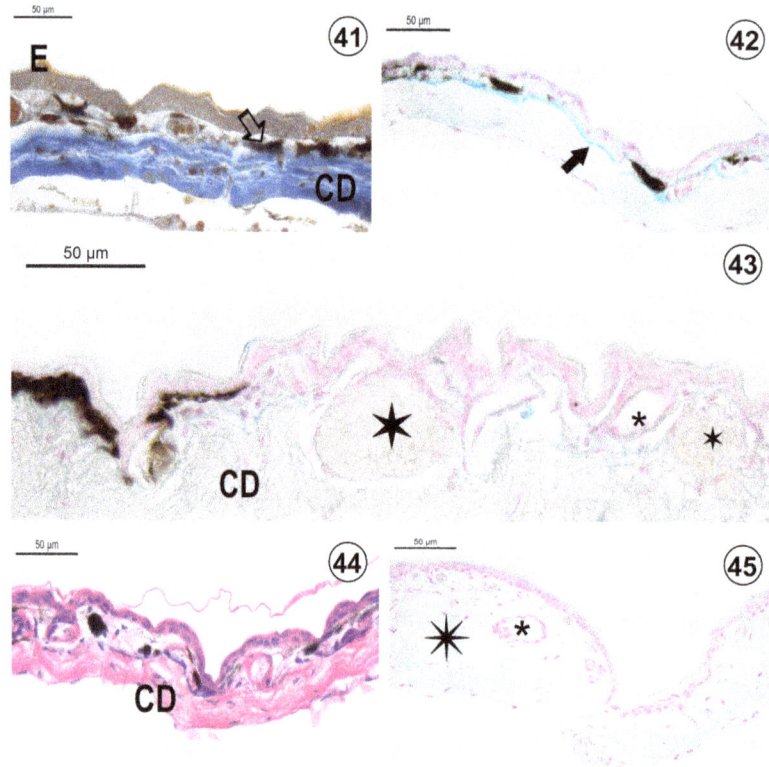

Figures 41–45. Light micrograph of the integument of *O. trapicheroi*: (41) Dorsal region (Mallory's trichrome staining); (42) Dorsal region (AB-method); (43) Ventrolateral region (AB-method); (44) Ventral region (HE-staining); (45) Ventral region (AB-method). Melanophores (⇨) occur in the spongious dermis of the dorsal integument, and in the ventrolateral region as isolated groups. They are absent in the ventral integument. In the dorsal region, the EK-layer (➡) is continuous and well stained by the AB-method. Isolated serous glands (★) occur in all integument regions. The apocrine glands with granular content (✳) are visualized in the ventrolateral integument, where they are more developed. In the ventral region, cutaneous elevations (✱) are also separated by grooves. The dermis contains several small blood vessels. E= epidermis; CD = compact dermis.

Table 6. Occurrence of the EK-layer in the integument of *Oloygon* spp. and *Scinax* spp. The EK-layer can occur as a continuous or discontinuous layer.

	EK-layer		
	Dorsal	Ventrolateral	Ventral
O. albicans	+/Ø	+/Ø	+/Ø
O. angrensis	+/Ø	+/Ø	+/å
O. flavoguttata	+/Ø	+/å	+/Ø
O. humilis	+/Ø	+/å	–
O. perpusilla	+/Ø	–	+/å
O. trapicheroi	+/Ø	–	–
O. v-signata	+/Ø	+/Ø	+/Ø
S. similis	+/Ø	+/Ø	–
S. hayii	+/å	–	–
S. x-signatus	+/Ø	+/Ø	–

(+) Present, (–) absent, (Ø) continuous, (å) discontinuous.

to release its secretory product to the extracellular space (Ham 1977, Kierzenbaum 2004).

Given that the secretory cell remains intact, according to the type of secretionproduced, the merocrine gland of mammals is classified as serous, mucous or mixed. Serous glands are essentially composed of serous cells with large spherical nuclei and an acidophilic cytoplasm that is individualized by the cell membrane. Mucous glands are composed by the mucus-secreting cells, which exhibit an irregular shape; the nuclei are basally located, and their cytoplasm is basophilic (through HE-staining), alcianophilic (through AB-staining to detect acid glycoconjugates), and/or PAS positive (detect neutral glycoprotein). Mixed glands are made up of both serous and mucous cells constituting the same secretory portion (Ham 1977, Kierszenbaum 2004, de Brito-Gitirana 2013).

In some anurans, serous and granular glands have been considered as the same type. Nevertheless, various subtypes of

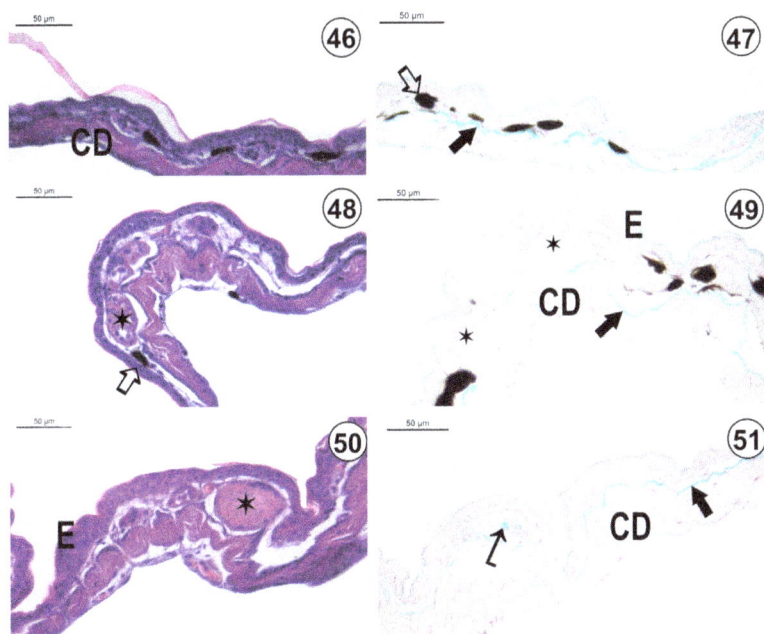

Figures 46–51. Light micrograph of the integument of *O. v-signata*: (46) Dorsal region (HE-staining); (47) Dorsal region (AB-method); (48) Ventrolateral region (HE-staining); (49) Ventrolateral region (AB-method); (50) Ventral region (HE-staining); (51) Ventral region (AB-method). Melanophores (⇨) occur in the spongious dermis that is poorly developed in the dorsal region. The EK-layer (➡) is a continuous layer in all integument regions. Apocrine glands (∗) with heterogeneous content occur in both ventrolateral and ventral regions. Serous glands are visualized in both dorsal and ventrolateral integument. Mixed glands (·) are observed in the ventral region. E = epidermis; CD = compact dermis.

serous glands with high morphological variability have been reported (Goniakowska-Witalinska and Kubiczek 1998, Delfino et al. 1999, Warburg et al. 2000, Nosi et al. 2002, Brunetti et al. 2012, Moreno-Gómez et al. 2014).

Actually, the granular gland described by Fox (1986a, b) and Duellman and Trueb (1994) is a kind of apocrine gland since its secretory product is a mixture of cytoplasm and secretory product. Furthermore, this glandular type is composed of a syncytium since the large mass of cytoplasm with many nuclei is not separated into individual cells by a cell membrane. In fact, the apocrine glands of anurans exhibit a high level of polymorphism.

In bufonids (de Brito-Gitirana and Azevedo 2005, de Brito-Gitirana et al. 2007, Almeida et al. 2007, Felsemburgh et al. 2009), ranids (Azevedo et al. 2006, Pelli et al. 2010), and leptodactylids (Felsemburgh et al. 2007, Gonçalves and de Brito-Gitirana 2008), apocrine granular glands occur scattered throughout the dorsal and ventral integument, exhibiting acidophilic granular secretion. In the integument of *S. similis*, the apocrine granular gland resembles those Ib type glands described for *Engystomops pustulosus* (Cope, 1864) (Delfino et al. 2015). In addition, apocrine glands with heterogeneous content occurred

in the integument of all species of Scinaxinae. Nevertheless, in *O. angrensis*, the secretory product contained peculiar basophilic granules with dense cores intermingled with acidophilic cytoplasm, and their histological features resembled those of type Ia glands described for *Pithecopus azureus* (Cope, 1862) (former *Phyllomedusa azurea*) (Nosi et al. 2002).

In all species, the apocrine gland included an acidophilic content mixed with slightly basophilic material. Although the secretory granules exhibited basophilic affinity, they revealed no alcianophilic reaction, demonstrating that their content had no glycoconjugate, and this material probably consists of glycolipids. On the other hand, lipids have been observed in cutaneous secretions of phyllomedusine (Blaylock et al. 1976), hylids *Litoria fallax* (Peter, 1880) and *L. peronii* (Tschidi, 1838) (Amey and Grigg 1995), African frogs (Withers et al. 1984), *Ranoidea australis* (Gray, 1842) (former *Cyclorana australis*) (Christian and Parry 1997), and the hylid *Ranoidea caerulea* (White, 1790) (former *Litoria caerulea*) (Warburg et al. 2000).

According to Warburg and co-workers (2000), lipid content represents the main adaptation of xeric-inhabiting arboreal frogs, enabling them to remain exposed throughout the year, even during dry seasons. Cutaneous lipids in tree frogs – *Phyl-*

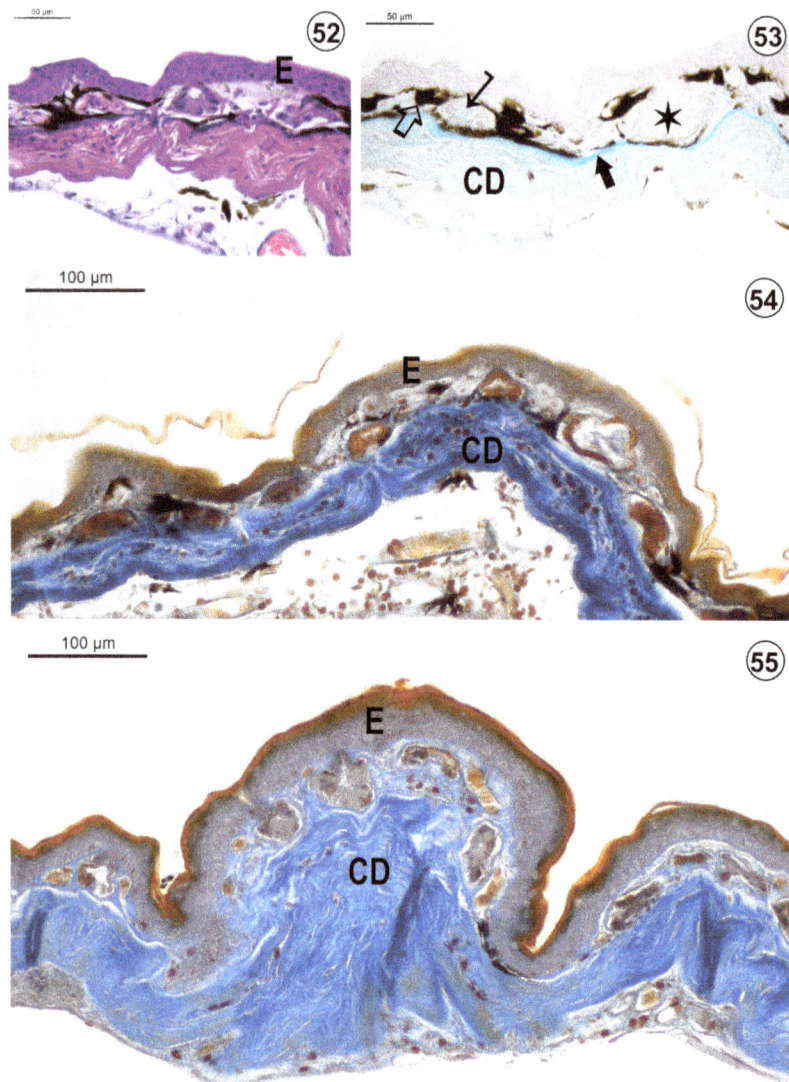

Figures 52–55. Light micrograph of the integument of *S. x-signatus*: (52) Dorsal region (HE-staining); (53) Dorsal region (AB-method); (54) Ventrolateral region (Mallory´s trichrome staining); (55) Ventral region (Mallory´s trichrome staining). The epidermis (E) is slightly ticker when compared to those of other hylids, as in the compact dermis (CD). Melanophores (⇨) are visulized in both dorsal and ventrolateral integument just beneath the epidermis. Serous glands are present in all regions; however, some of them show slightly alcianophilic content (→) in both ventrolateral and ventral regions. The apocrine glands (✳) with granular content occur in both dorsal and ventrolateral regions. The EK-layer (➡) is visualized in both dorsal and ventrolateral integument but is absent in the ventral integument.

lomedusa sauvagei Boulenger, 1882, *P. iherengii* Boulenger, 1885, *P. boliviana* Boulenger, 1902, *Ranoid gracilenta* (Peters, 1869), *R. caerulea* (White, 1790), *Polypedates maculatus* (Gray, 1830) – are a specialized adaptation to reduce dehydration in arid environments (Amey and Grigg 1995, Lillywhite 2004, Barbeau and Lillywhite 2005, Gomez et al. 2006). These lipids are spread over the body by complex self-wiping behavior to form an effective barrier that reduces evaporative water loss (Barbeau and Lillywhite 2005).

In this study, clusters of glands were observed in the ventrolateral integument of *O. albicans*, *O. angrensis*, *O. flavoguttata*, *S. hayii*. These glandular accumulations are probably present in a specialized region of the integument that provides special functions.

Clusters of tubuloalveolar alveoli in the ventral integument occurs in *Cycloramphus fuliginosus* (Gonçalves and de Brito-Gitirana 2008), being related to parental care. In *Rhinella icterica* (de Brito-Gitirana et al. 2007, Almeida et al. 2007) and in *R. ornata* (Felsemburgh et al. 2009), glandular aggregates constitute the parotoid gland, whose secretion is related to chemical defense against predators and parasites (Croce et al. 1973, Clarke 1997, Sakate and Lucas de Oliveira 2000).

In all hylids, examined in this study, melanophores occurred in the dorsal integument, but they did not always occur in the ventral integument. In contrast to melanophores, iridophores were visualized only in the dorsal region of *O. albicans*, *O. angrensis*, *O humilis*, and *S. similis*, while they occurred only in the ventral integument of *S. albicans*.

In all examined species, at least in the dorsal region of the integument, the Eberth-Katschenko (EK) layer was visualized as an acellular layer that was restricted to a region between the spongious and compact dermis. Moreover, the EK-layer was usually continuous in the dorsal integument, showing its typical basophilic and alcianophilic stainings, which were due to the glycoconjugate content. In *R. icterica* and *L. catesbeianus*, the EK-layer contained both dermatan sulfate and calcium, and occurred as scattered aggregates throughout the spongious dermis (Pelli et al. 2007, 2010). These mineral consists of calcium phosphate deposits (Katchburian et al. 2001). Moreover, calcium of the EK-layer was more concentrated in the dorsal integument of male toads, but no significant difference was detected in the integument of females (Azevedo et al. 2005). Elkan (1968) suggested that the absence of the EK layer in some anuran species may be correlated with the fixative type and storage time. Nevertheless, in this study, all hylids were fixed in the same manner, using the same fixative, and the presence or absence of the EK-layer varied according to the specimen and integument region. In addition, the EK-layer was absent in the ventral region of the integument of the species of *Scinax*, probably being a genus-specific feature. For some authors (Katchburian et al. 2001, Mangione et al. 2011), the EK-layer might be a remnant of an ancestral dermal skeleton. Toledo and Jared (1993) suggested that the calcium located in the EK layer participates in hydric balance, affecting the hydric absorption and retention. On the other hand, Azevedo and et al. (2007) demonstrated that hyaluronic acid (HA) occurs in the spongious dermis, suggesting that the entire spongious dermis acts as a hydric reservoir since HA, an important component of connective tissue matrices, is involved in promoting matrix assembly, tissue hydration and viscosity of some fluids (Laurent et al. 1996). However, the functional significance of the presence of calcium in the EK-layer remains unclear.

In *O. angrensis*, *O. flavoguttata*, *O. perpusilla*, *O. trapicheroi*, *S. similis* and *S. x-signatus*, cutaneous elevations were evident in the ventral integument, and were separated by a network of grooves. Cutaneous elevations were also noted in the ventral integument of *Hyla arborea* (Linnaeus, 1758) (Goniakowska-Witalinska and Kubiczek 1998). In *R. icterica* and in *Proceratophrys boiei* (Wied-Neuwied, 1824), *Proceratophrys laticeps* Izecksohn & Peixoto, 1981, *Proceratophrys appendiculata* (Günther, 1873) and *Odontophrynus americanus* (Duméril & Bibron, 1841), these elevations were separated by a network of grooves that probably acts as a distribution system of water from the ventral to dorsal surface of the integument (Azevedo and de Brito-Gitirana 2005, Felsemburgh et al. 2007). According to Lillywhite and Licht (1974), grooves can work as water distribution channels by a capillarity mechanism from one integument surface to another. Several authors have proposed that water distribution keeps the integument moist, protecting the animal against desiccation (Parakkal and Matoltsy 1964, Machin 1969, Duellman and Trueb 1994, de Brito-Gitirana and Azevedo 2005).

Although the usual patterns observed in the *Ololygon* and *Scinax* species, their integuments revealed histological characteristics. Thus, histological methods can be efficient to help characterize and differentiate of anuran integuments, thereby improving their taxonomy.

ACKNOWLEDGMENTS

We thank all colleagues for the help during fieldwork at Serra dos Órgãos National Park and the park staff for logistic support and permission for fieldwork. We also thank Conselho Nacional de Desenvolvimento Científico e Tecnológico (CNPq) and Fundação Carlos Chagas Filho de Amparo à Pesquisa do Estado do Rio de Janeiro (FAPERJ). TSS thanks support from CNPq (process 304374/2016-4).

REFERENCES

Almeida PG, Felsemburgh FA, Azevedo RA, de Brito-Gitirana L (2007) Morphological re-evaluation of the parotoid glands of *Bufo ictericus* (Amphibia, Anura, Bufonidae). Contributions to Zoology 763: 145–152.

Amey AP, Grigg GC (1995) Lipid-reduced evaporative water loss in two arboreal hylid frogs. Comparative Biochemistry and Physiology 111: 283–291. https://doi.org/10.1016/0300-9629(94)00213-D

Azevedo RA, Pelli AA, Ferreira-Pereira A, Santana ASJ, Felsemburgh F, de Brito-Gitirana L (2005) Structural aspects of the Eberth-Katschenko layer of *Bufo ictericus* integument: histochemical characterization and biochemical analysis of the cutaneous calcium (Amphibian, Bufonidae). Micron 361: 61–65. https://doi.org/10.1016/j.micron.2004.06.004

Azevedo RA, Santana ASJ, de Brito-Gitirana L (2006) Dermal collagen organization in *Bufo ictericus* and in *Rana catesbeiana* integument (Anuran, Amphibian) under the evaluation of laser confocal microscopy. Micron 37: 223–228. https://doi.org/10.1016/j.micron.2005.11.001

Bagnara JT, Taylor JD, Hadleu ME (1968) The dermal chromatophore unit. The Journal of Cell Biology 38: 67–79. https://doi.org/10.1083/jcb.38.1.67

Barbeau TR, Lillywhite HB (2005) Body wiping behaviors associated with cutaneous lipids in hylid tree frogs of Florida. The Journal of Cell Biology 208: 2147–2156. https://doi.org/10.1242/jeb.01623

Blaylock LA, Ruibal R, Platt-Aloia K (1976) Skin structure and wiping behavior of *Phyllomedusine* frogs. Copeia 2: 283–295. https://doi.org/10.1242/jeb.01623

Brizzi R, Delfino G, Pellegrini R (2002) Specialized mucous glands and their possible adaptive role in the males of some species of *Rana* (Amphibia, Anura). Journal of Morphology 254: 328–341. https://doi.org/10.1002/jmor.10039

Brizzi R, Delfino G, Pellegrini R (2002) Specialized mucous glands and their possible adaptive role in the males of some species of Rana (Amphibia, Anura). Journal of Morphology 254: 328–341. https://doi.org/10.1002/jmor.10039

Brunetti AE, Faivovich J, Hermida GN (2012) New insights into sexually dimorphic skin glands of anurans: the structure and ultrastructure of the mental and lateral lands in *Hypsiboas punctatus* (Amphibia: Anura: Hylidae). Journal of Morphology 273: 1257–1271. https://doi.org/10.1002/jmor.20056

Christian K, Parry D (1997) Reduced rates of water loss and chemical properties of skin secretions of the frog *Litoria caerulea* and *Cyclorana australis*. Australian Journal of Zoology 45: 13–20. https://doi.org/10.1071/ZO96046

Clarke BT (1997) The natural history of amphibian skin secretions, their normal functioning and potential medical applications. Biology Reviews 72: 365–379. https://doi.org/10.1111/j.1469-185X.1997.tb00018.x

Croce G, Giglioli N, Bolognani L (1973) Antimicrobial activity in the skin of *Bombina variegata pachypus*. Toxicon 11: 99–100. https://doi.org/10.1016/0041-0101(73)90159-1

Daly JW (1995) The chemistry of poisons in amphibian skin. Proceedings of the National Academy of Sciences 92: 9–13. https://doi.org/10.1073/pnas.92.1.9

de Brito-Gitirana L, Azevedo RA (2005) Morphology of *Bufo ictericus* integument. Micron 364: 532–538. https://doi.org/10.1016/j.micron.2005.03.013

de Brito-Gitirana L, Azevedo RA, Pelli AA (2007) Expression pattern of glycoconjugates in the integument of *Bufo ictericus* (Anura, Bufonidae): Biochemical and histochemical (lectin) profiles. Tissue and Cell 39: 415–421. https://doi.org/10.1016/j.tice.2007.08.002

de Brito-Gitirana L (2013) Coleção Conhecendo. Histologia dos Tecidos. Publit Soluções Editoriais, Rio de Janeiro, 252 pp.

de Brito-Gitirana L (2015) Coleção Conhecendo. Histologia: Sistema Tegumentar. Publit Soluções Editoriais, Rio de Janeiro, 59 pp.

Delfino G, Brizzi R, Feri L (1995) Chemical skin defense in *Bufo bufo*: an ultrastructural study during ontogenesis. Zoologische Anzeiger 234: 101–111.

Delfino G, Brizzi R, Alvarez BB, Gentili M (1999) Granular cutaneous glands in the frog *Physalaemus biligonigerus* (Anura, Leptodactylidae): comparison between ordinary serous and "inguinal" glands. Tissue and Cell 316: 576–586. https://doi.org/10.1054/tice.1999.0071

Delfino G, Drews RC, Magherini S, Malentacchi C, Nosi D, Terreni A (2006) Serous cutaneous glands of the Pacific tree-frog Hyla regilla (Anura, Hylidae): patterns of secretory release induced by nor-epinephrine. Tissue and Cell 38: 65–77. https://doi.org/10.1016/j.tice.2005.11.002

Delfino G, Giach F, Malentacchi C, Nosi D (2015) Ultrastrucutiral evidence of serous gland polymorphism in the skin of the tungara frog Engystomops pustulosus (Anura Leptodactylidae) Anatomical Record 298: 1659–1667. https://doi.org/10.1002/ar.23189

Duellman WR, Trueb L (1994) Biology of Amphibians. Johns Hopkins University Press, Baltimore.

Elkan E (1968) Mucopolysaccharides in the anuran defense against desiccation. Journal of Zoology 155: 19–53. https://doi.org/10.1111/j.1469-7998.1968.tb03028.x

Felsemburgh FA, Carvalho-e-Silva SP, de Brito-Gitirana L (2007) Morphological characterization of the anuran integument of the *Proceratophrys* and *Odontophrynus* genera (Amphibia, Anuran, Leptodactylidae). Micron 38: 439–445. https://doi.org/10.1016/j.micron.2006.06.015

Felsemburgh FA, Almeida PG, Carvalho-e-Silva SP, de Brito-Gitirana L (2009) Microscopical methods promote the understanding of the integument biology of *Rhinella ornata*. Micron 402: 198–205. https://doi.org/10.1016/j.micron.2008.09.003

Fox H (1986a) Epidermis. In: Bereiter-Hahn J, Matoltsy AG, Richards S (Eds) Biology of the Integument 2: Vertebrates. Springer-Verlag, Berlin, 78–110. https://doi.org/10.1007/978-3-662-00989-5_5

Fox H (1986b) Dermis. In: Bereiter-Hahn J, Matoltsy AG, Richards S (Eds) Biology of the Integument 2: Vertebrates. Springer-Verlag, Berlin, 111–149. https://doi.org/10.1007/978-3-662-00989-5_6

Frost DR (2016) Amphibian Species of the World: an Online Reference. American Museum of Natural History, New York. http://research.amnh.org/vz/herpetology/amphibia/index.php//Amphibia/Anura/Hylidae [Accessed: 04/05/2016]

Gomez NA, Acosta M, Zaidan F, Lillywhite HB (2006) Wiping behavior, skin resistance, and the Metabolic Response to dehydration in the Arboreal Frog *Phyllomedusa hypochondrialis*. Physiological and Biochemical Zoology 796: 1058–1068. https://doi.org/10.1086/507659

Gonçalves VF, de Brito-Gitirana L (2008) Structure of the sexually dimorphic gland of *Cycloramphus fuliginosus* (Amphibia, Anura, Cycloramphidae). Micron 39: 32–39. https://doi.org/10.1016/j.micron.2007.08.005

Goniakowska-Witalinska L, Kubiczek U (1998) The structure of the skin of the tree frog (*Hyla arborea* L.). Annals of Anatomy 180: 237–246. https://doi.org/10.1016/S0940-9602(98)80080-0

Greven H, Zanger K, Schwinger G (1995) Mechanical properties of the skin of *Xenopus laevis* (Anura, Amphibia). Journal of Morphology 224: 15–22. https://doi.org/10.1002/jmor.1052240103

Ham AW (1977) Histologia. Guanabara Koogan, Rio de Janeiro, 159 pp.

Jeckel AM, Saporito RA, Grant T (2015) The relationship between poison frog chemical defenses and age, body size, and sex. Frontiers in Zoology 12: 27. https://doi.org/10.1186/s12983-015-0120-2

Katchburian E, Antoniazzi MM, Jared C, Faria FP, Souza Santos H, Freymüller E (2001) Mineralized dermal layer of the Brazilian tree frog *Corythomantis greeningi*. Journal of Morphology 248: 56–63. https://doi.org/10.1002/jmor.1020

Kiernan JA (1990) Histological & Histochemical Methods: Theory and Practice. Pergamon Press, Frankfurt.

Kierszenbaum AL (2004) Histology and Cell Biology: An Introduction to Pathology. Mosby, New York.

Koyama H, Nagai T, Takeuchi H, Hillyard SD (2001) The spinal nerves innervate putative chemosensory cells in the ventral skin of desert toads, *Bufo alvarius*. Cell and Tissue 30: 185–192. https://doi.org/10.1007/s004410100370

Laurent TC, Laurent UBG, Fraser JRE (1996) The structure and function of hyaluronan: An over view. Immunology and Cell Biology 74: A1–A7. https://doi.org/10.1038/icb.1996.32

Lillie RD, Fullmer HM (1976) Histopathologic Technique and Practical Histochemistry. MacGraw-Hill Book Co, New York.

Lillywhite HB (2004) Plasticity of the water barrier in vertebrate integument. International Congress Series 1275: 283–290. https://doi.org/10.1016/j.ics.2004.08.088

Lillywhite HB, Licht PA (1974) Movement of water over toad skin: functional role of epidermal sculpturing. Copeia 1: 165–171. https://doi.org/10.2307/1443019

Machin J (1969) Passive water movements through skin of the toad *Bufo marinus* in air and in water. American Journal of Physiology 216: 1562–1568.

Mangione S, Garcia G, Cardozo OM (2011) The Eberth-Katschenko layer in three species of *Ceratophryines* anurans (Anura: Ceratophrydae). Acta Zoologica 921: 21–26. https://doi.org/10.1111/j.1463-6395.2009.00442.x

Mearow KS, Diamond J (1988) Merkel cells and the mechanosensitivity of normal and regenerating nerves in *Xenopus* skin. Neuroscience 26: 695–708. https://doi.org/10.1016/0306-4522(88)90175-3

Mills JW, Prum BE (1984) Morphology of the exocrine glands of the frog skin. American Journal of Anatomy 1711: 91–106. https://doi.org/10.1002/aja.1001710108

Moreno-Gómez F, Duque T, Fierro L, Arango J, Peckham X, Asencio-Santofimio H (2014) Histological Description of the Skin Glands of *Phyllobates bicolor* (Anura: Dendrobatidae) Using Three Staining Techniques. International Journal of Morphology 323: 882–888. https://doi.org/10.4067/S0717-95022014000300022

Mowry RW (1963) The special value of methods that color both acidic and vicinal hydroxyl groups in the histochemical study of mucins. With revised directions for the colloidal iron stain, the use of Alcian Blue G8X and their combinations with the periodic acid-Schiff reaction. Annals of the New York Academy of Sciences 106: 402–423. https://doi.org/10.1111/j.1749-6632.1963.tb16654.x

Nosi D, Terreni A, Alvarez BB, Delfino G (2002) Serous gland polymorphism in the skin of *Phyllomedusa hypochondrialis azurea* (Anura, Hylidae): response by different gland types to norepinephrine stimulation. Zoomorphology 121: 139–148. https://doi.org/10.1007/s004350100051

Parakkal PF, Matoltsy AG (1964) A study of the fine structure of the epidermis of *Rana pipiens*. The Journal of Cell Biology 20: 85–94. https://doi.org/10.1083/jcb.20.1.85

Pelli AA, Azevedo RA, Cinelli LP, Mourão PA, de Brito-Gitirana L (2007) Dematan sulfate is the major metachromatic glycosaminoglycan in the integument of the anuran *Bufo ictericus*. Comparative Biochemistry and Physiology Part B, Biochemistry Molecular Biology 146: 160–165. https://doi.org/10.1016/j.cbpb.2006.10.098

Pelli AA, Cinelli LP, Mourão PA, de Brito-Gitirana L (2010) Glycosaminoglycans and glycoconjugates in the adult anuran integument (*Lithobates catesbeianus*). Micron 41: 660–665. https://doi.org/10.1016/j.micron.2010.03.001

Rigolo JR, Almeida JA, Ananias F (2008) Histochemistry of skin glands of *Trachycephalus aff. venulosus* Laurenti, 1768 (Anura, Hylidae). Micron 391: 56–60. https://doi.org/10.1016/j.micron.2007.08.006

Sakate M, Oliveira LPC (2000) Toad envenoming in dogs: effects and treatment. Journal of Venomous Animals and Toxins 6: 46–58. https://doi.org/10.1590/S0104-79302000000100003

SBH (2016) Brazilian Amphibians: List of Species. Sociedade Brasileira de Herpetologia, available on line at: http://www.sbherpetologia.org.br/images/LISTAS/Lista_Anfibios 2016.pdf [Accessed: 30/06/2017]

Sullivan PA, Hoff KVS, Hillyard SD (2000) Effects of anion substitution on hydration behavior and water uptake of the red-spotted toad, *Bufo punctatus*: is there an anion paradox in amphibian skin? Chemical Senses 25: 167–172. https://doi.org/10.1111/j.1748-1716.2010.02200

Toledo RC, Jared C (1993) Cutaneous adaptations to water balance in amphibians. Comparative Biochemistry and Physiology 105A: 593–608.

Warburg MR, Rosenberg M, Roberts JR, Heatwole H (2000) Cutaneous glands in the Australian hylid *Litoria caerulea* (Amphibia, Hylidae). The Italian Journal of Anatomy and Embryology 201: 341–348. https://doi.org/10.1007/s004290050323

Withers PC, Hillman SS, Drewes RC (1984) Evaporative water loss and skin lipids of anuran amphibians. Journal of Experimental Biology 232: 11–17. https://doi.org/10.1002/jez.1402320103

Author Contributions: HMA, LBG and TSS participated equally in the preparation of this article.

Competing Interests: The authors have declared that no competing interests exist.

7

Age structure and growth of the rough scad, *Trachurus lathami* (Teleostei: Carangidae), in the Southeastern Brazilian Bight

Lygia C. Ruas[1], André M. Vaz-dos-Santos[2]

[1]*Programa de Pós-graduação em Aquicultura e Pesca, Instituto de Pesca. Avenida Francisco Matarazzo 455, Parque da Água Branca, 05001-970 São Paulo, SP, Brazil.*
[2]*Laboratório de Esclerocronologia, Departamento de Biodiversidade, Universidade Federal do Paraná. Rua Pioneiro 2153, Jardim Dallas 85950-000 Palotina, PR, Brazil.*
Corresponding author: André M. Vaz-dos-Santos (andrevaz@ufpr.br)

http://zoobank.org/87858997-8113-4717-A67A-673CD99E3D52

ABSTRACT. The rough scad, *Trachurus lathami* Nichols, 1920, is a small pelagic species distributed along the West Atlantic coast. It is most abundant in the Southern Brazil (28°30'–34°S) and in the Southeastern Brazilian Bight (SEBB, 22°–28°30'S). The rough scad is fished by purse seines, which main target is the Brazilian sardine, *Sardinella brasiliensis* (Steindachner, 1879). Age and growth are vital to understand the life cycle of a species, to fishery management and ecosystem modeling. This study aimed to assess the age and growth of *T. lathami*, to identify its age structure in the SEBB, and to evaluate what causes the wide differences among *Trachurus* species in terms of body size and growth parameters. Data available on *T. lathami* was attained between 2008 and 2010 from surveys at SEBB. A total of 278 whole otoliths of *T. lathami*, total length between 27 mm and 208 mm, were analyzed and compared with the only other source of otolith data, from 1975. Three blind readings were performed and assessed using traditional methods to study fish age and growth. Zero up to eight rings were found, each ring corresponding to one year in the life of an individual of this species. The von Bertalanffy growth model parameters were L_∞ = 211.90 mm and K = 0.319 year-1. The results of the analyses have shown similarities between 1975 and 2008-2010, indicating that the otolith development, the growth pattern and the age structure remained stable. *T. lathami* is the smallest species of *Trachurus* and it has the highest growth rates among them. This is probably related to the different temperatures where larvae/juvenile and adult grow, to the absence of a strong fishing pressure and to decadal population variability.

KEY WORDS. ECOSAR, otolith, sclerochronology, von Bertalanffy.

INTRODUCTION

In the Atlantic Ocean, the rough scad, *Trachurus lathami* Nichols, 1920, is a pelagic species distributed between the United States and North of Argentina, mainly on the continental shelf. It occurs between 50 and 100 m in depth, where it forms schools (Smith-Vaniz 2002). In Brazil, their largest concentrations have been recorded both in the Southeastern Brazilian Bight (SEBB, between 22° and 28°30'S) and in the Southern region (between 28°30' and 34°S) (Saccardo and Haimovici 2007). In these areas, the rough scad is an important fishery resource in purse seines, along with other important pelagic species such as the *Sardi-nella brasiliensis* (Steindachner, 1879) and *Opisthonema oglinum* (Lesueur, 1818) (Saccardo and Haimovici 2007, UNIVALI 2011).

In the SEBB, rough scad landings varied a lot until the end of the 1990's (Valentini and Pezzutto 2006), a period when this fish was caught in association with the Brazilian sardine (*S. brasiliensis*) (Fig. 1). When there were fewer sardines, purse seine fleets would capture and land the rough scad (Saccardo et al. 2005). After this period, sardine landings increased and rough scad landings dropped, without a clear association between them. Between 2005 and 2008, rough scad landings were around 700 t/year, with a later reduction to an average of 88 t (2009–2010) (MMA 2007a, b, 2008, UNIVALI 2009, 2011, Instituto de Pesca

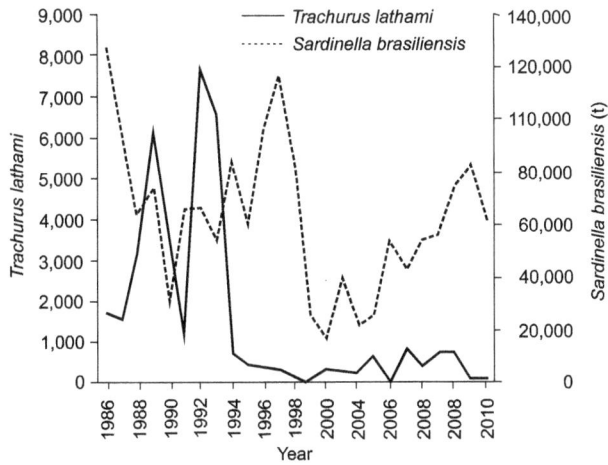

Figure 1. Commercial landings of *Trachurus lathami* and *Sardinella brasiliensis* in the Southeastern Brazilian Bight (SEBB). Data of 2008, 2009 and 2010 are restricted to states of Paraná and São Paulo that representing 95% of the total landings. Data source: Valentini and Pezzutto 2006, MMA 2007a, b, 2008 UNIVALI 2009, 2010, 2011, Instituto de Pesca 2013.

2013). Historically, this species has presented intense population fluctuations (Katsuragawa and Ekau 2003), oscillations that are also common to the other species of *Trachurus* Rafinesque, 1810, related to climate and environmental conditions, fishing pressure, biological elements (feeding, reproduction, growth, recruitment) and the combination of those (Arancibia and Neira 2002, Espino 2013, Geist et al. 2015).

There is a single stock of the rough scad at the SEBB (Saccardo and Haimovici 2007). Its life cycle and fishery (Saccardo 1987, Saccardo and Kasturagawa 1995, Saccardo et al. 2005); growth (Saccardo and Katsuragawa 1995); diet (Meneghetti and Alves 1971, Carvalho and Soares 2006); parasites (Braicovich et al. 2012); and larvae and juveniles (Katsuragawa and Matsuura 1992, Katsuragawa and Ekau 2003, Campos et al. 2010) have been investigated. Some biological aspects can be summarized from those studies, as follow. The rough scad breeds throughout the year, with a peak between October and December (springtime). The larvae are distributed all over the continental shelf. Individuals grow up to a total length of 200-250 mm (TL), with the first maturation at 115-132 mm TL, living up to 8–9 years. The rough scad is mainly a zooplankton feeder, despite some records of predatory behavior. The otoliths were studied in terms of age and growth (Saccardo and Katsuragawa 1995) and morphology (Siliprandi et al. 2014).

Studies on age and growth provide an essential tool to understand the biology and ecology of fish, providing a foundation for population dynamics assessments (King 2007). Growth parameters have multiple applications in fishery management (Beverton and Holt 1993, Froese and Binohlan 2000, Sponaugle 2010), and their estimates through the analysis of otoliths provide

precise and accurate results (Green et al. 2009). Several studies on age and growth, based on otoliths, have been conducted on different species of *Trachurus* (Webb and Grant 1979, Horn 1993, Karlou-Riga and Sinis 1997, Karlou-Riga 2000, Araya et al. 2001, Waldron and Kesrtan 2001, Kasapoglu and Duzgunes 2013, among others), mainly in view of their commercial relevance (Checkley et al. 2009). From these studies, it is possible to conclude that age and growth are very different among *Trachurus* species, and that otolith analysis is useful, allowing the comprehension of their biological patterns (Karlou-Riga 2000, Abaunza et al. 2003).

The significance of the small pelagic fisheries at SEBB led the Brazilian government to promote evaluation and monitoring programs of these resources. Although they were not continuous, the most recent initiative was the ECOSAR Program (Prospection and assessment of the sardine stock biomass in the Southeastern coast by the use of hydro-acoustic methods) carried out between 2008 and 2010 (Cergole and Dias Neto 2011). Even though the target species was *S. brasiliensis*, other representative species captured during the surveys, among which is *T. lathami*, were also accounted for. This study aimed to assess the age and growth of *T. lathami*, to identify its age structure in the Southeastern Brazilian Bight, and to evaluate what causes the wide differences among *Trachurus* species in terms of body size and growth parameters.

MATERIAL AND METHODS

Four survey cruises were carried out between 20 and 100 m deep with the OV Atlântico Sul by FURG during January-February 2008 (Summer), November 2008 (Spring), September-October 2009 (Spring) and February-March 2010 (Summer) in the Southeastern Brazilian Bight (22°–28°30′S). Transects in perpendicular profiles, oblique to the coast, were followed. The echo sounder (Simrad EK500) was employed and when shoals were detected, both pelagic trawling and purse seine were carried out. Details on the methodology are available in Rossi-Wongtschowski et al. (2014). *Trachurus lathami* were caught in 17 fishing operations (Fig. 2). On board, samples were frozen. In the laboratory, the total length (TL, mm) of individuals was measured and their *sagittae* otoliths were removed, washed, dried and stored in microtubes (FAO 1981).

Whenever possible, ten otoliths of *T. lathami* were selected from each survey by total length class (10 mm) (Araya et al. 2001). Images of the entire left otolith under water (Fig. 3) were obtained using a stereomicroscope coupled with an image analyzer, and the length (measurement of the horizontal projection of its ends in relation to the longer axis – OL, mm) and height (measurement of the vertical projection in relation to the higher axis of the structure – OH, mm) of the otolith were recorded. The weight of the otolith (OW, g) was obtained with an analytical balance. In order to describe otolith growth in relation to body growth, the allometric (potential) model (Huxley 1993) was used to fit regressions among the total length and the measurements of the otoliths (Vaz-dos-Santos 2015a). The adequacy of

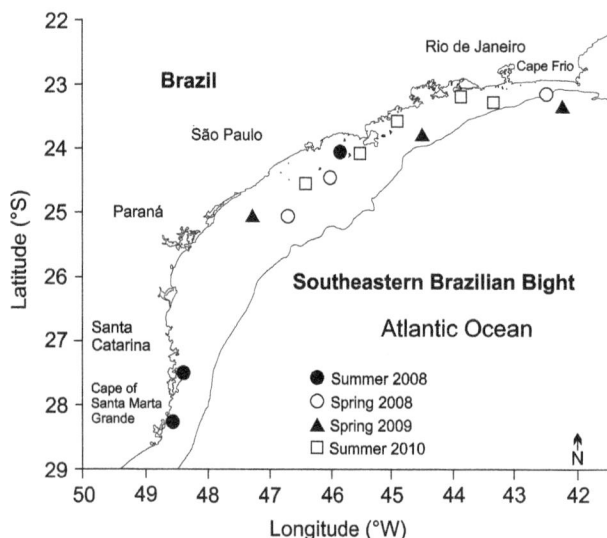

Figure 2. Map of the Southeastern Brazilian Bight (SEBB) showing the locations of fishing hauls with *Trachurus lathami* catches by period (n = 17, some points are overlapped in the map).

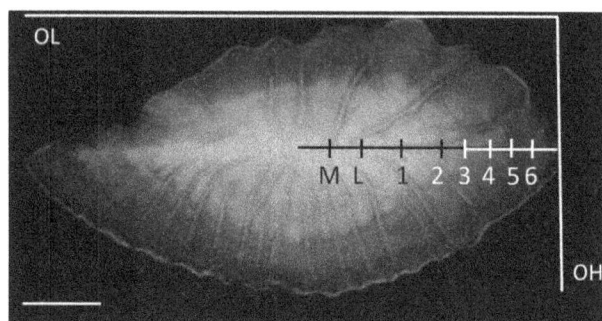

Figure 3. *Trachurus lathami*: external surface of a left otolith showing measurements (OL = otolith length and OH = otolith height) and ring analysis (M and L are rings formed before the first annual one; more details in the text). Scale bar: 1 mm.

the regressions was checked by the values of the coefficient of determination (r^2) and the residual analysis (Bervian et al. 2006, Vaz-dos-Santos and Rossi-Wongtschowski 2013).

In order to study age and growth, the annual growth zones of the otoliths (macrostructural analysis) were analyzed (Brothers 1987). Whole otolith images were used to identify the central opaque nucleus and, from this, in the posterior axis to the posterior edge (otolith radius, Ro, mm), the translucent zones. The complete and continuous translucent zones were counted and measured from the beginning of their formation (Fig. 3) (Saccardo and Katsuragawa 1995, Karlou-Riga and Sinis 1997, Waldron and Kerstan 2001). Three blind readings were carried out by the same reader, whose consistency was checked by the average percentage error (APE) (Beamish and Fournier 1981) and

the value of the coefficient of variation (CV) (Campana 2001). In order to check the consistency of the ring analysis, box plot and constancy analysis were applied (Vaz-dos-Santos 2015b). A box plot of the ring radius by the ring (category) was built in order to check whether there is overlap among measurements. After checking the assumptions, ANOVA followed by post-hoc Tukey test were performed to compare the averages of the radius. A constancy analysis (scatter plot between total length and the rings radius) was applied, and the linearity of each ring group was tested using regression analysis (except for rings 7 and 8, with a narrow range of total length). All statistical procedures followed Zar (2010) with 5% of significance. The ring counts and edge patterns allowed age attribution, since the formation of an annual ring had been previous validated, occurring between October and December (Saccardo and Katsuragawa 1995). Two rings are formed during the first year of life, before the first annual ring (Saccardo and Katsuragawa 1995). Whenever possible, these rings were recorded, but they were not used in any analysis.

The von Bertalanffy growth model parameters (VBGM) (L_∞, K, t_0) were estimated from observed lengths per age and by the average lengths per age using the least-squares iterative method (Aubone and Wöhler 2000). From the growth parameters, the inverse VBGM was used to estimate the age of the entire sample of *T. lathami*. Although the value of the theoretical age at zero length has been estimated, it was not used to estimate the age with the inverse VBGM. Next, an age length key (frequencies of individuals by age and TL class) was built and the space-time distributions by ages were analyzed in the surveyed periods using maps.

RESULTS

The total length of 1,312 *T. lathami* individuals sampled varied between 27 mm and 208 mm, resulting in bimodal distributions in each period surveyed, one of juveniles (20 to 60 mm) and a second mode composed of adults (>115 mm TL), which predominated in the seasons analyzed (Fig. 4). The few numbers of fish sampled between 100-140 mm TL resulted from the selectivity of the fishing gear employed.

Following the selection criteria, 282 otoliths of *T. lathami* were analyzed (it was not always possible to attain ten otoliths from each survey by TL class). The regressions among the otolith measurements (Table 1) represented properly the development of the structure (p < 0.001 in the three regressions, $0.970 < r^2 < 0.979$), thus enabling the use of the otoliths in the growth study. In the adjusted models (TL vs. OL, TL vs. OH, TL vs. OW), residuals did not show any bias. The best fit was presented by the TL vs. OL model ($r^2 = 0.979$), showing that the otolith length provides the best representation of fish growth, following the anterior-posterior axis.

After three readings, 278 otoliths were considered legible (98.6%) and four non legible (1.4%), which were not considered in the following analysis. The average percentage error among

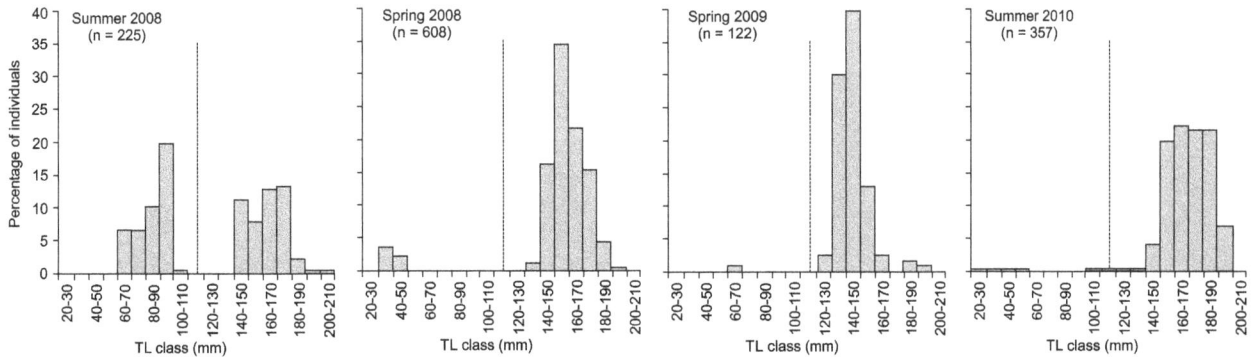

Figure 4. *Trachurus lathami*: length frequency distribution by period (dashed line indicates the average length of first maturity, L_{50} = 115 mm).

Table 1. *Trachurus lathami:* coefficients of potential regressions between total length (TL, mm) and otolith length (OL, mm), height (OH, mm) and weight (OW, g). (r^2) Coefficient of determination, n = 282.

Regression	a ± $IC_{95\%}$	b ± $IC_{95\%}$	r^2
TL vs. OL	0.041 ± 0.003	0.991 ± 0.017	0.979
TL vs. OH	0.050 ± 0.004	0.825 ± 0.017	0.970
TL vs. OW	$3.96. 10^{-8}$ ± $8.84. 10^{-9}$	2.524 ± 0.052	0.971

the three readings was 4.1%, and the coefficient of variation was 5.4%. Otoliths with up to eight rings were observed with high precision and consistency, revealed by the box plot and constancy analysis (Figs 5, 6). In relation to the box plot analysis, the average radius of rings was significantly different (ANOVA F = 1527.76, p < 0.001), except for rings 7 and 8 (p > 0.05). Linearity tested in the constancy analysis was not significant in all cases (p > 0.05). Fish with only two rings were not found. The age determination resulted in fishes with age zero up to eight years old.

The von Bertalanffy growth model parameters were L_{∞} = 211.90 mm, K = 0.319 year^{-1} and t_0 = -0.576 years for observed lengths per age, and L_{∞} = 206.31 mm, K = 0.336 year^{-1} and t_0 = -0.578 years for average lengths per age. When considering the maximum length obtained in the sampling (208 mm), the parameters estimated through the average lengths were disregarded, and the parameters fitted from observed lengths per age were adopted to estimate fish age (Fig. 7). The sample had fish from 0 to 8 years old or above (Table 2).

In relation to the space-time distributions, in the summer of 2008, the rough scad schools were concentrated closer to the coast at 27°–28°S (89%), with all ages represented, and a predominance of one-year old fish (Fig. 8), associated with the Cabo Frio upwelling (22°–23°S), around 100 m isobath. Another school was found at the continental shelf 24°–25°S (Fig. 9), similar to the summer of 2009 (Fig. 10). In the summer of 2010, the rough scad was more concentrated between 23°–24°S, and its school was composed mainly of 4-year old fish, or older (Fig. 11).

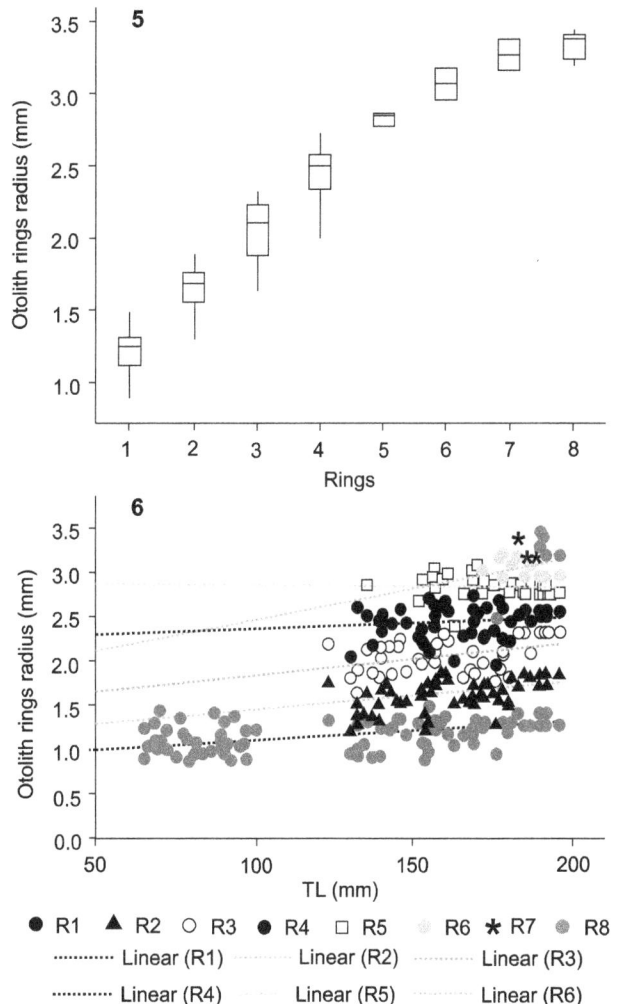

Figures 5–6. *Trachurus lathami:* (5) Box plot of ring radius (whiskers = minimum and maximum, box = interquartile range, bar = mean). (6) Constancy graph showing the position of rings radius (R) against total length (TL) (n = 278).

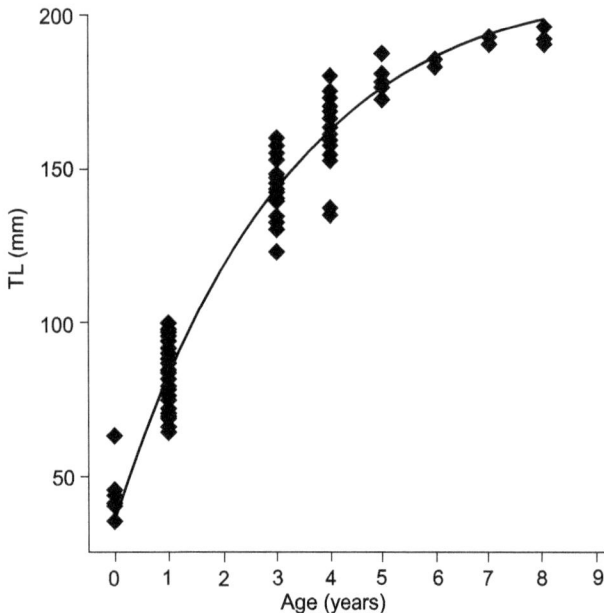

Figure 7. *Trachurus lathami:* von Bertalanffy growth curve fitted (line) to observed total lengths (points) by age (n = 278).

DISCUSSION

The growth parameters estimated for *T. lathami* in this study are the most recent after 1975 (Saccardo and Katsuragawa 1995). They originate from the only biological data available for this species in the Southwestern Atlantic (Cergole and Dias Neto 2011), and represent the population and the sampling adequately, due to the total length range analyzed. Previous information on the presence of up to 400 mm TL rough scads at the SEBB (Menezes and Figueiredo 1980) denotes very rare individuals, never recorded again. Historically, the longer total lengths recorded for the rough scad in the SEBB were 260 mm (Saccardo 1987), 207 mm (Figueiredo et al. 2002), 240 mm (Saccardo et al. 2005) and 261 mm (Bernardes et al. 2005). In all these studies, as well as in our results, individuals larger than 200 mm were seldom recorded. This indicates that *T. lathami* in the study area have smaller size structure (usually up to 180 mm SL) than in the Caribbean Sea (330 mm SL) (Smith-Vaniz 2002), but similar size structure to that found for the Uruguayan and Argentinean coasts (between 80-230 mm TL) (Cousseau and Perrotta 2004). Thus, the size structure analyzed in this work was representative of the stock, even in the absence of larger individuals.

The development of the otoliths of *T. lathami* had only been analyzed by Saccardo and Katsuragawa (1995), whose details are

Table 2. *Trachurus lathami:* age-length key in the Southeastern Brazilian Bight during 2008–2010: percentage (from the total) of individuals by length class and age; number of ages estimated through otolith readings and with the inverse von Bertalanffy Growth Model (VBGM).

Total length class (mm)	Age (years)									Otolith readings	Inverse VBGM	Total
	0	1	2	3	4	5	6	7	8+			
20–30	100%									–	2	2
30–40	100%									11	13	24
40–50	100%									15	–	15
50–60	100%									–	1	1
60–70		100%								16	–	16
70–80		100%								10	5	15
80–90		100%								10	13	23
90–100		100%								10	52	62
100–110			100%							1	1	2
110–120			100%							–	1	1
120–130			100%							3	1	4
130–140		7%	93%							18	27	45
140–150			100%							38	160	198
150–160			28%	72%						40	276	316
160–170				94%	6%					33	212	245
170–180					100%					30	171	201
180–190					25%	75%				27	84	111
190–200							80%		20%	16	14	30
200–210									100%	–	1	1
Total	42	116	10	328	458	244	83	24	7	278	1034	1312

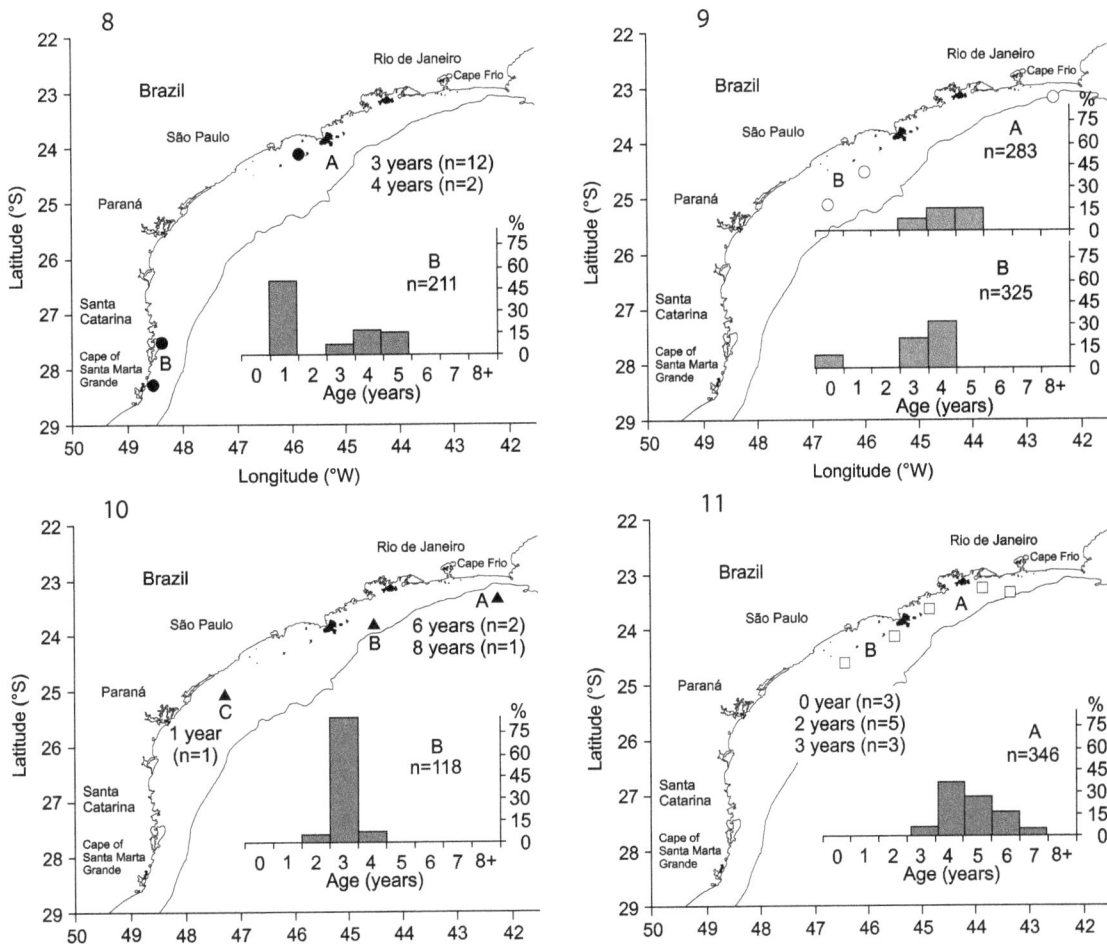

Figures 8–11. *Trachurus lathami*: space-time distribution of age groups in the Southeastern Brazilian Bight during (8) summer of 2008 (January-February), (9) spring of 2008 (November), (10) spring of 2009 (September-October) and (11) summer of 2010 (February–March).

available only in the unpublished thesis of Saccardo from 1980. Using these sources and Saccardo (1987), the visual inspection of scatter plots of the total length of individuals and otolith measurements (regression models were not adjusted in these former studies), it was possible to verify that otolith development has kept a similar pattern between the 1975 and 2008-2010. Namely, OL, OH and OW in relation to TL, respectively, showed similar values. In the present results, the allometric models and residual analysis did not show noticeable changes in the pattern of otolith development, indicating a single-phase growth (Bervian et al. 2006). Previously, it was evidenced a change in otolith thickness (two phases of otolith development) close to 130 mm TL (Saccardo and Katsuragawa 1995), due to the first gonadal maturation (Saccardo et al. 2005), which was not evidenced here probably due to the few numbers of individuals between 100-140 mm TL.

The quantitative elements (APE, CV, box plot, constancy analysis) used to evaluate otolith readings indicated that the

ring radius presented high precision and does not vary much (Campana 2001). Accuracy was also high, i.e., the position of the rings in the otoliths was similar to values reported previously (Saccardo and Katsuragawa 1995). These authors described the formation of two rings adjacent to the nucleus in 13% and 28% of the 1,908 otoliths analyzed, respectively. The material in this study hardly presented those rings and no clear patterns (at least in the macrostructural analysis); therefore, they were not taken into consideration. To investigate them, a microstructural analysis of the otoliths of *T. lathami* needs to be carried out.

The use of whole otoliths, sectioned otoliths or both, to count rings and to estimate age has been broadly discussed (Webb and Grant 1979, Karlou-Riga and Sinis 1997, Stewart and Ferrell 2001, Waldron and Kerstan 2001, Kerkich et al. 2013, Costa 2004, Dioses 2013, Goicochea et al. 2013, among others). Previous studies on *Trachurus* postulated that, when using whole otoliths, the following considerations need to be taken into account: (i) whole otoliths are suitable for estimating the age of fish smaller than 250 mm TL (Costa 2004); (ii) only complete, continuous and thicker rings should be counted (Dioses 2013); (iii) the precision (sensu Campana 2001) must be acceptable (Lyle et al. 2000); (iv) otoliths of adults must not show multi-ring formation (Karlou-Riga and Sinis 1997). These four considerations can be safely applied to our data and results. Actually, the analysis of whole otoliths in *Trachurus* is not recommended in older fish with thicker otoliths (Horn 1993, Karlou-Riga and Sinis 1997, Karlou-Riga 2000, Lyle et al. 2000). In the case of *T. lathami*, a smaller species than its congenerics, this was not an issue. Added to this, the present study was also looking for comparisons with the previously one (Saccardo and Katsuragawa 1995).

The lack of self-validation, which was not possible in view of our sample design, does not compromise our results. The previously adopted validation was correctly performed, based on 399 otoliths (21% of the total) with coincident ring analysis by Saccardo and Katsuragawa (1995). It was based on the percentage of the edge pattern (Panfili and Morales Nin 2002) and on the average lengths by age (Campana 2001), methods that are accepted for validation. Moreover, the resemblance found in the regressions and in the ring analysis between 1975 (Saccardo and Katsuragawa 1995) and 2008-2010 (present study) suggests that the development pattern of the otolith has remained stable. This could also be considered in the case of growth ring formation.

The growth parameters (L_∞, K) estimated with observed lengths per age were most suitable to describe the growth of *T. lathami*, since that obtained from the average lengths per age underestimated the maximum length values. The estimated theoretical age at zero length, is inconsistent in biological terms, since the rough scad hatches around 1.5 mm long (Katsuragawa and Ekau 2003), corresponding to seven days of age using the VBGM adjusted in this study. This value seldom has a biological meaning, since it would represent the zero length age whenever the growth continues following the same pattern described by the equation (Pauly 1984), which is not true. Several authors (Abaunza et al. 2003, Yankova et al. 2010, Yoda et al. 2014), working on *Trachurus*, also found inconsistent t_0 values due to the lack of small fish in their samples. In the present study, the

t_0 value led to overestimated ages, and it should not be used.

Comparison of the growth parameters obtained here with previous estimates in Brazil (Table 3) showed that the growth pattern and age structure of *T. lathami* have been stable (Saccardo and Katsuragawa 1995, Saccardo et al. 2005). However, the rough scad has historically presented seasonal displacements within the SEBB, resulting in variations in the space-time distribution of the individuals in each age (Saccardo 1987, Saccardo and Haimovici 2007). Except for the summer of 2008, the typical spring-summer pattern (September to March), when the rough scad schools usually concentrated between 23° and 26°S for spawning, was observed. Such movement coincides with the most productive period in terms of water column eutrophication (Castro et al. 2006) and biological productivity (Pires-Vanin et al. 1993), which favors the rough scad in terms of their feeding habits (Carvalho and Soares 2006) and conditions for larval growth (Katsuragawa and Matsuura 1992).

Comparison with the nine congeneric species, considering the parameters of VBGM for sex pooled and based on the total length available at the FishBase (Froese and Pauly 2015), confirmed that the age and growth of species of the genus are broadly different (Webb and Grant 1979, Araya et al. 2001). *Trachurus lathami* presented the lowest value of the maximum theoretical length ($L_\infty = 211.90$ mm), together with other lower estimates of this parameter for the species in Brazil (previously mentioned in the text). Although the inverse relationship between L_∞ and K is well known, some considerations about this can be done, as follows.

Worldwide, environmental conditions play an important role in the growth of *Trachurus* species (Geist et al. 2013, Sassa et al. 2014). In the FishBase (Froese and Pauly 2015), the growth parameters of other *Trachurus* species were associated with low water temperature (average = 14.4 °C, n = 50). This indicates that most of these species grow in cold waters, which also applies to most *T. lathami*. The temperature of the SEBB continental shelf is usually higher than 20 °C (Castro et al. 2006) and values lower than 20 °C are restricted to the upwelling areas, mainly in the spring and summer (October to March) (Braga and Niencheski 2006), when the spawning and initial growth of *T. lathami* larvae take place (Saccardo and Katsuragawa 1995). On the other hand,

Table 3. *Trachurus lathami:* growth parameters estimated and average total length at age for the species in the Southeastern Brazilian Bight.

Method	Growth parameters			Average length at age (mm)								
	L	K	t_0	TL_0	TL_1	TL_2	TL_3	TL_4	TL_5	TL_6	TL_7	TL_8
Otoliths (1975)[a]	258.97	0.160	-1.85	66	94	118	139	157	172	184	195	205
Back-calculation (1975)[a]	228.46	0.250	-0.56	30	74	108	135	155	172	184	194	202
Otoliths (1975)[b]	252.00	0.170	-1.73	64	94	118	139	157	172	184	195	204
ELEFAN (1997–1998)[b]	270.00	0.250	–	0	60	106	142	171	193	210	223	233
Otoliths (2008–2010)[c]	211.90	0.319	–	40	79	–	145	164	179	184	190	191
All sample (2008–2010)[c]	211.90	0.319	–	39	86	121	145	160	175	184	191	198
General average length[d]	–	–	–	48	81	114	141	161	177	188	198	205

Sources: [a]Saccardo and Katsuragawa 1995, [b]Saccardo et al. 2005, [c]present study, [d]average length of the data in the table

part of the adult stock can be found in deeper and colder waters (11.5–18.5 °C) (Haimovici et al. 2008). Larvae and juveniles grow in different environmental conditions of the adults, fact that may cause variations in the length and age structure, which could not be evaluated here.

Another factor that may explain the differences in body size and growth parameters (between 1975 and 2008-2010, and in relation to the congeneric species) is fishing. In Brazil, during the 1970s, the fishing pressure on *T. lathami* was quite strong (Saccardo and Katsuragawa 1995, Saccardo et al. 2005), but this has changed: since 1997, the species has been fished accidentally, as accessory fauna (Saccardo and Haimovici 2007). All other *Trachurus* species, on the other hand, are target species of fisheries, and are under an intense pressure (Araya et al. 2001). Usually, growth parameters are influenced by density dependent and independent causes (Cardoso and Haimovici 2011). In the present case, the reduction in body size and the increase in growth rates could be due to an increase in the density of the *T. lathami* population, which apparently has not been overfished in the period analyzed.

Decadal shifts in the populations of pelagic fish (Alheit et al. 2009), similar to the idea suggested by the visual aspect in Fig. 1 (although it only expresses the landings instead of abundance, density or biomass), deserve further inspection. The growth parameters estimated here are essential elements for the management of small pelagic species at SEBB, and this should not be ignored in future assessments. Besides, the fishing statistics in Brazil are limited and there is no tradition of sampling biological landings. The lack of biological data series is notorious. As long as such deplorable historic situation persists, basic and fundamental studies of auto-ecology like this will remain scanty. Undoubtedly, the next study on the age and growth of the *T. lathami* cannot wait 40 years.

ACKNOWLEDGMENTS

We thank IBAMA/ICMBio for funding ECOSAR surveys, Lauro S.P. Madureira (FURG) and their staff during surveys and all students who processed the biological samples. We express our gratitude to Maria Cristina Cergole, Suzana Anita Saccardo, Carmen Lúcia Del Bianco Rossi Wongtschowski, Antônio Olinto Ávila da Silva, Marcus Rodrigues da Costa and Teodoro Vaske Júnior. To CAPES for the scholarship granted to the first author. For referees and their valuable recommendations. The first author expresses his gratitude to CNPq due to the research grant 305403/2015-0.

REFERENCES

Abaunza P, Gordo L, Murta A, Eltink ATGW, Garc MT (2003) Growth and reproduction of horse mackerel, *Trachurus trachurus* (Carangidae). Reviews in Fish Biology and Fisheries 13: 27–61. https://doi.org/10.1023/A:1026334532390

Alheit J, Roy C, Kifani S (2009) Decadal-scale variability in populations. In: Checkley D, Alheit J, Oozeki Y, Roy C (Eds) Climate change and small pelagic fish. Cambridge University Press, Cambridge, 64–87. https://doi.org/10.1017/CBO9780511596681.007

Arancibia H, Neira S (2002) Does ENSO induce changes in recruitment of horse mackerel (*Trachurus symmetricus*) and in the long-term trend of the trophic level of fishery landings in central Chile? Investigaciones Marinas 30. https://doi.org/10.4067/S0717-71782002030100072

Araya M, Cubillos LA, Guzman M, Peñailillo J, Sepúlveda A (2001) Evidence of a relationship between age and otolith weight in the Chilean jack mackerel, *Trachurus symmetricus murphyi* (Nichols). Fisheries Research 51: 17–26. https://doi.org/10.1023/A:1026334532390

Aubone A, Wöhler OC (2000) Aplicación del método de máxima verosimilitud a la estimación y comparación de curvas de crecimiento de von Bertalanffy. INIDEP Informe Técnico 37: 1–21.

Beamish RJ, Fournier DA (1981) A method for comparing the precision of a set of age determinations. Canadian Journal of Fisheries and Aquatic Sciences 38: 982–983. https://doi.org/10.1139/f81-132

Bernardes RA, Rossi-Wongtschowski CLDB, Wahrlich R, Vieira RC, Santos AP, Rodrigues AR (2005) Prospecção pesqueira de recursos demersais com armadilhas e pargueiras na Zona Econômica Exclusiva da região Sudeste-Sul do Brasil. Instituto Oceanográfico, USP, São Paulo, 112 pp.

Bervian G, Fontoura NF, Haimovici M (2006) Statistical model of variable allometric growth: otolith growth in *Micropogonias furnieri* (Actinopterygii, Scianidae). Journal of Fish Biology 68: 196–208. https://doi.org/10.1111/j.1095-8649.2005.00890.x

Beverton RJH, Holt SJ (1993) On the dynamics of exploited fish populations. Chapman and Hall, London, 783 pp.

Braga ES, Niencheski LFH (2006) Composição de massas de água e seus respectivos potenciais produtivos na área entre o Cabo de São Tomé (RJ) e o Chuí (RS). In: Rossi-Wongtschowski CLDB, Madureira LSP (Eds) O ambiente oceanográfico da plataforma continental e do talude na região Sudeste-Sul do Brasil. Edusp, São Paulo, 161–218.

Braicovich PE, Luque JL, Timi J (2012) Geographical patterns of parasite infracommunities in the rough scad, *Trachurus lathami* Nichols, in the Southwestern Atlantic ocean. The Journal of Parasitology 98: 768–777. https://doi.org/10.1645/GE-2950.1

Brothers EB (1987) Methodological approaches to the examination of otoliths in aging studies. In: Summerfelt RC, Hall GH (Eds) Age and growth of fish. Iowa State University Press, Iowa, 319–330.

Campana SE (2001) Accuracy, precision and quality control in age determination, including a review of the use and abuse of age validation methods. Journal of Fish Biology 59: 197–242. https://doi.org/10.1006/jfbi.2001.1668

Campos PN, Castro MS, Bonecker ACT (2010) Occurrence and distribution of Carangidae larvae (Teleostei, Perciformes) from the Southwest Atlantic Ocean, Brazil (12-23°S). Journal of Applied Ichthyology 26: 920–924. https://doi.org/10.1111/j.1439-0426.2010.01511.x

Cardoso LG, Haimovici M (2011) Age and changes in growth of the king weakfish Macrodon atricauda (Günther, 1880) between 1977 and 2009 in southern Brazil. Fisheries Research 111: 177–187. https://doi.org/10.1016/j.fishres.2011.06.017

Carvalho MR, Soares LSH (2006) Diel feeding pattern and diet of rough scad Trachurus lathami Nichols, 1920 (Carangidae) from the Southwestern Atlantic. Neotropical Ichthyology 4: 419–426. https://doi.org/10.1590/S1679-62252006000400005

Castro BM, Lorenzetti JA, Silveira ICA, Miranda LB (2006) Estrutura termohalina e circulação na região entre o Cabo de São Tomé (RJ) e o Chuí (RS). In: Rossi-Wongtschowski CLDB, Madureira LSP (Eds) O ambiente oceanográfico da plataforma continental e do talude na região Sudeste-Sul do Brasil. Edusp, São Paulo, 11–120.

Cergole MC, Dias-Neto J (2011) Plano de gestão de uso sustentável de Sardinha verdadeira (Sardinella brasiliensis) no Brasil. Edições Ibama, Ministério do Meio Ambiente, Brasília, 242 pp.

Checkley D, Alheit J, Oozeki Y, Roy C (2009) Climate change and small pelagic fish. Cambridge University Press, Cambridge, 372 pp. https://doi.org/10.1017/CBO9780511596681

Costa AM (2004) Idade e crescimento do carapau (Trachurus trachurus L.) da costa portuguesa no período de 1992 a 1998. IPIMAR, Lisboa, 25 pp.

Cousseau MB, Perrotta RG (2004) Peces marinos de Argentina. Instituto Nacional de Investigaión y Desarrollo Pesquero, Mar del Plata, 167 pp.

Dioses T (2013) Edad y crecimiento del jurel Trachurus murphyi, (Nichols 1920) en el Perú. Revista Peruana de Biología 20: 45–52.

Espino M (2013) El jurel Trachurus murphyi y las variables ambientales de macroescala. Revista Peruana de Biología 20: 9–20. https://doi.org/10.15381/rpb.v20i1.2614

FAO (1981) Methods of collecting and analyzing size and age data for fish stock assessment. FAO Fisheries Circular 736: 1–100.

Figueiredo JL, Santos AP, Yamaguti N, Bernardes RA, Rossi-Wongtschowski CLDB (2002) Peixes da Zona Econômica Exclusiva do Sudeste e Sul do Brasil. Levantamento com rede de meia água. Edusp, São Paulo, 242 pp.

Froese R, Binohlan C (2000) Empirical relationships to estimate asymptotic length, length at first maturity and length at maximum yield per recruit in fishes, with a simple method to evaluate length frequency data. Journal of Fish Biology 56: 758–773. https://doi.org/10.1006/jfbi.1999.1194

Froese R, Pauly D (2015) FishBase. Available online at: http://www.fishbase.org, version 04/2015 [Accessed 24/06/2015].

Geist SJ, Ekau W, Kunzmann A (2013) Energy demand of larval and juvenile Cape horse mackerels, Trachurus capensis, and indications of hypoxia tolerance as benefit in a changing environment. Marine Biology 160: 3221–3232. https://doi.org/10.1007/s00227-013-2309-2

Geist SJ, Kunzmann A, Verheye HM, Eggert A, Schukat A, Ekau W (2015) Distribution, feeding behaviour, and condition of Cape horse mackerel early life stages, Trachurus capensis, under different environmental conditions in the northern Benguela upwelling ecosystem. ICES Journal of Marine Science 72: 543–557. https://doi.org/10.1093/icesjms/fsu087

Goicochea C, Mostacero J, Moquillaza P, Dioses T, Topiño Y, Guevara-Carrasco R (2013) Validación del ritmo de formación de los anillos de crecimiento en otolitos del jurel Trachurus murphyi Nichols 1920. Revista Peruana de Biología 20: 053–060. https://doi.org/10.15381/rpb.v20i1.2619

Green BS, Mapstone BD, Arlos G, Begg GA (2009) Tropical fish otoliths: information for assessment, management and ecology. Springer, New York, 313 pp. https://doi.org/10.1007/978-1-4020-5775-5

Haimovici M, Rossi-Wongtschowski CLDB, Bernardes RA, Fischer LG, Vooren CM, Santos RA, Rodrigues AR, Santos S (2008) Prospecção pesqueira de espécies demersais com rede de arrasto-de-fundo na Região Sudeste-Sul do Brasil. Instituto Oceanográfico, USP, São Paulo, 183 pp.

Horn PL (1993) Growth, age structure, and productivity of jack mackerels (Trachurus spp.) in New Zealand waters. New Zealand Journal of Marine and Freshwater Research 27: 145–155. https://doi.org/10.1080/00288330.1993.9516553

Huxley JS (1993) Problems of relative growth. The John Hopkins University Press, 2nd ed., Baltimore, 276 pp.

Instituto de Pesca (2013) Estatística Pesqueira. Sistema Gerenciador de Banco de Dados de Controle Estatístico de Produção Pesqueira Marítima, Instituto de Pesca/APTA/SAA/SP. Available online at: http://www.pesca.sp.gov.br/estatistica/index.php [Accessed: 20/04/2013]

Karlou-Riga C (2000) Otolith morphology and age and growth of Trachurus mediterraneus (Steindachner) in the Eastern Mediterranean. Fisheries Research 46: 69–82. https://doi.org/10.1016/S0165-7836(00)00134-X

Karlou-Riga C, Sinis A (1997) Age and growth of horse mackerel, Trachurus trachurus (L.) in the Gulf of Saronikos (Greece). Fisheries Research 32: 157–171. https://doi.org/10.1016/S0165-7836(97)00044-1

Kasapoglu N, Duzgunes E (2013) The relationship between somatic growth and otolith dimensions of Mediterranean horse mackerel (Trachurus mediterraneus) from the Black Sea. Journal of Applied Ichthyology 29: 230–233. https://doi.org/10.1111/jai.12019

Katsuragawa M, Ekau W (2003) Distribution, growth and mortality of young rough scad, Trachurus lathami, in the South-eastern Brazilian Bight. Journal of Applied Ichthyology 19: 21–28. https://doi.org/10.1046/j.1439-0426.2003.00335.x

Katsuragawa M, Matsuura Y (1992) Distribution and abundance of carangid larvae in the South-eastern Brazilian Bight during 1975–81. Boletim do Instituto Oceanográfico 40: 55–78. https://doi.org/10.1590/S0373-55241992000100005

Kerkich M, Aksissou M, Casal JAE (2013) Age and growth of the horse mackerel Trachurus trachurus (Linnaeus,1758) catches in the bay of M'diq (Mediterraneen coast of Morocco). IRACST - Engineering Science and Technology: An International Journal 3: 708–714.

King M (2007) Fisheries biology, assessment and management. Blackwell Publishing, 2nd ed., Oxford, 382 pp. https://doi.org/10.1002/9781118688038

Lyle JM, Krusic-Golub K, Morison AK (2000) Age and growth of jack mackerel and the age structure of the jack mackerel purse seine catch. Tasmanian Aquaculture and Fisheries Institute, Taroona, 49 pp.

Meneghetti JO, Alves CC (1971) Nota preliminar sobre o hábito alimentar de chicharro (Trachurus lathami, Nichols) e seu significado ecológico. Ciência e Cultura 23: 388–389.

Menezes NA, Figueiredo JL (1980) Manual de peixes marinhos do sudeste do Brasil. IV. Teleostei (3). Museu de Zoologia da Universidade de São Paulo, São Paulo, 96 pp.

MMA (2007a) Estatística da Pesca Brasil – 2005: grandes regiões e unidades da federação. Ministério do Meio Ambiente, Brasília, 105 pp.

MMA (2007b) Estatística da Pesca Brasil – 2007: grandes regiões e unidades da federação. Brasília, Ministério do Meio Ambiente, 113 pp.

MMA (2008) Estatística da Pesca Brasil – 2006: grandes regiões e unidades da federação. Ministério do Meio Ambiente, Brasília, 180 pp.

Panfili J, Morales Nin B (2002) Semi-direct validation. In: Panfili J, Pontual H, Troadec H, Wright PJ (Eds) Manual of fish sclerochronology. Éditions Ifremer, Brest, 463 pp.

Pauly D (1984) Fish population dynamics in tropical waters: a manual for use with programmable calculators. ICLARM, Manila, 325 pp.

Pires-Vanin AMS, Rossi-Wongtschowski CLDB, Aidar E, Mesquita HSL, Soares LSH, Katsuragawa M, Matsuura Y (1993) Estrutura e função do ecossistema de plataforma continental do Atlântico Sul brasileiro: síntese dos resultados. Publicação especial do Instituto Oceanográfico 10: 217–231.

Rossi-Wongtschowski CLDB, Vaz-dos-Santos AM, Siliprandi CC (2014) Checklist of the marine fishes collected during hydroacoustic surveys in the Southeastern Brazilian Bight from 1995 to 2010. Arquivos de Zoologia 45: 73–88. https://doi.org/10.11606/issn.2176-7793.v45iespp73-88

Saccardo SA (1987) Morfologia, distribuição e abundância de Trachurus lathami Nichols, 1920 (Teleostei: Carangidae) na região sudeste-sul do Brasil. Boletim do Instituto Oceanográfico 35: 65–95. https://doi.org/10.1590/S0373-55241987000100008

Saccardo SA, Cergole MC, Masumoto C (2005) Trachurus lathami. In: Cergole MC, Ávila-da-Silva AA, Rossi-Wongtschowski CLDB (Eds) Análise das principais pescarias comerciais da região Sudeste-Sul: dinâmica das principais espécies em exploração. Instituto Oceanográfico, USP, Série REVIZEE, São Paulo, 156–161.

Saccardo SA, Haimovici M (2007) Síntese sobre o chicharro (Trachurus lathami). In: Haimovici M (Ed.) A prospecção pesqueira e abundância de estoques marinhos no Brasil nas décadas de 1960 a 1990: levantamento de dados e avaliação crítica. Ministério do Meio Ambiente, Secretaria de Mudanças Climáticas e Qualidade Ambiental, Brasília, 233–237.

Saccardo SA, Katsuragawa M (1995) Biology of the rough scad Trachurus lathami, on the southeastern coast of Brazil. Scientia Marina 59: 265–277.

Sassa C, Takahashi M, Nishiuchi K, Tsukamoto Y (2014) Distribution, growth and mortality of larval jack mackerel Trachurus japonicus in the southern East China Sea in relation to oceanographic conditions. Journal of Plankton Research 36: 542–556. https://doi.org/10.1093/plankt/fbt134

Siliprandi CC, Rossi-Wongtschowski CLDB, Brenha MR, Gonsales SA, Santificetur C, Vaz-dos-Santos AM (2014) Atlas of marine bony fish otoliths (Sagittae) of Southeasthern-Southern Brazil Part II: Perciformes (Carangidae, Sciaenidae, Scombridae and Serranidae). Brazilian Journal of Oceanography 62: 28–100.

Smith-Vaniz WF (2002) Carangidae. In: Carpenter KE (Ed.) The living marine resources of the Western Central Atlantic. FAO, vol. 3, Rome, 1426–1468.

Sponaugle S (2010) Otolith microstructure reveals ecological and oceanographic processes important to ecosystem-based management. Environmental Biology of Fishes 89: 221–238. https://doi.org/10.1007/s10641-010-9676-z

Stewart J, Ferrell DJ (2001) Age, growth, and commercial landings of yellowtail scad (Trachurus novaezelandiae) and blue mackerel (Scomber australasicus) off the coast of New South Wales, Australia. New Zealand Journal of Marine and Freshwater Research 35: 541–551. https://doi.org/10.1080/00288330.2001.9517021

UNIVALI (2009) Boletim estatístico da pesca industrial de Santa Catarina – ano 2008: programa de apoio técnico e científico ao desenvolvimento da pesca no Sudeste e Sul do Brasil. Universidade do Vale do Itajaí, Centro de Ciências Tecnológicas da Terra e do Mar, Itajaí, 73 pp.

UNIVALI (2010) Boletim estatístico da pesca industrial de Santa Catarina – ano 2009 e panorama 2000–2009: programa de monitoramento e avaliação da atividade pesqueira industrial no Sudeste e Sul do Brasil. Universidade do Vale do Itajaí, Centro de Ciências Tecnológicas da Terra e do Mar, Itajaí, 97 pp.

UNIVALI (2011) Boletim estatístico da pesca industrial de Santa Catarina – Ano 2010. Universidade do Vale do Itajaí, Centro de Ciências Tecnológicas da Terra e do Mar, Itajaí, 59 pp.

Valentini H, Pezzuto PR (2006) Análise das principais pescarias comerciais da região Sudeste/Sul do Brasil com base na produção controlada do período 1986–2004. Instituto Oceanográfico, USP, São Paulo , 56 pp.

Vaz-dos-Santos AM (2015a) Métodos quantitativos aplicados ao estudo de otólitos. In: Volpedo AV, Vaz-dos-Santos AM (Eds) Métodos de estudos com otólitos: princípios e aplicações. CAFP-BA-PIESCI, Buenos Aires, 377–395.

Vaz-dos-Santos AM (2015b) Otólitos em estudos de idade e crescimento em peixes. In: Volpedo AV, Vaz-dos-Santos AM (Eds) Métodos de estudos com otólitos: princípios e aplicações. CAFP-BA-PIESCI, Buenos Aires, 303–333.

Vaz-dos-Santos AM, Rossi-Wongtschowski CLDB (2013) Length-weight relationships of the ichthyofauna associated with the Brazilian sardine, *Sardinella brasiliensis*, on the Southeastern Brazilian Bight (22°S-29°S) between 2008 and 2010. Biota Neotropica 13: 326–330. http://www.biotaneotropica.org.br/v13n2/en/abstract?short-communication+bn01613022013

Waldron ME, Kerstan M (2001) Age validation in horse mackerel (*Trachurus trachurus*) otoliths. ICES Journal of Marine Science 58: 806–813. https://doi.org/10.1006/jmsc.2001.1071

Webb BF, Grant CJ (1979) Age and growth of jack mackerel, *Trachurus declivis* (Jenyns), from Southeastern Australian Waters. Australian Journal of Marine and Freshwater Research 30: 1–9. https://doi.org/10.1071/MF9790001

Yankova MH, Raykov VS, Gerdzhikov DB, Frateva PB (2010) Growth and length-weight relationships of the horse mackerel, *Trachurus mediterraneus ponticus* (Aleev, 1956), off the Bulgarian Black Sea coast. Turkish Journal of Zoology 34: 85–92. https://doi.org/10.3906/zoo-0811-10

Yoda M, Shiraishi T, Yukami R, Ohshimo S (2014) Age and maturation of jack mackerel *Trachurus japonicus* in the East China Sea. Fisheries Science 80: 61–68. https://doi.org/10.1007/s12562-013-0687-5

Zar JH (2010) Biostatistical analysis. Pearson, 5th ed., New Jersey, 944 pp.

Author Contributions: LCR and AMVS contributed equally to this work.

Competing Interests: The authors have declared that no competing interests exist.

8

Redescription of *Malacomorpha cancellata* (Phasmatodea: Pseudophasmatidae): a geographically misplaced Neotropical species

Raphael Aquino Heleodoro[1], Ricardo Andreazze[2], José Albertino Rafael[1]

[1]Programa de Pós-graduação em Entomologia, Instituto Nacional de Pesquisas da Amazônia. Avenida André Araújo 2936, Petrópolis, 69067-375 Manaus, AM, Brazil.
[2]Departamento de Microbiologia e Parasitologia, Centro de Biociências, Universidade Federal do Rio Grande do Norte. Campus Universitário Lagoa Nova, Caixa Postal 1524, 59078-900 Natal, RN, Brazil.
Corresponding author: Raphael Aquino Heleodoro (raphaelnatal36@gmail.com)

http://zoobank.org/55315EC6-4C57-461A-A892-2A59B137B55C

ABSTRACT. *Olcyphides cancellatus* Redtenbacher, 1906 was described from Canton, China, in error. The species was transferred to *Pseudolcyphides* Karny, 1923, a genus that later on was synonymized with *Malacomorpha* Rehn, 1906. However, the name *P. cancellatus* was forgotten and was not mentioned in the publication where *Pseudolcyphides* was synonymized with *Malacomorpha* and thus was not transferred. Here the original geographical record is corrected and the species is transferred to *Malacomorpha*. The resulting new combination, *M. cancellata* **comb. nov.**, resulted from examination of specimens from state of Rio Grande do Norte, Brazil. In addition, species diagnosis, redescription of the female and the first description of male specimens, with comparative comments on other *Malacomorpha* species, are provided.

KEY WORDS. Entomology, Neotropics, phasmids, taxonomy.

INTRODUCTION

Olcyphides cancellatus Redtenbacher, 1906 was originally described based on a single female specimen from Canton, China. Its type-locality was mentioned with a question mark. Later, Weidner (1966) transferred *O. cancellatus* to *Pseudolcyphides* Karny, 1923, while failing to acknowledge that there was a problem with the type-locality. Conle et al. (2008) synonymized *Pseudolcyphides* with *Malacomorpha* in their revision of *Malacomorpha*, but failed to mention *Pseudolcyphides cancellatus* in their work, leaving the generic placement of the species in question. Hennemann et al. (2008) confirmed that the type-locality given for *O. cancellatus* was in fact in error and that the species is most likely from the Neotropical region.

This series of events involving *Pseudolcyphides cancellatus* resulted in three problems: (1) the species cannot be properly identified because its description is insufficient for diagnosis, (2) only one specimen is available for study and, and (3) the generic position of the species is uncertain.

In view of museum material identified as the species in question (from the state of Rio Grande do Norte, Brazil, after comparison with photographs of the holotype), we present an amended geographical distribution for it, a redescription of the female, the first description of the male, a new combination, information on copulatory behavior and host plant.

MATERIAL AND METHODS

This study is based on the examination of 72 specimens housed at Coleção Entomológica Adalberto Antonio Varela Freire, Universidade Federal do Rio Grande do Norte (CEAAVF/UFRN). The specimens are stored in vials containing 80% alcohol. Male specimens had their terminalia dissected and then macerated in 85% lactic acid heated at 120 °C for about two hours. The macerated piece was examined on concave slides with glycerin. After study, the genitalia was placed in a microvial with glycerin and stored together with the specimen. Genitalia terminology follows Helm et al. (2011). All measurements were taken with a digital caliper rule data of the shortest and longest specimen are given.

Photographs were taken with a Leica DFC500 digital camera fitted on a Leica MZ205 stereomicroscope connected to a computer loaded with the Leica Application Suite software. This software includes an Auto-Montage module (Syncroscopy software) used to combine multiple layers of photographs taken at different focus points into one photograph with greater depth of field. Label data are translated to English. Square brackets are used to indicate complementary data and semicolons are used to separate different specimens or groups of specimens.

The Brazilian specimens were compared with photographs of the holotype available at http://phasmida.speciesfile.org (Brock et al. 2016). The holotype is deposited in the Zoologisches Institut und Zoologisches Museum der Universität Hamburg (ZMUH), Germany. After study, part of the material was deposited at Coleção de Invertebrados do Instituto Nacional de Pesquisas da Amazônia (INPA) and Museu de Zoologia da Universidade de São Paulo (MZUSP). All specimens are in alcohol.

TAXONOMY

Malacomorpha Rehn, 1906

Malacomorpha Rehn, 1906; Zompro 2004 (key); Conle et al. 2008 (revision and synonymy); Brock et al. 2016 (catalogue).
Anisomorpha Gray, 1835; Bradley and Galil 1977 (taxonomy). Synonymized by Conle et al. (2008).

Type-species. *Malacomorpha androensis* Rehn, 1906 (Pseudophasmatidae): 113–114, fig. 2, by original designation.

Malacomorpha cancellata (Redtenbacher, 1906), comb. nov.

Figs 1–17, 19–30.

Olcyphides cancellatus Redtenbacher, 1906; Zompro 2002 (notes on holotype).
Pseudolcyphides cancellatus; Weidner 1966 (note on holotype); Hennemann et al. 2008 (geographical record); Brock et al. 2016 (world catalogue).

Diagnosis. Antenna with antennomeres alternating yellow and black from base to apex (Figs 4, 6). Pro- and mesonotum with three longitudinal and parallel black stripes (Figs 1, 7). Tegmina with white band between radial and medial veins (Figs 8, 14). Femora, tibiae and first tarsomeres alternating yellow and black from base to apex (Figs 3, 13–15, 19–21). Metasternum with black spots from apex of basal third to apex of median third (Figs 2, 9, 11, 15, 21). Sternites with irregular-shaped complex of spots at distal half (Figs 2, 10, 12, 15, 21).

Redescription. Females from Natal, Brazil, and holotype specimen (Figs 1–5, 7–10, 13–17, 30). Head. Frons and vertex dorsally smooth, yellow with black stripes (Fig. 13). Face yellow. Clypeus rectangular. Labrum U-shaped. Gena black (Figs 4, 14). Antennomeres covered by short setae, with color alternating yellow and black from base to apex (Figs 13–15). Ocellum present. Compound eye globose, black, dorsally with yellow stripes. Labial and maxillary palps yellow, covered by setae.

Thorax. Pronotum rectangular, 1.4 times longer than wide, rugged, yellow with three longitudinal and parallel black

stripes (Figs 1, 7, 13). Mesonotum rectangular, 1.5 times longer than pronotum (Figs 1, 7, 13). Propleuron black, smooth. Mesopleuron triangular, black, with yellow stripe at apex of apical third, rugged (Figs 4, 14). Prosternum rugged, yellow, anteriorly trapezoidal, posteriorly rectangular. Mesosternum smooth, rugged, yellow with black spots (Figs 2, 15). Metasternum smooth, yellow, with black spots from apex of basal third to apex of median third (Figs 2, 9, 15).

Legs. Covered by setae laterally. Coxae and trochanters black. Anterior femur black except yellow dorsally at basal third. Anterior tibia with basal third yellow, remaining black (Figs 13–15). Mid femur yellow at proximal half, with small black spot at base; distal half black (Figs 1, 14, 15). Mid tibia with basal third yellow, remaining black (Figs 13–15). Posterior femur yellow at basal and medial third, black at apical third (Figs 13–17). Tarsi with first tarsomeres alternating black and yellow; black from second to fifth tarsomeres (Figs 3, 13–15).

Wings. Tegmina black, elongated, two times longer than broad, with several longitudinal and transversal yellow veins; shoulder elevated, with white band between radial and medial veins (Figs 8, 14). Posterior wing six times longer than tegmina, costal area black with longitudinal and transversal yellow veins (Figs 13, 14); anal area reddish in live specimens and whitish in preserved specimens.

Abdomen. Laterally covered by small setae. Tergites 1–2 dorsally yellow, without spots; tergites 3–10 yellow with black spots apically. Tergites 2–7 dorsally rectangular, 1.5 times longer than wide. Tergites 8–9 dorsally trapezoidal, with arched spots at apex. Tergite 8 laterally rectangular, 1.3 times longer than high (Figs 5, 16). Tergite 9 laterally rectangular, 1.2 times higher than long (Figs 5, 16). Tergite 10 dorsally with straight basal margin, lateral margin convex and apical margin arched; laterally with basal and lateral margin straight, apical margin curved (Figs 5, 16). Sternites with irregular-shaped complex of spots at distal half (Figs 10, 15). Sternites 2–7 rectangular, two times longer than wide; sternite 2 yellow, gradually turning black from sternite 3 to sternite 7 (Fig. 15). Subgenital plate sword-shaped, with basal margin straight, lateral margin sinuous and apical margin acute; proximal half broad, gradually narrowing from base to apex of distal half; elongated, 1.5 times longer than broad; black with yellow spots, covered by setae (Fig. 17). Cercus slender, straight, narrowing from base to apex, covered by setae (Figs 16–17).

Measurements (mm). Body length 70.4–71.2; pronotum 3.9–4.1; mesonotum 9.4–10.0; anterior femur 11.0–11.8; anterior tibia 10.2–10.3; mid femur 8.4–8.5; mid tibia 6.7–6.8; posterior femur 12.3–12.5; posterior tibia 11.3–11.5.

Description of males (based on specimens from Natal, Brazil). Similar to female, but with shorter and slender body and with the following differences.

Thorax. Black stripes at pro- and mesonotum and white band in tegmina thinner. Mesosternum without black spots. Black spots of metasternum with lighter coloration. Abdomen. Tergite 8 dorsally quadrangular, laterally two times longer than high (Figs

Figures 1–5. *Malacomorpha cancellata* comb. nov. female holotype: (1) habitus, dorsal view; (2) idem, ventral view; (3) idem, lateral view; (4) head, pro- and mesothorax, lateral view; (5) apex of abdomen, lateral view. All photos are a courtesy of Paul D. Brock and all copyrights belong to the Zoologisches Institut und Zoologisches Museum der Universität Hamburg (ZMUH).

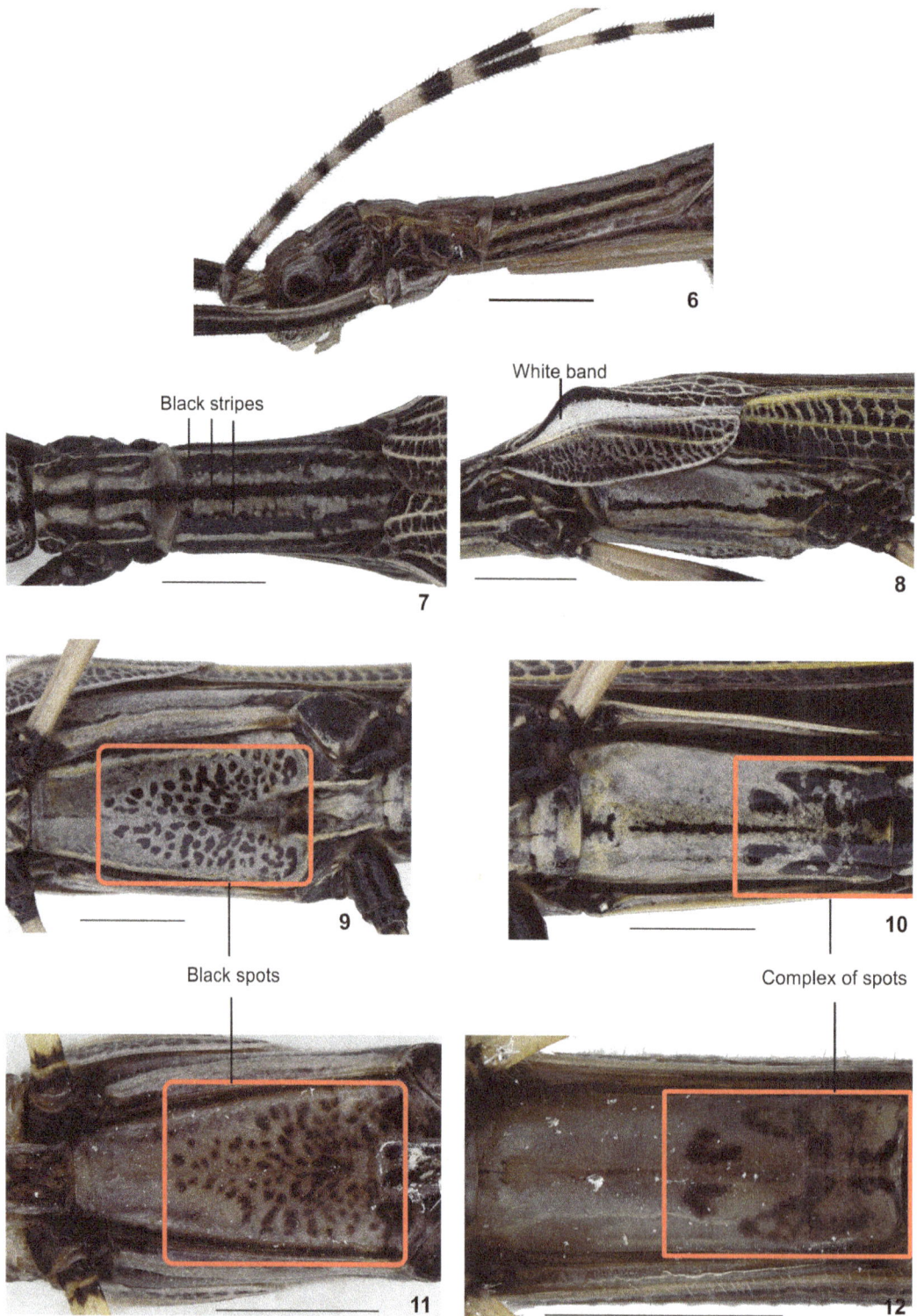

Figures 6–12. *Malacomorpha cancellata* comb. nov.: (6) male head, pro- and mesothorax, lateral view; (7) female pro- and mesothorax, dorsal view; (8) female meso-, metathorax and wings, lateral view; (9) female metathorax, ventral view; (10) female sternite 2, ventral view; (11) male metathorax, ventral view; (7) male sternite 2, ventral view. Scale bars: 6, 11, 12 = 2.0 mm, 7–10 = 2.5 mm.

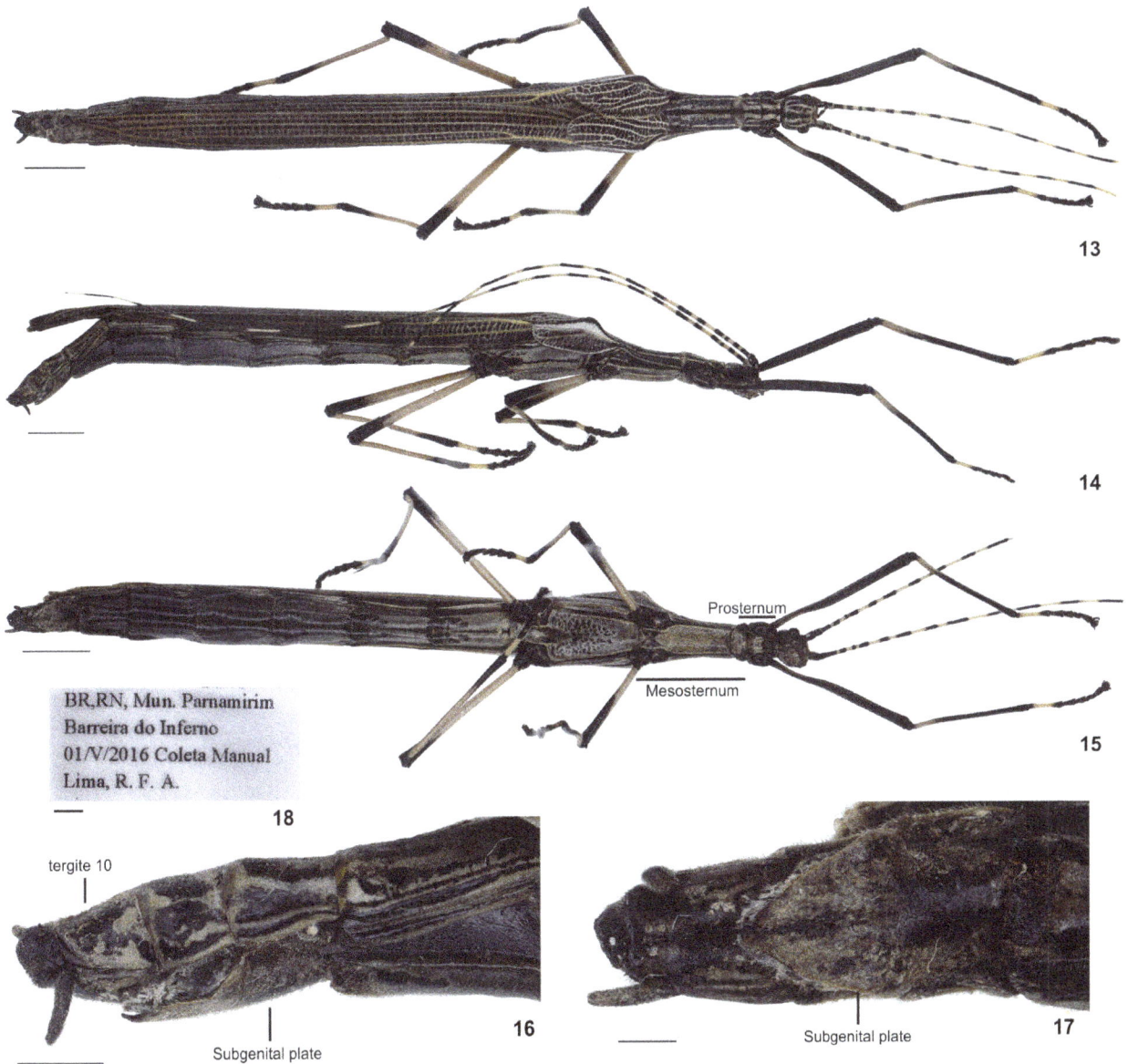

Figures 13–18. *Malacomorpha cancellata* comb. nov., female and specimen label: (13) habitus, dorsal view; (14) idem, lateral view; (15) idem, ventral view; (16) apex of abdomen, lateral view; (17) idem, ventral view; (18) specimen label. Scale bars: 13–15 = 5 mm, 16, 17 = 1 mm, 18 = 2 mm.

22–23). Tergite 9 dorsally and laterally trapezoidal (Figs 22–23). Tergite 10 dorsally rectangular, 1.4 times wider than long, with straight basal margin and lateral margin, apical margin slightly emarginated medially and projected laterally (Figs 22–24). Abdominal sterna yellow. Sternite 7 1.5 times longer than wide (Fig. 21). Sternite 8 with straight basal margin, lateral margin broadening towards apex, apical margin arched, convex (Fig. 24). Subgenital plate ventrally quadrangular, with straight basal margin, curved lateral margin, and straight apical margin; rugose, covered by setae (Fig. 24). Cercus robust, cylindrical, curved, covered by small setae (Figs 21–24). Vomer broad at basal third, narrowing at median third towards to apex; apex curved upwards (Fig. 25).

Genitalia. Connected to subgenital plate by two anterior points, pouch-like shaped and globose, mostly membranous, with some sclerotized parts (Figs 26–27), divided into two big lobes, dorsal and ventral. Dorsal lobe dorsally with small and scattered setae; dorsoapically with group of spines; antero-ventrally with sclerotized and sinuous right process of phallic

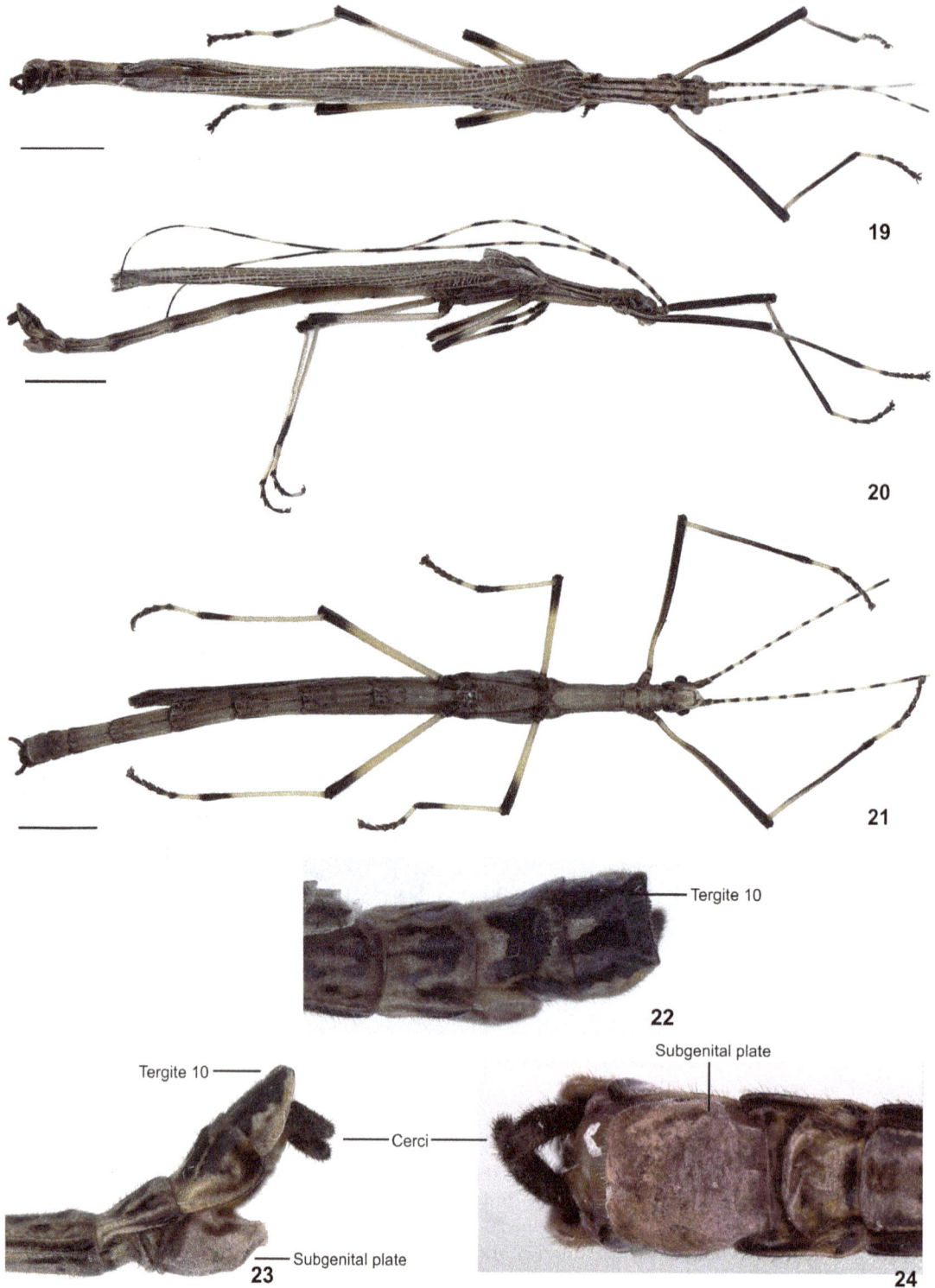

Figures 19–24. *Malacomorpha cancellata* comb. nov., male – specimen used: Parnamirim, Barreira do Inferno. 01.v.2016: (19) habitus, dorsal view; (20) idem, lateral view; (21) idem, ventral view; (22) apex of abdomen, dorsal view; (23) idem, lateral view; (24) idem, ventral view. Scale bars = 5 mm.

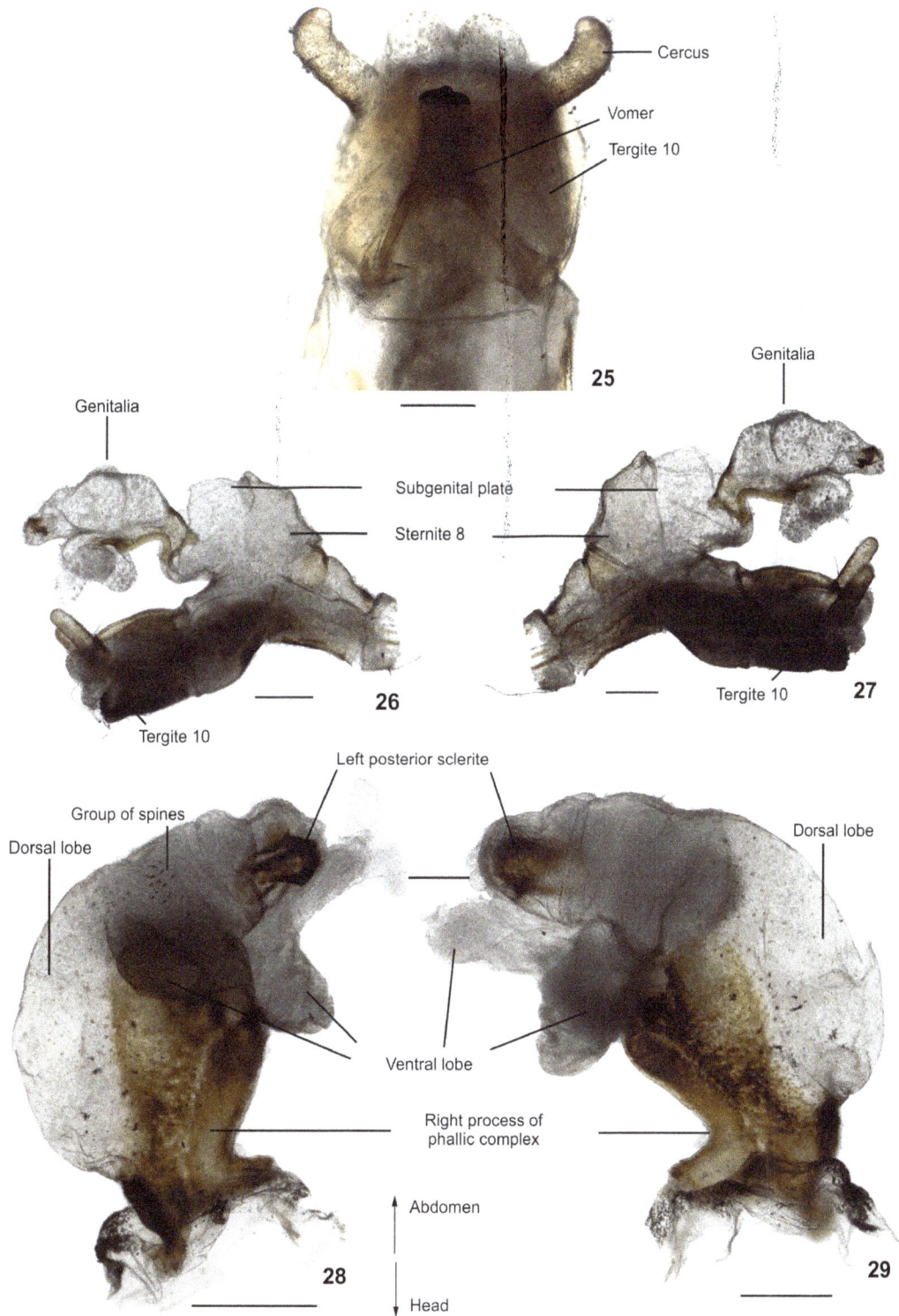

Figures 25–29. *Malacomorpha cancellata* comb. nov., male terminalia and genitalia – specimen used: Parnamirim, Barreira do Inferno. 01.v.2016: (25) apex of terga, ventral view; (26) Apex of abdomen with genitalia, right lateral view; (27) idem, left lateral view; (28) genitalia, left dorsolateral view; (29) idem, right dorsolateral view. Scale bars: 25, 28, 29 = 0.5 mm, 26, 27 = 1.0 mm.

Figure 30. *Malacomorpha cancellata* comb. nov., mating couple. Photo is a courtesy of Willianilson Pessoa.

complex; apically with well sclerotized, ellipsoidal left posterior sclerite (Figs 28–29). Ventral lobe divided into two ellipsoidal parts (Figs 28–29).

Measurements (mm). Body length 41.1–41.7; pronotum 1.8–2.0; mesonotum 3.4–3.7; anterior femur 8.8–9.1; anterior tibia 7.4–7.5; mid femur 5.7–5.8; mid tibia 5.3–5.4; posterior femur 8.2–8.3; posterior tibia 9.0–9.1.

Variations. The most noticeable variations are in the color of the antennomeres and legs, which can vary from dark to light yellow. Some male specimens have a lighter pigmentation of the thoracic and abdominal spots.

Material examined. Brasil, *Rio Grande do Norte*: Natal (Parque das Dunas Costeiras do Natal), 13.iv.1984, 29 females, Varela-Freire, A.A. leg. Collected manually on Ubaias [*Eugenia pyriformis* Cambess. (Myrtaceae)] (CEAAVF); same data but 3.vi.1984, 3 males, Varela-Freire, A.A. leg Collected manually on Ubaias (CEAAVF); same data but 3.vi.1986, 2 males, 2 females, A. A. Freire leg., Collected manually on Ubaia (INPA); same data but 05.iv.1989, 3males2females (INPA); Parnamirim (Barreira do Inferno), 19.iii.2009, 1 females, Oliveira, D.V. leg (INPA); Parnamirim (Coabinal), 25.iv.2009, 1 male, Magalhães, L.B. leg. (INPA); (Parque das Dunas), 15.vi.2010, 1 female, Soares, A.M. leg. (INPA); Goianinha (Usina Estivas), 30.iv.2011, 1 female Brito, M.M. leg. (INPA); same data but 26.iv.2011, 1male, Dantas, A.K. leg. (INPA); Parnamirim (Pium), 23.v.2015, 1 male, Silva, J. leg. (INPA); Baia Formosa (Mata Estrela), 14.iii.2016, 1 male, 1 female, Silva, G.M. leg., collected manually on dune (INPA); Natal (UFRN,

Mata dos Saguis), 20.iii.2016, 1 female, Garcia G.S. leg. (INPA); Parnamirim (Barreira do Inferno), 1.v.2016, 1female, Silva, F. leg. (INPA); idem, 1male, Coutinho, J.R.S. leg. (INPA); same data but 1 female, Lima, L. leg. (INPA); Natal (Parque das Dunas, zona inter-dunal), 03.vi.1984, 5 males, 5 females, A.A. Freire, leg., collected manually on Ubaia (MZUSP); Parnamirim, 30.iv.2009, 2 females, Oliveira, D.V. leg. (MZUSP); Natal [actually Parnamirim] (Barreira do Inferno), 11.vi.2009, 1 male, 1 female, Damião Valdenor de Oliveira leg. (MZUSP); Parnamirim, 29.iv.2010, 2females, Oliveira, C.A.S. leg. (MZUSP); Parnamirim, 2.v.2010, 1female, Lima, G.R.R. leg. (MZUSP); Natal, (Ponta Negra), 11.vi.2011, 1female, Bezerra, A.M. leg. (MZUSP); Parnamirim (Barreira do Inferno), 04.vii.2014, 01 female, Medeiros, A.G.N. leg. (MZUSP).

Biological information. A couple of individuals in copula were photographed in a forest fragment near the city of Natal (Fig. 30). The male was positioned adjacently to the female, the body turned almost upside down, clasping the female with its terminalia. The copula lasted approximately 45 minutes.

Known distribution. Brazil, *Rio Grande do Norte*: Natal, Baía Formosa, Goianinha, Parnamirim.

Remarks. The name of the genus comes from the Greek *Malakós* (soft, gentle) and *Morphé* (shape, form) and ends with the suffix "a", indicating that it is a feminine name. Thus, the specific name also has to be feminine. "Cancellatus" comes from the Latin and means latticed, being a masculine word. Therefore, we changed the specific name to "cancellata" to agree with the gender of the genus.

Conle et al. (2008) did not mention this species in their revision of *Malacomorpha* and hence they did not transfer it from *Pseudolcyphides* to *Malacomorpha*. Since they did not propose a new combination, the generic placement of this species remained questionable (i.e., whether it should be *Pseudolcyphides* or *Malacomorpha*). In addition, it is important to highlight that the specific name "cancellatus" had never been combined with *Malacomorpha* in a scientific publication before.

Although Brock et al. (2016) already used the combination "*Malacomorpha cancellatus*", it was used in an online digital database that does not constitute a formal scientific publication, since it is not registered with the Official Register of Zoological Nomenclature (ZooBank). According to the International Code of Zoological Nomenclature, this violates Article 8.5.3, which mandates that a valid electronic publication: "be registered in the Official Register of Zoological Nomenclature (ZooBank) (see Article 78.2.4) and contain evidence in the work itself that such registration has occurred." Thus, we are here treating this combination as new.

Malacomorpha cancellata comb. nov. has a striking and unique coloration pattern. In addition, it can be distinguished from the wingless *M. androsensis* Rehn, 1906, *M. bastardoae* Conle et al., 2008, *M. guamuhayaense* Zompro and Fritzsche, 2008, *M. jamaicana* (Redtenbacher, 1906), *M. macaya* Conle et al., 2008, *M. multipunctata* Conle et al., 2008, *M. obscura* Conle et al., 2008, and *M. sanchezi* Conle et al., 2008 by the presence of wings. Among winged species it differs from *M. cyllara* (Westwood, 1859) by having a wider and quadrangular subgenital plate (rounded in *M. cyllara*); from *M. hispaniola* Conle et al., 2008 by having longer and ellipsoidal tegmina (short and rectangular in *M. hispaniola*).

DISCUSSION

This is the first record of *Malacomorpha* from Brazil and South America. Thus it is an important record that increases the range of distribution from the Bahamas, Cuba, Hispaniola and Jamaica (Conle et al. 2008). It also strengthens the idea proposed by Hennemann et al. (2008) that this is definitively a Neotropical genus.

So far, the species has only been found in Brazil in the state of Rio Grande do Norte, in Atlantic Forest with traces of Caatinga and Dune vegetation. *M. cancellata* comb. nov. has been repeatedly collected since 1984 at Parque Estadual Dunas de Natal (a conservational urban forest fragment, located in Natal city, capital of Rio Grande do Norte) feeding on *Eugenia pyriformis* (Cambess.), commonly known in northeastern Brazil as "Ubaia" or "Uvaia". Parque Estadual Dunas de Natal is close to the sea and mostly harbors Atlantic Forest, but it also has traces of Caatinga and Dune vegetation. *M. cancellata* comb. nov. has also been reported from Dune areas inside and outside the park.

This habitat information is valuable because the Atlantic Forest and Caatinga biomes, as well as Dune environments, are critically endangered in Brazil. Less than 2% of the total area of the Caatinga is currently protected by conservational units (Tab-arelli et al. 2000), and there is no reliable estimate of how much deforested the biome is. Concerning the Atlantic Forest, less than 15% of the original forest survived (Ribeiro et al. 2009). There are no reliable sources on how much of the Dune vegetation is being protected by conservational units, nor is it known how much of it has been already deforested. Hence, it is possible to assume that *M. cancellata* comb. nov. is also endangered, especially if it has become a specialist species in these biomes. Future studies are needed to evaluate whether *M. cancellata* comb. nov. is endangered or not. Lastly, this is the first record of the host plant and copulatory habits of *M. cancellata* comb. nov., as well as the first description of the male genitalia of *Malacomorpha*.

ACKNOWLEDGEMENTS

To Alberto da Silva Neto, Diego Matheus de Mello Mendes, and João Rafael Oliveira for support in the laboratory; to Conselho Nacional de Desenvolvimento Científico e Tecnológico (CNPq) for funding our research; to Fundação de Amparo à Pesquisa do Estado do Amazonas (FAPEAM) and Programa de Apoio a Núcleos de Excelência (PRONEX) grant 016/2006, process 1437/2007; to Willianilson Pessoa for the use of his photo; to Paul D. Brock and to Zoologisches Institut und Zoologisches Museum der Universität Hamurg (ZMUH) for allowing us to use his photos of the holotype. *In memoriam* to Adalberto Antonio Varela Freire.

REFERENCES

Bradley JC, Galil BS (1977) The taxonomic arrangement of the Phasmatodea with keys to the subfamilies and tribes. Proceedings of the Entomological Society of Washington 79: 176–204.

Brock PD, Büscher T, Baker E (2016) Catalog of stick and leaf insects of the world. v. 5.5. http://phasmida.speciesfile.org [Accessed: 16/09/2016]

Conle OV, Hennemann FH, Perez-Gelabert DE (2008) Studies on neotropical Phasmatodea II: Revision of the genus *Malacomorpha* Rehn, 1906, with the descriptions of seven new species (Phasmatodea: Pseudophasmatidae: Pseudophasmatinae). Zootaxa 1478: 1–64.

Helm CS, Treulieb K, Werler S, Bradler S, Klass KD (2011) The male genitalia of *Oxyartes lamellatus* – phasmatodeans do have complex phallic organs (Insecta: Phasmatodea). Zoologischer Anzeiger 250: 223–245. https://doi.org/10.1016/j.jcz.2011.04.005

Hennemann FH, Conle OV, Zhang W (2008) Catalogue of the Stick and Leaf-insects (Phasmatodea) of China, with a faunistic analysis, review of recent ecological and biological studies and bibliography (Insecta: Orthoptera: Phasmatodea). Zootaxa 1735: 1–77.

Redtenbacher J (1906) Die insektenfamilie der Phasmiden. I. Phasmidae Areolate. Leipzig, 108 pp.

Rehn JAG (1906) The Orthoptera of the Bahamas. Bulletin of the New York Museum 22: 113–115.

Ribeiro MC, Metzger JP, Martensen AC, Ponzoni FJ, Hirota MM (2009) The Brazilian Atlantic Forest: How much is left, and how is the remaining forest distributed? Implications for conservation. Biological Conservation 142: 1141–1153. https://doi.org/10.1016/j.biocon.2009.02.021

Tabarelli M, Silva JMC, Santos AMM, Vicente A (2000) Análise de representatividade das unidades de conservação de uso direto e indireto na Caatinga: análise preliminar. In: Silva JMC, Tabarelli M (Eds) Workshop avaliação e identificação de ações prioritárias para a conservação, utilização sustentável e repartição de benefícios da biodiversidade do bioma Caatinga. Petrolina, Pernambuco. www.biodiversitas.org.br./caatinga [Accessed 13/09/2016]

Weidner H (1966) Die Entomologischen Sammlungen des Zoologischen Staatsinstituts und Zoologischen Museums, Hamburg. Mitteilungen aus dem Hamburgischen Zoologischen Museum und Institut 63: 209–264.

Zompro O (2002) Catalogue of type material of the insect order Phasmatodea at the Zoologisches Museum der Universität Hamburg (Insecta: Orthoptera: Phasmatodea). Mitteilungen aus dem Hamburgischen Zoologischen Museum und Institut 99: 179–201.

Zompro O (2004) Revision of the genera of the Areolatae, including the status of *Timema* and *Agathemera* (Insecta, Phasmatodea). Abhandlungen des Naturwissenschaftlichen Vereins Hamburg 37: 1–327.

Author Contributions: RAH, RA, and JAR identified and described the specimens. RAH, RA, and JAR wrote the paper. RAH dissected the specimens.

Competing Interests: The authors have declared that no competing interests exist.

9

Ecological and reproductive aspects of *Aparasphenodon brunoi* (Anura: Hylidae) in an ombrophilous forest area of the Atlantic Rainforest Biome, Brazil

Laura Gomez-Mesa[1,2], Juliane Pereira-Ribeiro[1], Atilla C. Ferreguetti[1], Marlon Almeida-Santos[1], Helena G. Bergallo[1], Carlos F. D. Rocha[1]

[1]*Departamento de Ecologia, Universidade do Estado do Rio de Janeiro. Rua São Francisco Xavier 524, PHLC sala 220, Maracanã, 20550-019 Rio de Janeiro, RJ, Brazil.*
[2]*Programa de Biología, Universidad CES-EIA, Calle 10 A, No. 22 – 04, Medellín, Colombia.*
Corresponding author: Juliane Pereira-Ribeiro (julianeribeiro25@gmail.com)

http://zoobank.org/9DD60207-1773-4A32-A0FD-25973BCB4F61

ABSTRACT. Presented is the first information on the ecological and reproductive aspects of the treefrog, *Aparasphenodon brunoi* Miranda-Ribeiro, 1920, living in ombrophilous forest areas of the Atlantic Rainforest, Brazil. We recorded the species' daily activity and over the course of a year, population density during the year, microhabitat usage, diet, and some reproductive features (quantity, diameter and mean mass of oocytes, mean reproductive effort of female). Field sampling was conducted monthly from June 2015 to July 2016. Searches for treefrogs were systematic, using visual encounter surveys along 14 plots RAPELD long term research modules established in the forest. For each captured individual, we recorded the hour, microhabitat used, and perch height. The diet of the population was ascertained based on 15 individuals collected outside the study plot areas. Treefrogs used seven different types of microhabitats in the forest but the preferred microhabitats were tree-trunks and lianas. The amount of accumulated rainfall and air temperature interacted to explain the number of *A. brunoi* individuals active throughout the year. The reproductive strategy for females of this comparatively large arboreal frog in the ombrophilous forest is to produce clutches with a large number (900.8 ± 358.1) of relatively small-sized eggs. We conclude that in the ombrophious forest of the Vale Natural Reserve, *A. brunoi* is a nocturnal arboreal treefrog active throughout the year but activity increases during the wet season as a result of increased precipitation. In the forest, treefrogs tend to perch mainly on tree-trunks and lianas about 1 m above ground, where it feeds preferably on relatively large bodied arthropod prey. When living in the ombrophilous forest of the Atlantic rainforest, *A. brunoi* may change some features of its ecology (e.g. marked difference in the use of bromeliads) compared to when living in restinga habitats.

KEY WORDS. Casque-headed frog, ecological aspects, ecology, habitat use.

INTRODUCTION

Bruno's casque-headed frog, *Aparasphenodon brunoi* Miranda-Ribeiro, 1920, is endemic to the Brazilian Atlantic Rainforest (Haddad et al. 2013), occurring from the south of Bahia, southward to the state of São Paulo, along the states of Espírito Santo, Minas Gerais and Rio de Janeiro, and can be found mostly in restinga habitats of this Biome (Carvalho 1939, Feio et al. 1998, Argôlo 2000, Mollo Neto and Teixeira Jr 2012, Ruas et al. 2013, Haddad et al. 2013, Oliveira and Rocha 2014). Restinga is a typical environment of the Brazilian coast, which is characterized by sand dune formations (Rizzini 1997).

Although this species has been considered as decreasing in population size (Rocha et al. 2004), information regarding its ecology is still scarce. Along its distribution range, the relatively low information available providing aspects on its ecology comes from areas of restinga habitats (Teixeira et al. 2002, Mesquita et al. 2004, Sluys et al. 2004, Wogel et al. 2006, Haddad et al. 2013), with no information available on the ecology of the species found within ombrophilous forest environments of the Biome.

The occurrence of *A. brunoi* in forests is comparatively less frequent, with most records being composed of lists for the studied area (e.g. Feio et al. 1998, Silva-Soares et al. 2010, Almeida and Gasparini 2014). In one of the largest remnants of forests in the Atlantic Rainforest Biome, the Vale Natural Reserve (VNR) in the state of Espírito Santo in southeastern Brazil, *A. brunoi* was reported to occur (Almeida and Gasparini 2014) but with no local information regarding its ecology in this forested environment.

Recently, Jared et al. (2015) provided new biological information on this species by identifying highly toxic cutaneous secretions, and well-developed delivery mechanisms for both *A. brunoi* and *Corythomantis greeningi* Boulenger, 1896. The cutaneous secretions of these species present proteolytic and fibrinolytic action, as well as hyaluronidase that promotes the diffusion of toxins. These secretions are associated with a delivery mechanism which consists of using bony spines, located in the skull, that pierce the skin and inject the toxin into the predator (Jared et al. 2015). This unique defense mechanism makes *A. brunoi* a particularly important treefrog because it has a high toxicity, which is 25 times higher than that found in vipers of the genus *Bothrops*. Thus, aspects of their ecology are important for a better understanding of their biology in different environments.

Considering the known differences in the structural habitat among restinga and ombrophilous forest, we would expect that differences in aspects of the ecology of this frog could arise when living in the ombrophilous forest. In this context, we aimed to contribute information on such a unique species by analyzing the ecological and reproductive aspects of *A. brunoi* in the ombrophilous environment of the VNR over a one year period. We specifically addressed the following questions: i) What is the daily activity of *A. brunoi* and what is their activity throughout the year? ii) Which are the preferred microhabitats used by *A. brunoi* in the forest? iii) What is the vertical range of *A. brunoi* when perching in their habitat? iv) What prey composes the treefrog's diet and which prey items make up the majority of the diet? v) What is the overall morphometrics (mean quantity, diameter, and mass) of *A. brunoi* oocytes? vi) What is the average female reproductive effort for *A. brunoi*?

MATERIAL AND METHODS

The study was carried out in the Vale Natural Reserve (19°06′45″S, 40°03′03″W), located in Linhares and Jaguaré municipality, north of Espírito Santo, Southeastern Brazil. The reserve consists of approximately 23,500 ha and is one of the largest and most important remnants of the Atlantic Rainforest Biome (MMA 2000). The regional is tropical, rainy, and warm with mean annual rainfall of 1,214.6 mm and a mean annual temperature of 24.3 °C (Kierulff et al. 2014). The reserve is covered by a mosaic of habitats with four main vegetation types: Ombrophilous forest ("Tabuleiro"), a dense forest with trees reaching ca. 40 m height; the Sandy soil forest ("Mussununga forest"), which follows the cordons of sandy soils with trees of comparatively lower height and shrubs that allow most sunlight to reach the ground; the Permanent or Seasonally flooded forest (composed locally by swamps, and lowland and riparian forests) which are associated to water bodies that differ structurally; and

the Natural grasslands ("Campos nativos"), which are open fields that emerge as enclaves in the forest, covered by herbaceous and shrubby vegetation forming thickets (Fig. 1, Peixoto et al. 2008). In this study, we sampled only the ombrophilous forest vegetation type.

Field sampling was done monthly from June 2015 through July 2016, including months from dry (April to September) and rainy (October to March) seasons in the area. Sampling was carried out during diurnal (11:00 am to 05:00 pm) and nocturnal (06:00 pm to 11 pm) periods in order to identify activity patterns of the treefrogs. Frog sampling was conducted in 14 plots using the RAPELD sampling method (Magnusson et al. 2005), distributed proportionally along four modules. This sampling method consisted of permanent and standardized plots of 250 m extensions each, following contour lines of the ground at a distance of 1 km between each.

Treefrogs were captured on transects along plots of the module using visual encounter and acoustic surveys (Crump and Scott 1994) simultaneously by two observers. Each plot was sampled five times but only once per month. Before sampling each plot, we measured air temperature (°C) and relative air humidity (%) using a thermohygrometer. During each transect, the plot was carefully inspected by the observers, looking for treefrogs in the leaf-litter, on trees, branches, bushes, fallen logs or other microhabitats when present. We were careful to record only active individuals (e.g. moving, foraging) to include in our activity analysis. All individuals were found and sampled within five meters from the center of each side of the plot (totaling a strip of 10 m wide). For each captured individual we recorded the hour, date, plot number, microhabitat used, and the height above ground (in cm) that the individual was perched when first sighted. At the end of each sampling period, air temperature and relative humidity were measured and recorded again. Data on daily rainfall (in mm) in the area of the Reserve were obtained from the meteorological station at the VNR.

We estimated treefrog density in the area (ind/ha), based on the area of each plot (250 m extension x 10 m wide = 2500 m²), calculating the total searched area of plots [considering all plots transected/month x 2500 m² extension = total area searched (in m²)] and divided the number of individuals of *A. brunoi* per month. Then, we calculated and compared mean treefrog density among the months of dry and rainy seasons using Student t-Test. We used Multiple Regression Analysis to evaluate the effects of temperature and accumulated rainfall of the sampling days for *A. brunoi* density and Simple Regression Analysis to evaluate the effect of humidity throughout the sampled days on *A. brunoi* density in the respective month of the sampling. We obtained the accumulated temperature and humidity in the sampled period from the averages of the values measured in the plots and obtained precipitation estimates of the sampled period from the sum of the accumulated precipitation that occurred during the sampling period. The analyses were conducted using the software R – 3.3.1.

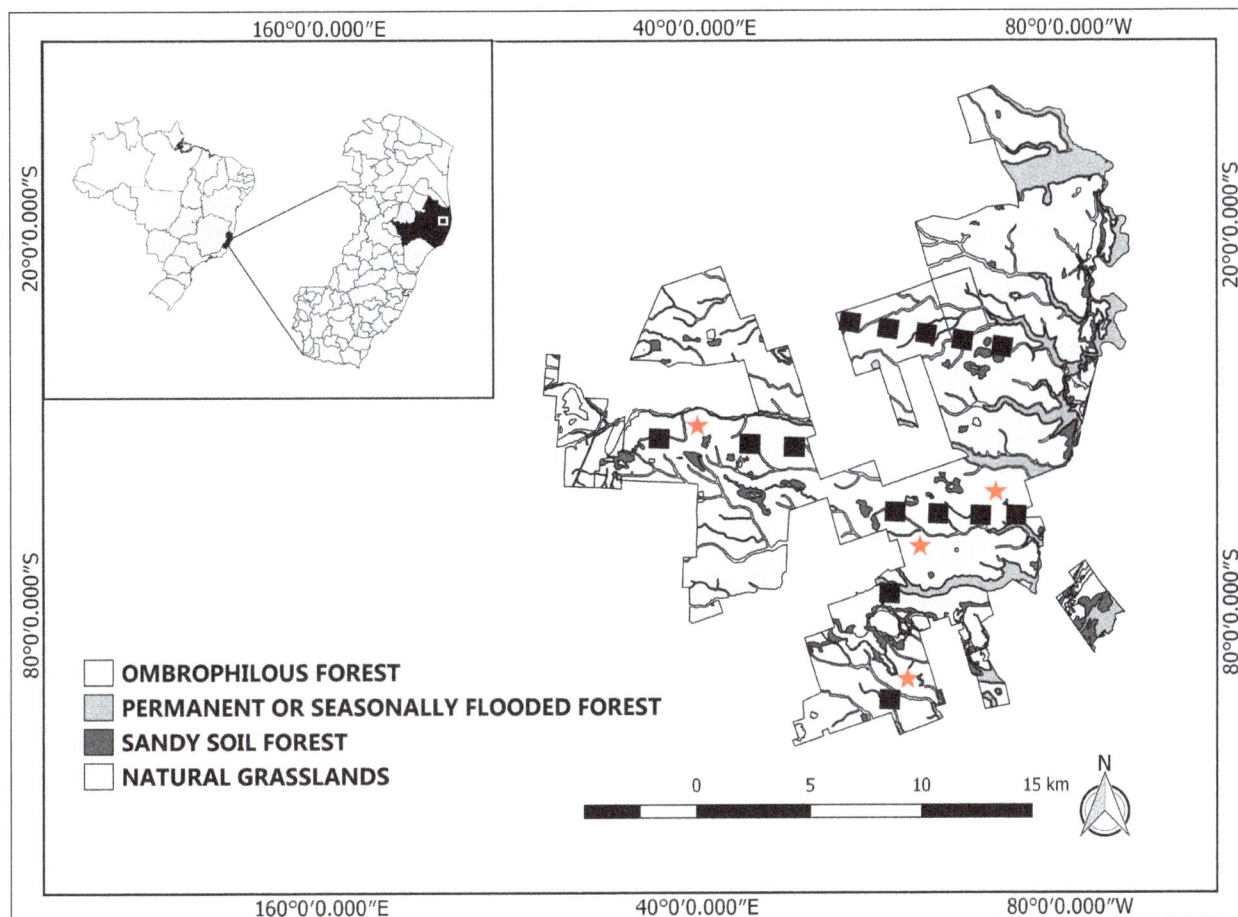

Figure 1. Location of the Vale Natural Reserve, north of Espírito Santo, southeastern Brazil, showing the vegetation types present in the reserve and the location of the plots (black squares) and collection sites of individuals for diet analysis and reproductive aspects (stars).

We estimated the frequency of the different microhabitats used by the treefrog in order to identify those preferentially used. We measured the distribution of heights that individuals were found to identify the range and preferred height that *A. brunoi* perched in the forest. We found some individuals occupying the hole of PVC tubes (used to demarcate the plots in the forest), but we did not consider such records for microhabitat usage estimates because they constituted artificial microhabitats.

We analyzed diet composition in 15 individuals collected outside areas of the plots, in order to avoid interference with our density estimates within plots. The frogs were euthanized with a topical anesthetic gel (lidocaine 5%), fixed in 10% formalin solution and preserved in 70% alcohol (IBAMA license 46327-4). The treefrog specimens were deposited in the Museu Nacional, Rio de Janeiro (MNRJ). We measured snout-vent length (SVL) and jaw width (JW) of the frogs using a Vernier Caliper (to the nearest 0.1 mm) and weighed them using a Pesola dynamometer (to the nearest 0.1 g). Individuals were dissected

to determine the sex and to analyze stomach contents under a stereomicroscope. Animal prey items were measured (to the nearest 0.1 mm) using a Vernier Caliper and categorized to the taxonomic level of order (or family in the case of ants). Unidentified arthropod remains were grouped in a separate category "unidentified arthropod remains" (U.A.R). We measured the length (L) and width (W) of each prey item and its volume (V) was estimated using the ellipsoid formula: $V = 4/3\varpi (L/2) (W/2)^2$ (Dunham 1983). Diet composition was estimated in terms of number (N), volume (V, in mm^3) and frequency (F) of occurrence (percentage of stomachs containing a particular prey category) of each prey type in the stomachs. We estimated an index of relative importance (I_x) of each prey category in the diet which represents the sum of percentages of the number, volume and frequency of each prey type: [$I_x = (\%N + \%V + \%F)$] (Powell et al. 1990). The effect of frog mouth width (in mm) on the volume (in mm^3) of the largest prey consumed was estimated by simple Regression Analysis.

For each gravid female, we recorded the mass of each ovary using an electronic scale (precision of 0.001 g), counted the total number of mature oocytes in both ovaries and measured the diameter of ten ovarian oocytes from each individual female using digital calipers (to the nearest 0.1 mm). We estimated female reproductive effort by dividing the total mass (g) of eggs by the total female body mass (g) including egg mass (Prado et al. 2000). We did not perform a regression analyses between female body size (SVL mm) and the respective number of oocytes due to the small sample size of gravid females (N = 4). We calculated an egg diameter effort index and a number of oocytes effort index by dividing the mean egg diameter and the mean number of oocytes, respectively, by the mean SVL. The results are represented as the mean ± SD and the range of data, the smallest and the largest number. Voucher specimens of A. brunoi are deposited in the Museu Nacional do Rio de Janeiro (MNRJ) under the voucher numbers Linhares, MNRJ 91008, 91009, 91010, 91011, 91012, 91013, 91014, 91015, 91016, 91017, 91018, 91019, 91020, 91021, 91022.

RESULTS

We recorded a total of 77 individuals of A. brunoi, all of them by visual encounters, in the ombrophilous forest. Treefrogs occurred in all months throughout the study, except July and September 2015, when no individuals were found. The abundance of individuals varied consistently throughout the year, with most individuals active during the rainy season (October to March) (n = 62; 80.6% of all individuals recorded) compared with that in the dry season (April to September) (N = 15; 19.4%). The estimated density of A. brunoi for the area varied markedly between the months of dry season (x = 1.79 + 1.90 ind/ha) and months of rainy season (x = 8.06 + 2.81 ind/ha) (t-test = 5.247, t = <0.001, n = 12). The relationship between the accumulated rainfall of the sampling period in each month (p = 0.01) and temperature (p = 0.04) was significantly related with corresponding density of individuals active in that particular month (Multiple Regression Analysis; F_{2-9} = 10.328; R^2 = 0.695; p = 0.005; n = 12; Density = -15.156+0.115*Rainfall+0.636*Temperature) (Figs 2, 3). In contrast, the accumulated humidity of the sampling period in each month was not significantly related to the corresponding density of active individuals in that particular month ($F1_{-10}$ = 0.083; R^2 = 0.008; p = 0.77; n = 12).

In relation to daily activity, the first individuals were found active at dusk, from 06:00 pm, period at which we registered the highest number of individuals. Then, the number of active individuals decreased steadily till 22:00. Before 06:00 pm and after 10:00 pm no individuals of A. brunoi were found (Fig. 4).

In relation to use of the microhabitat, A. brunoi (n = 51) used seven different types of microhabitats in the forest, tree-trunks (51%; n = 26) and lianas (23.5%; n = 12) being the most frequent microhabitats used (Fig. 5, Suppl. material 1: Table S1). Individuals were found perched from 10 to 500 cm above the ground with a median perch height of 40 cm (first quartile = 20 cm, third quartile = 100 cm, n = 77).

We analyzed the stomach content of 15 A. brunoi individuals (8 males and 7 females) (x = 54,4 ± 6,8; 46.0–66.0 mm SVL; 5.5–46.0 g) and, of these, three stomachs were empty. Aparasphenodon brunoi consumed nine different types of prey in its diet (Table 1). The most representative were Orthoptera (20%) and Acari (16%). However, Phasmatodea dominated in volume

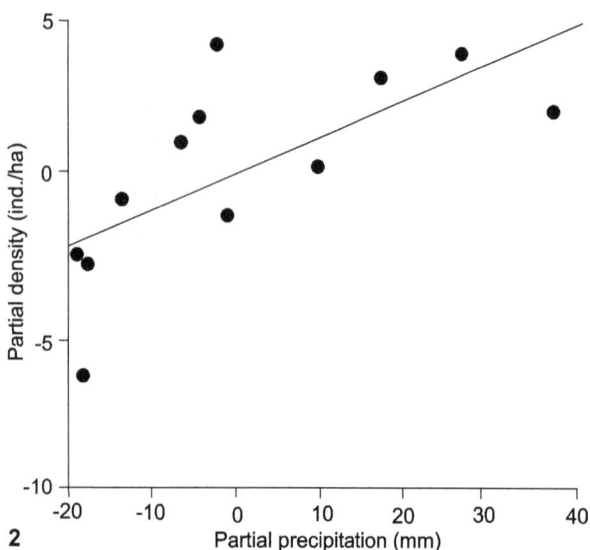

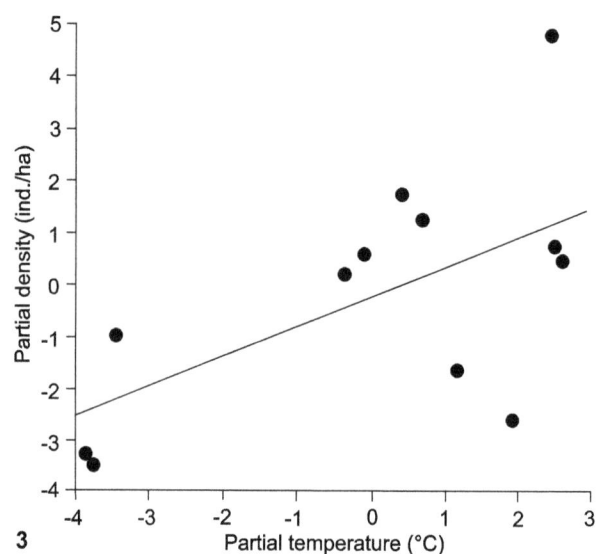

Figures 2–3. Results of Multiple Regression Analysis between (2) the accumulated rainfall of the sampling period in each month (June 2015 – July 2016) and (3) temperature with corresponding density of active individuals of Aparasphenodon brunoi in the Vale Natural Reserve, municipality of Linhares, Espírito Santo, Southeastern Brazil (Density = -15.156+0.115*Rainfall+0.636*Temperature).

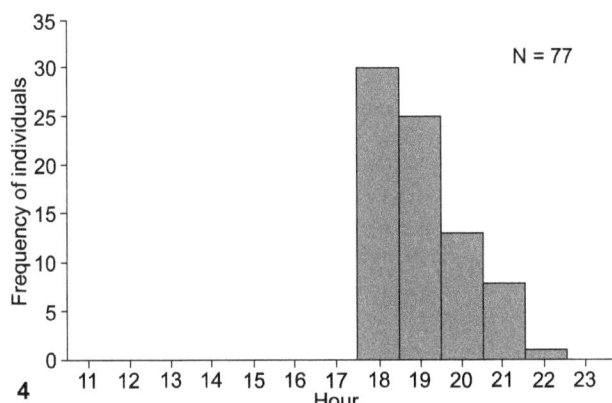

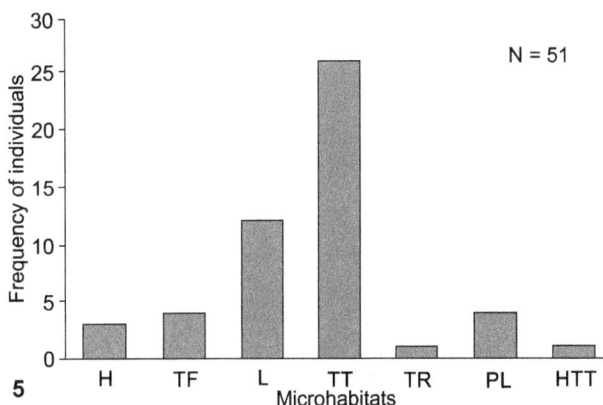

Figures 4–5. Activity and microhabitat use of *Aparasphenodon brunoi*: (4) Number of individuals of *A. brunoi* (N = 77) recorded between 11:00 am and 11:00 pm in transects in the Vale Natural Reserve (VNR), municipality of Linhares, Espírito Santo, Southeastern Brazil). (5) Use of natural microhabitats by individuals of *A. brunoi* (N = 51) in the VNR. (H) On herbaceous plant, (TF) on a fallen tree trunk, (L) on liana, (TT) on a tree trunk, (TR) on a tree root, (PL) on a palm leaf, or in a (HTT) hollow in a tree trunk.

Table 1. Diet of *Aparasphenodon brunoi* at the Vale Natural Reserve, north of Espírito Santo, Brazil. Number (N), volume (V mm³), frequency (F) and Importance index (Ix) of prey categories. U.A.R = unidentified arthropod remains.

Gut Contents	N (%)	V (%)	F (%)	Ix
Arachnida				
Aranae	2 (8)	314.3 (5.48)	1 (6.67)	20.15
Acari	4 (16)	0.06 (0.001)	3 (20)	36.00
Insecta				
Orthoptera	5 (20)	1008.57 (31.51)	3 (20)	71,51
Phasmatodea	1 (4)	1822.21 (31.75)	1 (6.67)	42.42
Isoptera	1 (4)	0.30 (0.01)	1 (6.67)	10.68
Lepidoptera	1 (4)	1117.61 (19.47)	1 (6.67)	30.14
Hymenoptera (ants)	1 (4)	0.20 (0.004)	1 (6.67)	10.67
Coleoptera	1 (4)	157.72 (2.57)	1 (6.67)	13.24
Larvae	2 (8)	245.04 (4.27)	2 (13.33)	25.6
Plant Remains	7 (28)	53.12 (0.93)	7 (46.67)	75.6
U.A.R	–	220.293 (3.84)	–	–
Total	25 (100)	5739.42 (100)	21	–

(31.7%) followed by Orthoptera (31.5%), and Orthoptera (20%) and Acari were the most frequent, with the same percentage (20%) and Larvae represented 13.3% of frequency (Table 1). In relation to the index of relative importance (I_x), Orthoptera dominated (71.51%) followed by Phasmatodea (42.42%). The relationship between mean prey volume and treefrog JW was significantly related ($F_{1-13} = 8.388$; $R^2 = 0.392$; p = 0.012; n = 15).

The mean number of oocytes (± 1 SD) per female was 900.8 ± 358.1 (535–1338; n = 4), with 455.8 ± 218.4 oocytes in the left ovary (183–641) and 445 ± 182.5 oocytes in the right ovary (320–715; n = 4). The mean oocyte diameter was 1.36 ± 0.13 mm (1.22–1.52 mm; n = 40) and mean volume was 1.34 ± 0.38 mm³. The average total oocyte mass was 1.8 ± 0.7 g (0.8–2.4_ g; n = 4) (Suppl. material 2: Table S2). The average female reproductive effort was 7.0 ± 2.0% (4–9%, n = 4). The eggs diameter effort index was 2% and the number of oocytes effort index was 1.4 oocytes/mm.

DISCUSSION

Our data indicated that *A. brunoi* is an essentially nocturnal treefrog species as most anurans (Duellman and Trueb 1994), which is active in the first hours of the night. The species is active throughout the year but with a higher intensity from November to March, coinciding with the rainy season in the area. The density of individuals during the wet season was considerably higher (4.7 times higher) than that recorded during the dry season. In fact, the only months which we did not find active individuals (July and September) were months of the dry season in the area. *A. brunoi* activity (80.6%) occurred mostly during the wet season. Also, the amount of rainfall accumulated during sampling period interacted with temperature to explain the number of individuals of *A. brunoi* active throughout the year. All these data together reinforce that annual activity of *A. brunoi* predominates during the wet season in the ombrophilous forest. Although in the Restinga environment of Praia da Neves (Espírito Santo state) Teixeira et al. (2002) recorded some individuals of this species in some months of both dry and wet seasons. At the restinga of Barra de Maricá (Rio de Janeiro state) activity of this treefrog was restricted to months of the rainy season (Britto-Pereira et al. 1988). This tendency of increased activity during the rainy season in the restinga environment may result from the fact that this season provides more source of humidity to *A. brunoi*. The importance of humidity to favor activity in many frog species is well known and mostly results from their permeable skin (and associated risks of desiccation) and the need of humidity for reproduction (Duellman and Trueb 1994). This may be the reason that in restinga habitats, an environment characterized by low availability of free water and high temperatures (Silva et al. 2011), *A. brunoi* tend to live relatively restricted to bromeliads using the water stored inside the tank of these plants as source of moisture (Britto-Pereira et al. 1988, Teixeira et al. 2002). In restinga habitats, during diurnal period (the hottest one), individuals of *A. brunoi* remain inside the tank of bromeliads whereas nocturnally they remain frequently outside these plants, on leaves of bromeliads and some other plants and trunks (Sazima and Cardoso 1980, Schineider and Teixeira 2001, Mesquita et al. 2004, Sluys et al. 2004). Conversely, in the ombrophilous forest of the VNR, our data indicated that *A. brunoi* used seven different types of microhabitats where it perches preferentially about 1 m above ground, the preferred microhabitats for this species in the forest were tree trunks and lianas and coincidently, not bromeliads. We believe that the increased moisture and milder environmental temperatures of the ombrophilous forest favors a wider range of microhabitats and allows the frogs to remain active throughout the year. Interestingly, from the 77 *A. brunoi* individuals recorded during our study, not a single frog was associated with bromeliads, although these plants are frequent and abundant in the VNR, especially in the non-forest vegetation types (Siqueira et al. 2014, pers. obs. of authors). This is suggestive that the observed differences in the use of bromeliads as microhabitats among forested and restinga environments may be associated with the role of these plants as a source of moisture for these treefrogs in restinga habitats, a question that remains to be investigated.

Our data indicated that in the ombrophilous forest *A. brunoi* is an arthropod predator, with Orthoptera, Phasmatodea, and Lepidoptera being the most important prey items in its diet. In restinga areas, a diet composed by arthropods with consumption of Insecta, Arachnida and Myriapoda has aslo been reported (Teixeira et al. 2002, Mesquita et al. 2004, Sluys et al. 2004). The relatively wide array of prey types consumed indicates that *A. brunoi* is not a selective species and the consumption of relatively active prey can indicate ambush foraging behavior in this species. This can be advantageous to some species that have low foraging velocity and thus, a comparatively lower intensity than active foraging species (Strussman et al. 1984). The inclusion of some relatively large preys as Orthoptera, Phasmatodea and Lepidoptera in the diet can be suggestive of a specific energetic need by this treefrog. Probably the relative preference of arthropods of a relatively large size may be related to the energetic balance of this relatively large-bodied arboreal treefrog (costs and benefits of ingesting large prey). This idea is also supported by our data that shows that about 40% of the variation in prey volume (size) ingested was explained by frog mouth size, an indicative that as the treefrog increases in size tend to prey on larger prey (volumes), probably to keep a positive energy balance. Also, a positive relationship between frog mouth size and ingested prey size is expected for preda-

tors which do not chew their prey and are limited by gape size (Lima and Moreira 1993, Maia-Carneiro et al. 2013). The low amount of plant matter in the frog's diet probably corresponds to plant parts ingested during attempts to capture prey, as has been suggested also for the diet of many other frog species (e.g. Martins et al. 2010, Machado et al. 2016).

Reproductive data showed that ovigerous females of *A. brunoi* produce on average about 900 oocytes per reproductive event with a reproductive effort of about 7%. This relatively high number of oocytes produced by reproductive event may be related to the relatively large body size of this arboreal treefrog. The diameter of each egg was relatively small (x = 1.36 ± 0.13 mm) when compared to the diameters of the eggs of other treefrogs, such as *Hypsiboas faber* (1.92 ± 0.15 mm), *Aplastodiscus eugenioi* (2.31 ± 0.22 mm) and *Phasmahyla gutatta* (2.40 ± 0.25 mm) (Hartmann et al. 2010), and may result from a trade-off between number and size of oocytes produced in which females produce large clutches with smaller eggs, a reproductive strategy for female *A. brunoi* in the ombrophilous forest (an interesting issue to compare to populations in restinga environments).

We conclude that *A. brunoi* is a nocturnal arboreal treefrog, active throughout the year but having increased activity during the wet season, resulting from the large amount of rain. In the forest, the treefrog tend to perch mainly on tree-trunks and lianas about 1 m above ground (instead of using predominantly bromeliads as in restingas), where it feeds preferably on relatively large bodied arthropod prey. The reproductive strategy for females of this comparatively large arboreal treefrog in the ombrophilous forest is to produce a large amount of relatively small-sized eggs. When living in the ombrophilous forest *A. brunoi* may change some features of its ecology (e.g. marked difference in the use of bromeliads) compared to when living in restinga habitats. Our study is the first to gather information on the ecology of this treefrog in ombrophilous forest of the Atlantic Rainforest Biome of Brazil.

ACKNOWLEDGMENTS

This study is part of the results of the "PPBio Mata Atlântica", a program of the Ministry of Science, Technology and Innovation (MCTI). The authors benefitted from grants provided to HGB (process 307715/2009-4 and 457458/2012-7) and to CFDR (304791/2010-5, 470265/2010-8 and 302974/2015-6) from Conselho Nacional de Desenvolvimento Científico e Tecnológico (CNPq) and through "Cientistas do Nosso Estado" Program from FAPERJ to CFDR (processes E-26/102.765.2012 and E-26/202.920.2015) and to HGB (process E-26/103.016.2011). LGM thanks to the program "Becas Iberoamérica, Estudiantes de Grado. Santander Universidades". JPR and ACF received fellowships for master and PhD respectively from Coordenação de Aperfeiçoamento de Pessoal de Nível Superior (CAPES). The ICMBio provided the permit for the development of the study (46327-3) and the Vale Natural Reserve the permit to research in the Reserve.

REFERENCES

Almeida AP, Gasparini JLR (2014) Anfíbios na Reserva Natural Vale, Linhares, Espírito Santo, Brasil. Ciência & Ambiente 49: 211–235.

Argôlo AJS (2000) Geographic distribution. *Aparasphenodon brunoi*. Herpetological Review 31: 108.

Britto-Pereira MC, Cerqueira R, Silva HR, Caramashi U (1988) Anfíbios anuros da restinga de Barra de Maricá, RJ: levantamento e observações preliminares sobre a atividade reprodutiva das espécies registradas. Anais do V Seminário Regional de Ecologia, Universidade de São Carlos, São Carlos, 295–306.

Carvalho AL (1939) Nota previa sobre os habitos de uma interessante 'perereca' bromelicola do litoral SE brasileiro (*Aparasphenodon brunoi* Mir. Rib. 1920). O Campo 10: 25–26.

Crump ML, Scott-Jr NJ (1994) Visual encounter surveys. In: Heyer WR, Donnelly MA, Mcdiarmid RW, Hayek LAC, Foster MS (Eds) Measuring and monitoring biological diversity: standard methods for amphibians. Smithsonian Institution Press,Washington, DC, 84–92.

Duellman WE, Trueb L (1994) Biology of amphibians. The Johns Hopkins University Press, Baltimore.

Dunham AE (1983) Realized niche overlap, resource abundance, and intensity of interspecific competition. In: Huey RB, Pianka ER, Schoener TW (Eds) Lizard Ecology: studies of a model organisms. Harvard University Press, Cambridge, 261–280. https://doi.org/10.4159/harvard.9780674183384.c15

Feio RN, Braga UML, Wiederhecker H, Santos PS (1998) Anfíbios do Parque Estadual do Rio Doce (Minas Gerais). Universidade Federal de Viçosa, Instituto Estadual de Florestas, Viçosa.

Haddad CFB, Toledo LF, Prado CPA, Loebmann D, Gasparini JL, Sazima I (2013) Guia dos anfíbios da Mata Atlântica: diversidade e biologia. Anolis Books, São Paulo.

Hartmann MT, Hartmann PA, Haddad CFB (2010) Reproductive modes and fecundity of an assemblage of anuran amphibians in the Atlantic rainforest, Brazil. Iheringia, Série Zoologia, 100: 207–215. https://doi.org/10.1590/S0073-47212010000300004

Jared C, Mailho-Fontana PL, Antoniazzi MM, Mendes VA, Barbaro KC, Rodrigues MT, Brodie ED (2015) Venomous Frogs Use Heads as Weapons. Current Biology 25: 2166–2170. https://doi.org/10.1016/j.cub.2015.06.061

Kierulff MCM, Avelar LHS, Ferreira MES, Povoa KF, Bérnils RS (2014) Reserva Natural Vale: História e aspectos físicos. Ciência & Ambiente 49: 7–40.

Lima AP, Moreira G (1993) Effects of prey size and foraging mode on the ontogenetic change in feeding niche of *Colostethus stepheni* (Anura: Dendrobatidae). Oecologia 95: 93–102. https://doi.org/10.1007/BF00649512

Machado AO, Wink G, Dorigo TA, Rocha CFD (2016) Diet, Diel Activity Pattern, Habitat Use, and Reproductive Effort of *Hylodes nasus* (Anura: Hylodidae) in One of the World's Largest Urban Parks (Tijuca National Park), Southeastern Brazil. South American Journal of Herpetology 11: 127–135. https://doi.org/10.2994/SAJH-D-16-00004.1

Magnusson WE, Lima AP, Luizão R, Luizão F, Costa FRC, Castilho CVD, Kinupp VF (2005) RAPELD: A modification of the Gentry Method for biodiversity surveys in long-term ecological research sites. Biota neotropica 5: 1–6. https://doi.org/10.1590/S1676-06032005000300002

Maia-Carneiro T, Kiefer MC, Sluys MV, Rocha CFD (2013) Feeding habits, microhabitat use, and daily activity period of *Rhinella ornata* (Anura, Bufonidae) from three Atlantic rainforest remnants in southeastern Brazil. North-Western Journal of Zoology 9: 157–165.

Martins ACJS, Kiefer MC, Siqueira CC, Van Sluys M, Menezes VA, Rocha CFD (2010) Ecology of *Ischnocnema parva* (Anura: Brachycephalidae) at the Atlantic rainforest of Serra da Concórdia, state of Rio de Janeiro, Brazil. Zoologia 27: 201–208. https://doi.org/10.1590/S1984-46702010000200007

Mesquita DO, Costa GC, Zatz MG (2004) Ecological aspects of the casque-headed frog *Aparasphenodon brunoi* (Anura, Hylidae) in a Restinga habitat in southeastern Brazil. Phyllomedusa: Journal of Herpetology 3: 51–59. https://doi.org/10.11606/issn.2316-9079.v3i1p51-59

MMA (2000) Avaliação e Ações Prioritárias para a Conservação da Biodiversidade da Mata Atlântica e Campos Sulinos. Ministério do Meio Ambiente, SBF, Brasília.

Mollo Neto A, Teixeira-Jr M (2012) Checklist of the genus *Aparasphenodon* Miranda Ribeiro, 1920 (Anura: Hylidae): Distribution map, and new record from São Paulo state, Brazil. Check List 8: 1303–1307. https://doi.org/10.15560/8.6.1303

Oliveira JCF, Rocha CFD (2014) Journal of coastal conservation: a review on the anurofauna of Brazil's sandy coastal plains. How much do we know about it? Journal of Coastal Conservation 19: 35–49. https://doi.org/10.1007/s11852-014-0354-8

Prado CPA, Uetanabaro M, Lopes FS (2000) Reproductive strategies of *Leptodactylus chaquensis* and *L. podicipinus* in the Pantanal, Brazil. Journal of Herpetology 34: 135–139. https://doi.org/10.2307/1565249

Peixoto AL, Silva IM, Pereira OJ, Simonelli M, Jesus RM, Rolim SG (2008) Tabuleiro Forests North of the Rio Doce: Their Representation in the Vale do Rio Doce Natural Reserve, Espírito Santo, Brazil. In: Thomas WW, Britton EG (Eds) The Atlantic coastal forest of Northeastern Brazil. The New York Botanical Garden Press, New York, 313–348.

Powell R, Parmerlee JS, Rice MA (1990) Ecological observations on *Hemidactylus brooki haititanus* Meerwarth (Sauria: Gekkonidae) from Hispaniola. Caribbean Journal of Science 26: 67–70.

Rizzini CT (1997) Tratado de fitogeografia do Brasil: aspectos ecológicos, sociológicos e florísticos. Âmbito Cultural Edições, Rio de Janeiro.

Rocha CFD, Carvalho-e-Silva S, Van Sluys M (2004) *Aparasphenodon brunoi*. The IUCN Red List of Threatened Species 2004: e.T55298A11277104. https://doi.org/10.2305/IUCN.UK.2004.RLTS.T55298A11277104.en [Accessed: 10/07 2016]

Ruas DS, Mendes CVM, Del-Grande ML, Solé M (2013) *Aparasphenodon brunoi* Miranda-Ribeiro, 1920 (Anura: Hylidae): Distribution extension and geographic distribution map for Bahia state, Brazil. Check List 9: 858–859. https://doi.org/10.15560/9.4.858

Sazima I, Cardoso AJ (1980) Notas sobre a distribuição de *Corythomantis greeningi* Boulenger, 1896 e *Aparasphenodon brunoi* Miranda-Ribeiro, 1920 (Amphibia, Hylidae). Iheringia, Série Zoologia, 55: 3–7.

Schineider JAP, Teixeira RL (2001) Relacionamento entre anfíbios anuros e bromélias da restinga de Regência, Linhares, Espírito Santo, Brasil. Iheringia, Série Zoologia, 91: 41–48. https://doi.org/10.1590/S0073-47212001000200005

Silva HRD, Carvalho ALGD, Bittencourt-Silva GB (2011) Selecting a Hiding Place: Anuran Diversity and the use of Bromeliads in a Threatened Coastal Sand Dune Habitat in Brazil. Biotropica 43: 218–227. https://doi.org/10.1111/j.1744-7429.2010.00656.x

Silva-Soares T, Hepp F, Costa PN, Luna-Dias C, Gomes MR, Carvalho-e-Silva AMPT, Carvalho-e-Silva SP (2010) Anfíbios anuros da RPPN Campo Escoteiro Geraldo Hugo Nunes, Município de Guapimirim, Rio de Janeiro, Sudeste do Brasil. Biota Neotropica 10: 225–233. https://doi.org/10.1590/S1676-06032010000200025

Sluys MV, Rocha CFD, Hatano FH, Boquimpani-Freitas L, Marra RV (2004) Anfíbios da restinga de Jurubatiba: composição e história natural. In: Pesquisas de longa duração na Restinga de Jurubatiba: ecologia, história natural e conservação. RiMa, São Carlos, 165–178.

Siqueira GS, Kierulff MCM, Alves-Araújo A (2014) Florística das plantas vasculares da Reserva Natural Vale, Linhares, Espírito Santo, Brasil. Ciência e Ambiente 49: 67–129.

Strussman C, Vale MBR, Meneghini MH, Magnusson WE (1984) Diet and foraging mode of Bufo marinus and Leptodactylus ocellatus. Journal of Herpetology 18: 138–146. https://doi.org/10.2307/1563741

Teixeira RL, Schineider JAP, Almeida GI (2002) The occurrence of amphibians in bromeliads from a Southeastern Brazilian restinga habitat, with special reference to *Aparasphenodon brunoi* (Anura, Hylidae). Brazilian Journal of Biology 62: 263–268. https://doi.org/10.1590/S1519-69842002000200010

Wogel H, Weber LN, Abrunhosa PA (2006) The tadpole of the casque-headed frog, *Aparasphenodon brunoi* Miranda-Ribeiro (Anura: Hylidae). South American Journal of Herpetology 1: 54–60. https://doi.org/10.2994/1808-9798(2006)1[54:TTOTCF]2.0.CO;2

Author Contributions: LGM, JPR and ACF collected the data; LGM, JPR and MAS did the analysis of diet and reproductive aspects; HGB helped with the statistical analysis; CFDR and all authors participated in the writing and review of the manuscript.

Competing Interests: The authors have declared that no competing interests exist.

10

The breeding biology, nest success, habitat and behavior of the endangered Saffron-cowled Blackbird, *Xanthopsar lavus* (Aves: Icteridae), at an Important Bird Area (IBA) in Rio Grande do Sul ,Brazil

Luciane R. da Silva Mohr[1]**, Eduardo Périco**[1]**, Vanda S. da Silva Fonseca**[1,2]**, Alexsandro R. Mohr**[2,3]

[1]*Programa de Pós-graduação em Ambiente e Desenvolvimento, Museu de Ciências Naturais, Setor de Ecologia e Evolução, Centro Universitário Univates. Rua Avelino Tallini 171, 95900-000 Lajeado, RS, Brazil.*
[2]*Bioimagens Consultoria Ambiental. Rua Felicíssimo de Azevedo 1352, 90540110 Porto Alegre, RS, Brazil.*
[3]*Graduação em Biologia, Universidade de Santa Cruz do Sul. Avenida Independência 2293, 96816-501 Santa Cruz do Sul, RS, Brazil.*
Corresponding author: Luciane Rosa da Silva Mohr (lu.mohr@hotmail.com)

http://zoobank.org/68508FE5-9679-43E5-83AC-B2369249B68A

ABSTRACT. The Saffron-cowled Blackbird, *Xanthopsar flavus* (Gmelin, 1788), is a globally vulnerable icterid endemic to grasslands and open areas, and a priority species for research and conservation programs. This contribution provides information on the population size, habitat, behavior, breeding biology and nest success of *X. flavus* in two conservation units (CUs) in Viamão, state of Rio Grande do Sul, Brazil: the Environmental Protection Area Banhado Grande, and the Wildlife Refuge Banhado dos Pachecos, classified as an "Important Bird Area". Searches for *X. flavus* were carried out mainly in open areas, the type of habitat favored by the species. Outside the breeding season individual behavior was recorded by the *ad libitum* method; during the breeding season, selected *X. flavus* pairs were observed following the sequence sampling method. The research areas were visited once a month, totaling approximately 530 hours of observations (September 2014 to June 2016) over 84 days, which included two breeding seasons. The species was observed across all months (not necessarily within the same year) and several *X. flavus* flocks were encountered, some with more than one hundred individuals (range = 2-137). Additionally, the behavior and feeding aspects, habitat use and breeding information on *X. flavus* were recorded. Two breeding colonies were found, and eleven nests were monitored. The estimated nesting success was 10% in Colony 1, but zero in Colony 2, where all eggs and nestlings were predated. Saffron-cowled Blackbirds were recorded in mixed flocks, mostly with *Pseudoleistes guirahuro* (Vieillot, 1819), *P. virescens* (Vieillot, 1819) and *Xolmis dominicanus* (Vieillot, 1823), the last also a globally endangered species. The collected information highlights the importance of CUs for the maintenance of *X. flavus* populations in the region. Maintenance of proper areas for feeding and breeding is necessary and urgent. Information from current research is being employed in the management plan of the Wildlife Refuge Banhado dos Pachecos in which *X. flavus* is one of the conservation target-species.

KEY WORDS. Conservation, habitat degradation, natural history, vulnerable species.

INTRODUCTION

The Saffron-Cowled Blackbird, *Xanthopsar flavus* (Gmelin, 1788), is an endemic species of the grasslands of southern South America (Collar et al. 1992). The species is of high priority for conservation and research (Stotz et al. 1996, Fontana et al. 2013). This bird depends on heterogeneous areas of natural grassland, and uses different habitats for feeding and nesting (Fonseca et al. 2004, Fraga 2005, Azpiroz et al. 2012). It is currently listed as vulnerable at the regional, national and global levels (State Decree 51.797/2014, ICMBio/2014, IUCN 2016), largely due to the destruction and degradation of its habitat, and the consequent disruption of its biological cycle.

The geographical distribution of the Saffron-Cowled Blackbird includes southern Brazil (states of Santa Catarina and Rio Grande do Sul), southern Paraguay, Uruguay and northeastern

Argentina (Collar et al. 1992, Azpiroz 2000, Dias and Maurício 2002, Fonseca et al. 2004, Fraga 2005, Birdlife International 2016). According to Collar et al. (1992), Saffron-Cowled Blackbird populations are decreasing throughout their distribution range, and for this reason it is important to protect the areas that still remains adequate to its life-cycle (Azpiroz 2000, Azpiroz et al. 2012). The world population of the species is estimated at 10,000 individuals, at the most (Birdlife International 2016).

In Brazil, information on the species' natural history and conservation comes mainly from studies and observations carried out in the extreme north-eastern corner of the country (Belton 1994, Fonseca et al. 2004, Krüger and Petry 2010, Petry and Krüger 2010, Moura 2013) and on the southern coastal plain of the state of Rio Grande do Sul (Dias and Maurício 2002). Due to its status as a vulnerable species, new information on the species at other sites is both relevant and important. The current study provides data on the size of the Saffron-Cowled Blackbird population, habitat, behavior, breeding biology and nest success from an Important Bird Area (IBA) and its immediate vicinity, on the internal coastal plain of Rio Grande do Sul, Brazil.

MATERIAL AND METHODS

The study area lies in the municipality of Viamão, within the coastal plain of the state of Rio Grande do Sul, southern Brazil (Fig. 1). It comprises the "Wildlife Refuge Banhado dos Pachecos (WRBP)" (30°05'45.43"S, 50°51'46.38"W), and the "Environmental Protection Area of Banhado Grande (EPABG)" and its immediate vicinity. The study areas lie within an ecotone zone between the Atlantic Rain Forest and Pampas biome. Sandbank vegetation, pioneer shrub-tree woods, swampy areas with Cyperaceae and high grass, flooded fields, dry marshes, pastures and areas with anthropic activities (mainly livestock and rice fields) occur within the limits of WRBP and EPABG (Accordi and Hartz 2006).

EPABG, a Sustainable Conservation Unit of approximately 133,000 ha, was established in 1998 (State Decree 38.971/1998). WRBP is an Integral Protection Conservation Unit, with 2,543.46 ha, established in 2002. No environmentally degrading anthropic activity is allowed there (State Decree 41.559/2002). The preserved areas protect the region's wetlands and water sources. The WRBP is classified as an "Important Bird Area" (IBA), highly relevant for bird conservation, particularly for species that depend on dense marshes and wet grasslands, including endangered species (Bencke et al. 2006). The research in the two protected areas was authorized by the Department of Environment (SEMA) of the state of Rio Grande do Sul (Authorization 01/2015).

The areas for the study of the Saffron-Cowled Blackbirds (SCB) were selected based on the results of Accordi and Barcellos (2006), Accordi and Hartz (2006), and on observations in the region during January 2014, when 12 SCBs and a bird couple with nest and two nestlings were reported in the EPABG. Month-ly reports on SCB occurrences (from September 2014 to June 2016) were the result of research within the WRBP, and on the main and secondary roads in the area surrounding the EPABG. Roads were traversed by car at low speed (<10 km/h) and on foot. Research efforts focused on marshy areas and grasslands characterized by *Eryngium* sp. L. (Apiaceae), Cyperaceae and Asteraceae which may be used for breeding and in fields or pasturelands, with low vegetation, used for feeding.

When a SCB was detected, the number of individuals in the flock was counted, and the associated species, or species that interacted with SCBs, were identified. The birds were taxonomically identified and quantified by direct observation, or identified by their calls. Classification followed Piacentini et al. (2015). After the counting of individuals, each area was observed for 60 minutes to investigate individual arrivals and their behavior. Outside the breeding season, all SCB occurrences were noted, along with associated behaviors (Franchin et al. 2010). During the breeding season both *ad libitum* and sampling sequences methods were used (Franchin et al. 2010). In the latter case, selected pairs were monitored. However, whenever the birds seemed uncomfortable with the presence of the observer, monitoring was temporarily interrupted. Areas with SCB occurrences were visited once a month, totaling approximately 530 hours of field observations during 84 days (between September 2014 and June 2016), including two breeding seasons, across 22 months. Observations were conducted in the morning, during the day, and in the afternoon.

In the breeding seasons, two nesting areas were found and named "Nesting Colony 1" and "Nesting Colony 2". After the first report of breeding activity, the colonies were assessed for four consecutive days, after which nests were visited every three days (with one exception in each area when four days had passed). Once the nesting areas were abandoned, each nest's internal and external diameter, internal nest depth, external nest height and nest height from the ground were measured. All plant species supporting nests were identified and their heights were measured. All plants more than 60 cm high within a 1 m radius of the nest were also identified (Fonseca et al. 2004, modified), since they could support nests or predators. The tallest and the shortest plants (including leaves and inflorescences when present) were measured. Variables for each nesting colony were compared with a t-test using the statistical program Past 3.13 (Hammer et al. 2001).

Since no nest was found either under construction or during egg-laying, egg and nestling numbers are reported, but clutch size was not. To avoid research activities causing any reproductive loss, eggs and nestlings were not measured, and no specimens were captured or marked. To make comparisons possible, reproductive success was presented in two ways: first based on fledged chicks/eggs or nestlings rates so as to be comparable with some previous studies (Fraga et al. 1998, Dias and Mauricio 2002), and by apparent success (number of successful nests/total number of nests) to be comparable with others (Moura 2013). In the current study, nests were considered successful when at least one chick fledged.

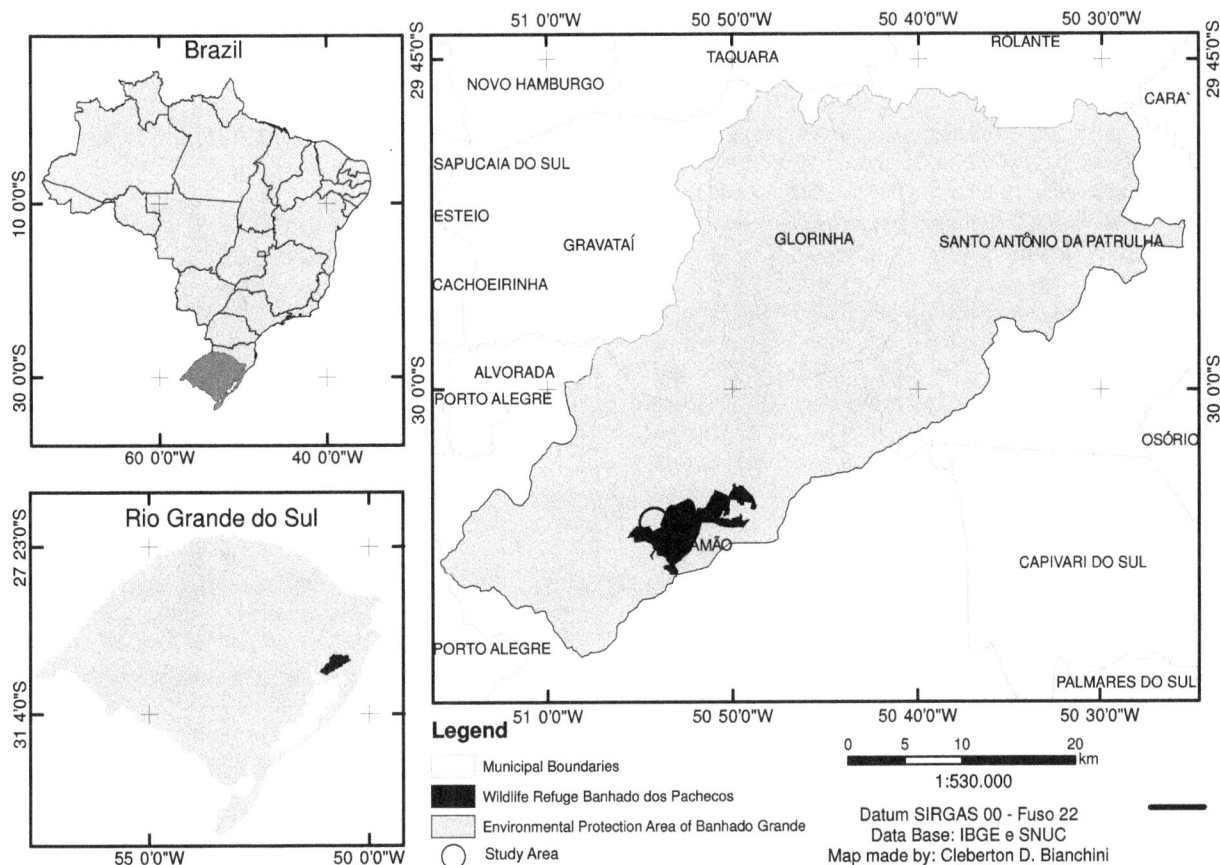

Figure 1. Map of study area, with the Wildlife Refuge Banhado dos Pachecos (WRBP) and, in its immediate vicinity, the Environmental Protection Area of Banhado Grande (EPABG), in the state of Rio Grande do Sul, Brazil.

RESULTS

SCBs in the WRBP mainly occupied a landscape characterized by fields and wetlands dominated by *Eryngium* sp. and *Typha domingensis* (Pers.) (Typhaceae). In the immediate vicinity of EPABG *X. flavus* was reported in dry field areas, mainly with *Schizachyrium microstachyum* (Desv. ex Ham.) Roseng., B.R. Arrill. and Izag. (Poaceae), in wet fields with *Eryngium* sp. occupied by livestock and horses, and in rice fields.

SCB were not encountered in the study areas in January and May, 2015, and in February, 2106. In the non-breeding season (during the austral autumn), the largest SCB flock (137 individuals) was reported in March 2015. Other sizeable flocks were reported in February 2015, January, April and May 2016, with 69, 96, 75 and 70 individuals, respectively. All reports occurred in areas close to the WRBP within the EPABG. During the first breeding season, the largest SCBs flock (16 individuals) was reported (November and December 2014) at a Nesting Colony 1 and a flock of 21 SCBs outside the nesting colony, on the edge of WRBP. In the second breeding season (November and December 2015) the largest flocks observed were 10 SCBs in Nesting Colony 2 and with 35 SCBs in a flock which was feeding within an adjacent area outside the Nesting Colony.

The habitats most commonly used by *X. flavus* for feeding were fields covered with short grasses and rice fields during the early maturation stages and post-harvest period. SCBs were mainly observed feeding on insects and insect larvae, usually foraging for them on the ground, while probing soil and vegetation. While some individuals were feeding on the ground, others took the role of sentinels, perched on eucalyptus *Eucalyptus* sp. L'Hér. (Myrtaceae) and maricá trees *Mimosa bimucronata* (DC.) Kuntze (Fabaceae). During the breeding season, two males, while acting as sentinels at the nest, were observed capturing flying insects and eating them in a manner similar to flycatchers.

In the non-breeding season, when SCBs were foraging in intraspecific flocks, two or three males took the role of sentinels and, through vocalization, appeared to warn off the other individuals to move away when researchers or possible predators came close. Sentinels rested on the ground and perched on trees (*Eucalyptus* sp. and *M. bimucronata*). In the non-breeding season, SCB individuals were seen in mixed flocks with *Pseudoleistes guirahuro* (Vieillot, 1819) (Icteridae), *P. virescens* (Vieillot, 1819) (Icteridae), *Agelasticus thilius* (Molina, 1782) (Icteridae), *Sicalis luteola* (Sparrman, 1789) (Thraupidae), *Chrysomus ruficapillus* (Vieillot, 1819) (Icteridae), *Zenaida auriculata* (Des Murs, 1847) (Columbidae) and *Xolmis dominicanus* (Vieillot, 1823) (Tyrannidae). Within these *P. guirahuro* and *P. virescens* acted as sentinels.

SCB were recorded with the Black-and-White Monjitas (BWM), *X. dominicanus*, only in the autumn and winter, in April, May, June, July and August. During these months, the largest BWM gatherings respectively comprised of 2, 15, 20, 15 and 15 individuals. Largest mixed flocks of SCBs and BWMs were, respectively, 14 and 36 in June 2015; 15 and 15 in July 2015; 70 and 15 in May 2016; and 30 and 12 in June 2016.

In October, during the two breeding seasons evaluated, and approximately one month prior to the discovery of nests, SCB males were displaying to females in aerial courtship flights. Females in Nesting Colony 1 were reported inspecting *Eryngium* sp., while males perched close by. In the breeding season, SCB individuals were reported close to *Tyrannus savanna* Daudin, 1802 (Tyrannidae), *Progne tapera* (Vieillot, 1817) (Hirundinidae) and *Sturnella superciliaris* (Bonaparte, 1850) (Icteridae). During territorial disputes and defense, SCBs attacked individuals of the same and of other species, including *P. virescens* and *P. guirahuro*, and each defended a territory some 3 m-radius around the nest. On one occasion, a nesting pair in Nesting Colony 2 joined a mixed flock of *P. virescens* and *P. guirahuro* to feed in a field area approximately 30 m from the colony. Several times, SCB gathered together with specimens of *P. guirahuro*, *P. virescens*, *T. savanna*, *P. tapera* to drive off such predators as *Circus buffoni* (Gmelin, 1788) (Accipitridae), *C. cinereus*, Vieillot, 1816 (Accipitridae), *Milvago chimango* (Vieillot, 1816) (Falconidae) and *Falco femoralis* Temminck, 1822 (Falconidae).

In the first breeding season assessed, there were six nests (1-6) in Nesting Colony 1, five of which were measured (Table 1). One nest was damaged by water, making it impossible to take the full set of measurements. It should be underscored that in the same area, in January 2014, a pair with a nest containing two nestlings was recorded. During the second breeding season, five nests (7-11) were found and measured in Nesting Colony 2 (Table 1). Colonies were separated by some 3 km from one another. Nesting Colony 1 was characterized by small, near-isolated, clusters of *Eryngium* sp., whilst Nesting Colony 2 was composed of isolated *Eryngium* sp. individuals and a small dense cluster of these plants, although the nests lay at the edge of the site, in a matrix of otherwise low-lying grazed grassland. All nests in Nesting Colony 1 were attached to *Eryngium* sp., the dominant plant in the region, and were, on average, 51 cm from the ground. Only one nest (Nest 1) was fixed on the tallest plant in the area, even though this was only 35 cm from the ground. In Nesting

Table 1. Mean ± standard deviation (SD) of measurements of Saffron-Cowled Blackbirds' nests.

Measures (in cm)	Nesting Colony 1			Nesting Colony 2		
	N	Mean ± SD	Range	N	Mean ± SD	Range
Height of the nests' supporting plants	6	105.17 ± 23.80	84.0–147.0	5	89.8 ± 21.88	58.0–113.0
Height of tallest plants*	6	140.67 ± 17.10	122.0–164.0	5	96.8 ± 24.59	61.0–125.0
Height of shortest plants*	6	83.00 ± 9.70	72.0–98.0	5	46.8 ± 23.05	11.0–63.0
Nest height above the ground	6	51.33 ± 13.10	35.0–68.0	5	46.4 ± 18.65	21.0–73.0
Nest internal depth	5	5.24 ± 0.15	5.1–5.4	5	5.7 ± 0.51	5.1–6.3
Nest external height	5	12.38 ± 0.57	11.6–13.0	5	11.82 ± 1.58	10.5–14.5
Nest internal diameter	5	6.32 ± 0.35	5.9–6.7	5	6.88 ± 0.52	6.4–7.6
Nest external diameter	5	10.92 ± 0.31	10.4–11.2	5	11.67 ± 0.5	11.1–12.3

*Leaves and inflorescence.

Colony 1, the closest nests were 5 m apart, whilst the most distant were 52 m apart. In Nesting Colony 2, the closest nests were separated by 4.5 m, and the most distant by 70 m. One nest was fixed only to a *Scirpus* sp. L. (Cyperaceae); three nests were fixed to *Scirpus* sp. and *Eryngium* sp. simultaneously, and one nest was attached only to grass. The latter was closest to the ground. No nest was fixed to the tallest plant within the area surrounding the nests. In Nesting Colony 2, nest diameters and depths were greater than those of Nesting Colony 1 (Table 1), but there was no significant variation in all nest measurements between the two colonies during the two breeding seasons under analysis (p > 0.05). Although the variables mean heights of the tallest plant and mean height of the shortest plant around the nests differed significantly between the two nesting colonies analyzed (p < 0.05 for both), the mean height of the plants supporting the nests were not significantly different between sites (p > 0.05).

Based on the number of eggs and nestlings observed in the two colonies (2-5), the average number of eggs per nest was calculated as 3.6 (± 0.84). One egg was laid per day and eggs hatched after 11-12 days (n = 1). Nestling lifespan (the time it took from egg-hatching to fledging) was analyzed following Azpiroz (2000). Nest 1 was discovered on November 28th, 2014 with four nestlings, aged one or two days. The nestlings vanished between the 9th and 12th day of life: it is possible that they flew out of the nest, but not probable. Since those nestlings were too young and were unable to take long flights, it would be expected that, if they were alive and well, they would be found nearby, but this did not happen.

Nest 2 was discovered on November 28th, 2014, with four one-day-old nestlings. They vanished during the night on the fourth day. This nest was the first monitored nest to be preyed upon. It was the farthest from the others, but the closest to the ground, on the edge of an irrigation canal. Paw tracks of the pampas fox *Lycalopex gymnocercus* (G. Fischer, 1814) (Canidae) were found nearby. Nest 3 was discovered on November 29th, 2014 with two eggs. Then, on the following days, two more eggs were laid, one on each day. Nestlings were seen exiting the nest on the 9-10th day. This was sooner than the 12th day in the results of Azpiroz (2000). This accelerated fledging may owe to the intense heat on the nest, since it was exposed to the sun.

Nest 4 was discovered ready, but empty, on November 29th, 2014. It remained empty and it was impossible to say whether there had been eggs and/or nestlings in it, or whether it was abandoned right after it was built. Nest 5 was discovered on November 29th, 2014 with three nestlings, which were observed till the 9th day. They were not seen after this, even though they could fly out of the nest, albeit not for long flights. They were not seen within the area close to the nest. Nest 6 was found on November 30th, 2014 when it contained four eggs and a nestling aged one or two days. The latter was observed during the following eight days but was not seen after the 11th day. It may have flown out of the nest since it was not seen in the immediate vicinity. The eggs were neither predated nor removed from the nest. They still had not hatched on the last day of observations, on January 14th, 2015.

Based on the ratio of successful nests (at least one chick produced)/total number of nests evaluated for Nesting Colony 1, five out of six nests were successful (83%). On the other hand, only two fledgling birds at different stages of development were reported between December 10th, 2014 and December 23th, 2014 in the vicinity of Colony 1. Since they were only capable of hopping and short flights around the area, it is possible that they were the fledglings from Nests 1 or 5 and 3. If reproductive success is, following Fraga et al. (1998), calculated as the number of fledged chicks/eggs laid or nestling observed, then our estimates are close to 10%. In fact, out of the 20 possible individuals observed that could have developed in the Colony, only these two were observed.

Eight well-developed juvenile SCBs and two adult females were reported at another site, on the edge of the WRBP, on November 30th, 2014. They were perched on *M. bimucronata* and then flew together to the interior of the WRBP. At another area, on the edge of the WRBP, a pair was seen feeding four well-developed fledglings accompanied by a male helper. While the male flew in search of food, the female and the helper attacked a low-flying *C. buffoni* together. The four fledglings hid among cattail brushes (*T. domingensis*). In addition, a male with two juveniles were seen close by.

Although two pairs of Shiny Cowbirds, *Molothrus bonariensis* (Gmelin, 1789) (Icteridae), had been seen in Nesting Colony 1 in October, they were not observed there during the breeding period. Also, parasitism by this species on SCB nests was not observed. During a two-hour morning observation on Nest 2, a SCB pair was observed feeding four nestlings every 12 minutes on average (range 4-30 min). During the observation period of one hour and fifteen minutes, in the late afternoon, the average interval of feeding the nestlings of *X. flavus* was 12.3 minutes

(range 10–13), with only the female engaged in this activity. During this period, the male was seen engaging in territorial defense in the nesting area.

During the second breeding season, Nest 7 was found on November 13th, 2015 with three eggs; the nest had two nestlings and one egg on November 21th, 2015; there were three nestlings on November 22th. Four days later, all the nestlings had been preyed upon. Nest 8 was discovered on November 13th, 2015 with three eggs, but only two eggs remained on November 15th. On November 21th, there was a nestling in the nest, but it was not seen again after November 26th. The pair associated with this nest was also not observed after this date. Nest 9 was seen between November 14th and November 16th with four eggs, but it was empty on November 18th. Traces of eggshells were seen on the ground close to the nest. Nest 10 was recorded in the morning (10 a.m.) of November 15th with four eggs, but it was empty in the afternoon (04:15 p.m.). The eggs were likely preyed upon during the day, but there were no traces of eggshells. Nest 11 was observed on November 15th with two eggs; two nestlings were recorded on November 21th; there was only one nestling on November 26th, and the nest was empty on December 1st. The fledgling may have flown out of the nest, but it was not seen in the vicinity.

Apparently, in Colony 2 there were two successful nests out of five (40%). However, we have almost certainly overestimated the reproductive success of the parent birds, since the fledged and juvenile offspring were not observed in the area close to the nests. It is possible that successful breeding did not occur at all in the colony. Beyond those attempts related above, the colonies were not observed making any new breeding endeavors during the study period.

DISCUSSION

Habitat

The habitats used by individual SCBs were the same as those recorded in other studies (Belton 1994, Azpiroz 2000, Dias and Maurício 2002, Fonseca et al. 2004). SCBs engaged in activities such as resting, concealment and feeding of young mainly in wetlands containing *Eryngium* sp. and *T. domingensis*. *X. flavus* feeds in low grassland areas. As there is no active land management at WRBP, the grasslands have a high percentage of tall vegetation, making them inappropriate as SCB feeding areas. As a result, the birds must fly long distances to find suitable feeding areas. In some of our observations, adults had to fly between 0.6 and 1km to obtain food.

The lack of established SBC breeding areas at WRBP may be due to the distances between potential nesting sites and feeding areas within the protected area. In the immediate vicinity of EPABG, SCBs were frequently reported feeding in areas with rice crops, in dry fields and in wet fields with *Eryngium* sp. In the fields with *Eryngium* sp., there were also livestock and horses.

Although these herbivores impair the growth of plants that SBCs use for nest building, they keep the vegetation stature low, which is appropriate for the foraging of *X. flavus*.

Number of Blackbird

According to the Birdlife International (2016), there are approximately 10,000 SCB individuals within the four countries where this bird is distributed. Reports from Uruguay indicate flocks averaging 60 specimens (1-135) (Azpiroz 2000), whereas in Paraguay, flocks are composed of 30-50 individuals, but up to 250 specimens were observed feeding in wetlands and rice fields (Esquivel et al. 2007), and some 300 SCBs were recorded at an area where soybean and corn were being cultivated (Codesido and Fraga 2009). Flocks of up to 240 specimens have been reported from Argentina (Fraga 2005).

In Brazil, flocks with more than 70 individuals have been reported in Bom Jesus (RS) and Lages (SC) (Fontana et al. 2008). In the natural higher-altitude grasslands of Rio Grande do Sul, at a locally known as "Campos de Cima da Serra", flocks of up to 30 individuals have been recorded (Fonseca et al. 2004), with up to 100 individuals in breeding colonies, and 300 birds outside the breeding season (Moura 2013). Flocks with approximately 60 individuals have been reported from the southern coastal plain of Rio Grande (Dias and Maurício 2002). The occurrence of *X. flavus* in the region under study (EPABG and WRBP) has been known for at least fifteen years (Accordi and Hartz 2006). In the past, flocks with 52 specimens, with estimates of no more than 100 specimens, were recorded (Bencke et al. 2003). Therefore, the frequent observations of flocks with fewer individuals during the current study, flocks ranging between 50 and 90 specimens on 18 occasions, and a flock of 137 specimens on a single occasion, are important findings, since they show that the species is present in the region and demonstrate the importance of landscape heterogeneity for the conservation of the species.

During the two breeding seasons in this study, the number of individuals was seldom greater than the number of couples engaged in building and taking care of their nests. In contrast, during the summer and the beginning of autumn, when juvenile birds were present, the most numerous flocks were reported. Since breeding success was very low and no more than a single nesting attempt was observed per colony per year, the occurrence of juvenile SCBs in the study area indicates that other colonies are extant in the region and that at least a portion of the population undertakes annual movements. According to Fraga et al. (1998), flocks may make irregular dispersive movements and may be highly mobile in the non-breeding season, which accounts for variations in the number of individuals.

Foraging habitats

The most common foraging habitats of the birds observed in this study were areas of low-growing native grassland or with rice stubble. This is consistent with observations made in other locations (Belton 1994, Dias and Maurício 2002, Fraga 2005, Bencke et al. 2003). In the current study, SCBs were seen perching on exotic and native trees on which they rested during feeding intervals. Fraga et al. (1998) also observed it in Argentina. In the breeding season, two males were observed capturing insects in the air and eating on them, which is rather unusual. Since they were defending their territory and females were absent, we believe that this is a way by which they can eat and guard the site at the same time.

Social behavior and interactions with other species

When it comes to interspecific interactions, SCBs interacted most frequently with *P. guirahuro* and *P. virescens*, while foraging outside the breeding season. Similar interactions were also reported by Fraga et al. (1998), Azpiroz (2000) and Dias and Maurício (2002). Frequently, the marshbirds acted as sentinels and benefitted the SCBs as they are ground-based feeders. In the non-breeding season, two or three SBCs acted as the sentinels in the intra-specific flocks, perching either on the ground or on trees.

Between April and August, we often observed SCBs feeding with the BWM, *X. dominicanus*. This is also an endangered species, in the same category as the SCB. In fact, SCBs were always observed following BWMs during foraging. This association is beneficial since BWMs play the role of sentinels (Fraga 2005, Kruger and Petry 2010), and since the two species differ in their foraging modes, they are not in competition with each other. Frequently, BWMs were more conspicuous than the SCBs in mixed flocks. Their association has been defined as a proto-co-operation relationship (a non-mandatory ecological relationship in which both species benefit) (Bencke et al. 2003). This type of interaction was observed in Brazil (Dias and Maurício 2002, Fonseca et al. 2004, Mohr et al. 2012), and elsewhere (Azpiroz 2000, Fraga et al. 1998). BWMs occur in the region only during the austral autumn and winter (Accordi and Hartz 2006), and show seasonal dispersion movements. The significant numbers of BWMs encountered shows that the study region is also important to the life cycle of this vulnerable species. The current study found almost three times as many BWM specimens than previously reported for the WRBP region (Bencke et al. 2003).

We observed SCBs defending small areas around the nests, but even when these birds were close to the colonies, they did not defend their feeding areas. They joined flocks of individuals from other species to mob and drive off *C. cinereus*, *C. buffoni* (several times), *M. chimango* and *F. femoralis*. Other studies also have described SCBs and other species chasing predators away (Fraga 2005, Fraga et al. 1998).

Breeding biology and nest success

The characteristics of the nests of *X. flavus* were similar to those described by Azpiroz (2000). The nests were fixed to plants such as *Eryngium* sp. and *Scirpus* sp., as reported for other sites (Fraga et al. 1998, Azpiroz 2000, Dias and Maurício 2002). We did not observe nests on *Baccharis* sp. L. (Asteraceae) and *Ludwigia* sp. L. (Onagraceae), even though in the highland grasslands of Rio Grande do Sul those two plants are used by SCBs to build nests (Fonseca et al. 2004, Moura 2013).

All measurements of internal and external diameter, internal depth and external height of nests obtained in Uruguay (Azpiroz 2000) were greater than in the current research, and in Fonseca et al. (2004), in Rio Grande do Sul, Brazil. Contrastingly, mean height of nest attachment from the ground was greater in the current study than in the data of Azpiroz (2000), but less than in Fonseca et al. (2004). Overall, the mean height of the nest-supporting plants was greater than reported by Fonseca et al. (2004).

Mean heights of the tallest and shortest plants surrounding the nest and the mean height of plants on which the nests were fixed were smaller for the colony evaluated during the second breeding season, albeit not significantly. The isolated nests near the ground, and not covered by vegetation were the first to be preyed upon. The height of the supporting plant does not seem to be a selection factor for nest building, since no nest was fixed on the tallest of the plants in the studied area around the nests.

The laying of one egg per day by SCBs was reported by Azpiroz (2000) and Moura (2013). The mean number of eggs laid was lower than reported by Moura (2013) for the highland fields of Rio Grande do Sul. The four unhatched eggs in Nest 6 were not preyed upon, which may owe to the fact that this nest was the highest from the ground, with the tallest *Eryngium* sp. surrounding it. Some fledglings survived long enough to leave the nest, but they would not have been sufficiently well developed to fly. Additionally, they were not seen near the nests. Two nestlings were recorded jumping out of the nest prior to being fully-fledged. It is possible that this premature nest fledging was due to the heat, since the nest was unprotected from the sun. Females were frequently reported feeding nestlings, while males were on territorial duty. This was also reported by Moura (2013).

Comparing the rate of fledged chicks/eggs or nestlings observed, the estimated breeding success for Nesting Colony 1 (10%) was higher than the 8.4% in Argentina, calculated by Fraga et al. (1998), but lower than for Uruguay (42.9%: Azpiroz 2000) and the southern coastal plain of the state of Rio Grande do Sul, Brazil (Dias and Maurício 2002). Dias and Maurício (2002) estimated a breeding success between 31.8% and 36.3% in the southern region of the state.

In the highland fields of Rio Grande do Sul, and using the apparent reproductive success, Moura (2013) reported that 19 of 47 nests produced nestlings (41%). Regarding to the apparent success reported in the current analysis, the rate of Colony 2 was (40%), close to the above, whilst the rate of Colony 1 was much higher (83%). In these cases, the apparent success based on nests that produced at least one chick, seems to overestimate reproductive success in the colony, since it is not possible to know the fate of the chicks. This is pertinent since in Colony 1 only two juveniles were observed, and none in Colony 2. It is likely that any other juveniles raised in the study colonies would have been seen close-by, since they were not able to undertake long flights.

It is highly probable that the Nesting Colony 1 area has not been used again for breeding, since the landowner removed approximately 60% of the *Eryngium* sp. late in September 2015. Petry and Krüger (2010) observed that, after a burning event that destroyed the vegetation in a wetland used for breeding, SCBs took three years to return to the site. If the Nesting Colony 1 area is not severely altered again, it will perhaps be used in the future by *X. flavus*, as several SCB specimens have been reported using the area for resting and foraging throughout the year.

Although two Shiny Cowbird pairs (*M. bonariensis*) were seen at Nesting Colony 1 on October 2014, parasitism of SCB nests was not recorded. While at some locations no nest parasitism is reported even when the Shiny Cowbird is commonly observed (Belton 1994, Dias and Maurício 2002), its parasitism has been reported in other areas of the state of Rio Grande do Sul (Moura 2013). Nest parasitism by this species has been considered one of the main drivers of low breeding success of the SCB in Uruguay and Argentina (Fraga et al. 1998, Azpiroz 2000).

Predation on nestlings and eggs was higher at Nesting Colony 2, where, for example, four eggs from the same nest were preyed upon on the same day. It was there that the nest placed nearest to the ground and built on grasses might have facilitated the predatory activities. Traces of eggshells were discovered near another nest; possible predators of eggs and nestlings may have been the colubrid snake *Philodryas patagoniensis* (Girard, 1858) (Dipsadidae), or the pampas fox, observed close to the colony area, among others.

Fledgling SCBs from Nests 8 and 11 were sufficiently old to fly off the nest but they were not seen in the area. Normally, juveniles remain for almost a month within the nest area after their first flight (Azpiroz 2000). Since juveniles were not reported in the area close to the nests, lack of breeding success in the colony is suspected. However, the 14 SCB juveniles in the area at the edge of WRBP in November 2014 and the 15 juvenile specimens seen in the EPABP in January 2016 indicate that there are other breeding colonies of *X. flavus* in the region. Due to the development stage of these specimens and to reports in the areas under analysis, it is not likely that they belonged to the assessed colonies.

Conservation

Overall, the Birdlife International (2016) considers *X. flavus* as being in rapid and continuing decline. The main causes of this are destruction and degradation of its natural habitats, mainly the transformation of natural grasslands into monocultural fields, and the drainage of wetlands. The SCB is no longer found at traditional sites for the species in Rio Grande do Sul, such as Novo Hamburgo, Guaíba and others (Belton 1994, Bencke et al. 2003).

Within the area analyzed, the main threats in areas adjacent to the WRBP are the transformation of open fields (used for cattle) into rice fields and the removal of the shrubs and *Eryngium* sp., which SCBs use to support their nests. The "cleansing" of the fields is a culturally common practice in the state; it is considered to provide more space for livestock. According to the legislation that accompanied its establishment, EPABG should be sustainable and preserve the Gravataí River basin and associated wetlands. However, few wardens are allocated for its protection and environmental degradation is constantly on the increase. On the positive side, there are no livestock and agricultural activities at the WRBP, an area that is completely protected. As a rule, open natural grassland communities within conservation areas in southern Brazil are few and small (less than 0.5%) (Overbeck et al. 2007). This is also the case of the WRBP, where the few areas of natural grassland lack management and therefore are prone to invasion by shrubs and woodlands. Given their preference for rough grassland for nesting, SCBs at WRBP may have areas appropriate for nesting, but not for long term feeding. This may have been the cause of their movement to areas adjacent to the EPABG. The WRBP has extensive and continuous wetlands, with plants that are appropriate for the breeding of *X. flavus*, but they could not be reached due to the swampy features of the land. It is considered likely that this area is used as a dormitory by SCBs. Despite the evidence (observed juveniles) of other breeding colonies near the study area, such other sites were not found. Although there are reports of more than one hundred individuals in the study area, the population may be isolated from other flocks in the state, favoring inbreeding and the loss of genetic variability (Frankham et al. 2003).

Xanthopsar flavus seems to be adapted to cattle-breeding and agriculture environments, although they require open fields and wetlands with vegetation for nest construction (Fraga et al. 1998, Dias and Maurício 2002, Fonseca et al. 2004). Like many other endangered birds, *X. flavus* and *X. dominicanus* have great mobility and the delimitation of conservation units for the protection of these species is rather difficult (Fraga et al. 1998). In fact, *X. dominicanus*, like *X. flavus*, also build their nests in wetland areas, and the fact that their dispersion movements include the areas under analysis indicates the existence of other nearby areas where the species can breed. However, the distances that these species may travel remain unknown, and the possible location of these sites is difficult to estimate.

Like Petry and Krüger (2010), we would like to emphasize the relevance of conservation units that include areas of open natural grassland and to highlight the importance of the preservation of such areas with wetlands outside the CUs. Since 1998, a 380 household-strong settlement, called "Filhos de Sepé", has been established in the EPABG surrounding the WRBP. All reports of *X. flavus* from the EPABG have come from within the settlement (including the rice fields). Farmers from Filhos de Sepé are not allowed to use pesticides on their crops. This will certainly minimize the impacts of agricultural activities

in the region and contribute to the maintenance of biodiversity, including the food items consumed by SCBs.

Subsidies for livestock production coupled with the preservation of native fields similar to the "Alianza del Pastizal" in Pampa biome countries, payment for environmental services, establishment and implementation of conservation units and increase in research activities will certainly assist the conservation of the species and the preservation of their environment (Develey et al. 2008, Fontana et al. 2013). Genetic and demographic analyses and investigation of regional movements should be prioritized when planning further studied on the conservation biology of *X. flavus*. More measures are required to protect the environmental mosaic needed for the maintenance of *X. flavus*, *X. dominicanus* and other endangered plant and animal species in the region. Information derived from current research is already being used in the preparation of a Management Plan for WRBP in which *X. flavus* is one of the target species for conservation.

ACKNOWLEDGEMENTS

We are grateful to Centro Universitário Univates and to Coordenação de Aperfeiçoamento de Pessoal de Nível Superior (CAPES) for the scholarship granted to Luciane Mohr. Thanks are also due to Enerplan Group and Logos/Bioimagens Consultoria Ambiental for subsidizing several practical activities undertaken; to the farmers of the Assentamento Filhos de Sepé who allowed us access to their farms; to André Osório, the manager of WRBP and to Cleberton Bianchini, Jonas John, Samuel Gaedke, Camila Schmidt, Rafael Dalssotto, Ricardo Stertz, Manfred Ramminger and Andrea F. Steffens for helping with data collection. Adrian Barnett helped with the English.

REFERENCES

Accordi IA, Barcellos A (2006) Composição da avifauna em oito áreas úmidas da Bacia Hidrográfica do Lago Guaíba, Rio Grande do Sul. Revista Brasileira de Ornitologia 14: 101–115.

Accordi IA, Hartz SM (2006) Distribuição espacial e sazonal da avifauna em uma área úmida costeira do sul do Brasil. Revista Brasileira de Ornitologia 14: 117–135.

Azpiroz A (2000) Biología y conservación del Dragón (*Xanthopsar flavus*) en la reserva de biosfera Bañados del Este. Documentos de Trabajo 29: 1–32.

Azpiroz AB, Isacch JP, Dias RA, Di Giacomo AS, Fontana CS, Palarea CM (2012) Ecology and conservation of grassland birds in southeastern South America: a review. Journal of Field Ornithology 83: 217–246. https://doi.org/10.1111/j.1557-9263.2012.00372.x

Belton W (1994) Aves do Rio Grande do Sul: distribuição e biologia. São Leopoldo, Editora Unisinos, 584 pp.

Bencke GA, Fontana CS, Dias RA, Maurício GN, Mähler Jr JKF (2003) Aves. In: Fontana CS, Bencke GA, Reis RE (Orgs) Livro vermelho da fauna ameaçada de extinção no Rio Grande do Sul. Porto Alegre, Editora PUCRS, 189–479.

Bencke GA, Maurício GN, Develey PF, Goerck JM (2006) Áreas Importantes para a Conservação das Aves no Brasil. Parte I – Estados do Domínio da Mata Atlântica. São Paulo, SAVE Brasil, 494 pp.

BirdLife International (2016) Saffron-cowled Blackbird *Xanthopsar flavus*. http://www.birdlife.org/datazone/species/factsheet/22724673; http://www.museum.lsu.edu/~Remsen/SACCBaseline.html [Accessed: 04/07/2016]

Codesido M, Fraga R (2009) Distributions of threatened grassland passerines of Paraguay, Argentina and Uruguay, with new locality records and notes on their natural history and habitat. Ornitología Neotropical 20: 585–595.

Collar NJ, Gonzaga LP, Krabbe NK, Madroño Nieto A, Naranjo LG, Parker III TA, Wege DC (1992) Threatened birds of the Americas: the ICBP/IUCN Red Data Book. (Third edition, part 2). Cambridge, ICBP/IUCN Red Data Book, 1150 pp.

Develey PF, Setubal RB, Dias RA, Bencke GA (2008) Conservação das aves e da biodiversidade no bioma Pampa aliada a sistemas de produção animal. Revista Brasileira de Ornitologia 16: 308–315.

Dias RA, Maurício G (2002) Natural history notes and conservation of a Saffron-cowled Blackbird *Xanthopsar flavus* population in the southern coastal plain of Rio Grande do Sul, Brazil. Bird Conservation International 12: 255–268. https://doi.org/10.1017/S0959270902002162

Esquivel A, Velázquez MC, Bodrati A, Fraga R, Del Castillo H, Klavins J, Clay RP, Madroño A, Peris SJ (2007) Status of the avifauna on San Rafael National Park, one of the last large fragments of Atlantic Forest in Paraguay. Bird Conservation International 17: 301–317. https://doi.org/10.1017/S095927090700086X

Fonseca VS, Petry MV, Fonseca FL (2004) A new breeding colony of the Saffron-cowled Blackbird (*Xanthopsar flavus*) in Rio Grande do Sul, Brazil. Ornitologia Neotropical 15: 133–137.

Fontana CS, Rovedder CE, Repenning M, Gonçalves ML (2008) Estado atual do conhecimento e conservação da avifauna dos Campos de Cima da Serra do sul do Brasil, Rio Grande do Sul e Santa Catarina. Revista Brasileira de Ornitologia 16: 281–307.

Fontana CS, Dias RA, Maurício GN (2013) *Xanthopsar flavus*. In: Serafini PP (Org.) Plano de ação nacional para a conservação dos passeriformes ameaçados dos campos sulinos e espinilho. Brasília, ICMBio, 120–125.

Fraga R (2005) Ecology, behavior and social organization of Saffron-cowled blackbirds (*Xanthopsar flavus*). Ornitologia Neotropical 16: 15–29.

Fraga R, Casañas H, Pugnali G (1998) Natural history and conservation of the endangered Saffron-cowled Blackbird *Xanthopsar flavus* in Argentina. Bird Conservation International 8: 255–267. https://doi.org/10.1017/S095927090000191X

Franchin AG, Júnior OM, Del-Claro K (2010) Ecologia Comportamental: métodos, técnicas e ferramentas utilizadas no estudo de aves. In: Von Matter S, Straube FC, Accordi I, Piacentini V, Candido-Jr JF (Orgs) Ornitologia e Conservação: Ciência Aplicada, Técnicas de Pesquisa e Levantamento. Rio de Janeiro, Technical Books, 281–293.

Frankham R, Ballou JD, Briscoe DA (2003) Introduction to conservation genetics. Cambridge, Cambridge University Press.

Hammer O, Harper DA, Ryan PD (2001) Paleontological statistics software package for education and data analysis (PAST). Paleontologia Electronica 4: 1–9.

ICMBio (2014) Lista das espécies terrestres e mamíferos aquáticos ameaçados de extinção do Brasil. Portaria MMA n° 444, de 17 de dezembro de 2014. http://www.icmbio.gov.br/portal/especies-ameacadas-destaque [Accessed: 21/06/2016]

IUCN (2016) The IUCN Red List of Threatened Species. http://www.iucnredlist.org [Accessed: 15/07/2016]

Krüger L, Petry MV (2010) Black-and-white monjita (*Xolmis dominicanus*) followed by the saffron-cowled blackbird (*Xanthopsar flavus*): statistical evidence. Ornitologia neotropical 21: 299–303.

Mohr RS, Fonseca VS, Perico E, Mohr AR (2012) Interações ecológicas de *Xanthopsar flavus* (Aves: Icteridae), uma espécie prioritária para a conservação, em uma nova área de ocorrência no Bioma Pampa, RS. In: Deble ASO, Deble LP, Leão ALS (Orgs) Bioma Pampa: Ambiente x Sociedade. Bagé, Ediurcamp, 96–103.

Moura EJT (2013) Biologia reprodutiva do veste-amarela (*Xanthopsar flavus*, Gmelin 1788) nos Campos de Cima da Serra, Sul do Brasil. Master's Degree dissertation, São Leopoldo, RS, Brazil: Universidade do Vale do Rio dos Sinos. http://www.repositorio.jesuita.org.br/handle/UNISINOS/3133 [Accessed: 01/07/2016]

Overbeck GE, Müller SC, Fidelis A, Pfadenhauer J, Pillar VD, Blanco CC, Boldrini II, Both R, Forneck ED (2007) Brazil's neglected biome: The South Brazilian Campos. Perspectives in Plant Ecology, Evolution and Systematics 9: 101–116. https://doi.org/10.1016/j.ppees.2007.07.005

Petry MV, Krüger L (2010) Frequent use of burned grasslands by the vulnerable Saffron-cowled Blackbird *Xanthopsar flavus*: implications for the conservation of the species. Journal of Ornithology 151: 599–605. https://doi.org/10.1007/s10336-009-0489-9

Piacentini VQ, Aleixo A, Agne CE, Maurício GN, Pacheco JF, Bravo GA, Brito GRR, Nakas LN, Olmos F, Posso S, Silveira LF, Betini GS, Carrano E, Franz I, Lees AC, Lima LM, Pioli D, Schunck F, Amaral FR, Bencke GA, Cohn-Haft M, Figueiredo LFA, Straube FC, Cesari E (2015) Annotated checklist of the birds of Brazil by the Brazilian Ornithological Records Committee. Revista Brasileira de Ornitologia 23: 91–298.

State Decree 38.971, de 23 de outubro de 1998. Cria a Área de Proteção Ambiental do Banhado Grande e dá outras providências. Governo do Estado do Rio Grande do Sul, RS, Brasil. http://www.icmbio.gov.br/cepsul/images/stories/legislacao/Decretos/1998/dec_rs_38971_1998_uc_apa_banhadogrande_rs.pdf [Accessed: 05/05/2016]

State Decree 41.559, de 24 de abril de 2002. Cria o Refúgio de Vida Silvestre Banhado dos Pachecos e dá outras providências. Governo do Estado do Rio Grande do Sul, RS, Brasil. http://

www.al.rs.gov.br/legis/M010/M0100099.ASP?Hid_Tipo=TEX-TO&Hid_TodasNormas=1235&hTexto=&Hid_IDNorma=1235 [Accessed: 05/05/2016]

State Decree 51.797, de 09 de setembro de 2014. Declara as Espé-cies da Fauna Silvestre Ameaçadas de Extinção no Estado do Rio Grande do Sul. http://www.al.rs.gov.br/legis/M010/M0100099. ASP?Hid_Tipo=TEXTO&Hid_TodasNormas=61313&hTex-to=&Hid_IDNorma=61313 [Accessed: 11/05/2016].

Stotz D, Fitzpatrick JW, Parker III TA, Moskovits DK (1996) Neo-tropical birds: ecology and conservation. Chicago, University of Chicago.

Author Contributions: LRSM and ARM conducted the field work; LRSM wrote most of the paper input from all other authors, whom also contributed to the data analysis.

11

Taxonomy of Xylographhellini (Coleoptera: Ciidae) from the Australian and Oriental regions with descriptions of new species of *Scolytocis* and *Xylographella*

Igor Souza-Gonçalves[1,2], Cristiano Lopes-Andrade[2]

[1]*Programa de Pós-Graduação em Ecologia, Departamento de Biologia Geral, Universidade Federal de Viçosa. 36570-900 Viçosa, MG, Brazil.*
[2]*Laboratório de Sistemática e Biologia de Coleoptera, Departamento de Biologia Animal, Universidade Federal de Viçosa. 36570-900 Viçosa, MG, Brazil.*
Corresponding author: Igor Souza-Gonçalves (igao_bio@yahoo.com.br)

http://zoobank.org/55B5AE5A-84AB-4963-BF0B-8C141703995D

ABSTRACT. Xylographhellini beetles occur mainly in lands of the Southern Hemisphere. However, the taxonomy of Australian and Oriental species is incipient. The tribe comprises four genera, of which *Scolytocis* Blair, 1928 and *Xylographella* Miyatake, 1985 were recently redescribed and reported from Australia but without descriptions of new species. Here, three new species of Xylographhellini are described: *Scolytocis australimontensis* **sp. n.** from Australia, with smooth interspaces of pronotal punctures; *Scolytocis insularis* **sp. n.** from the Pohnpei Island (Caroline Islands, Micronesia), with microstriated interspaces of pronotal punctures; and *Xylographella frithae* **sp. n.** from Australia, with six raised keels in elytral declivity. *Scolytocis samoensis* Blair, 1928, type species of the genus, is recorded from Guam (Mariana Islands, Micronesia) and redescribed. Keys for the *Scolytocis* and *Xylographella* occurring in the Australian and Oriental regions are also provided.

KEY WORDS. Australia, ciid, Ciinae, Micronesia, minute tree-fungus beetles

INTRODUCTION

Xylographhellini (Ciidae: Ciinae) comprises four genera in two subtribes (Lopes-Andrade 2008): *Xylographella* Miyatake, 1985 and *Scolytocis* Blair, 1928 in Xylographhellina; and *Syncosmetus* Sharp, 1891 and *Tropicis* Scott, 1926 in Syncosmetina. Xylographhellina occur in the Australian (New Zealand), Oriental (Philippines, Fiji and Samoa) and Palearctic (Japan) regions, throughout the Neotropical region, and in the Chinese transition zone (Lopes-Andrade 2008, Lopes-Andrade and Grebennikov 2015). Syncosmetina occur in the Ethiopian region (islands of the Indian Ocean), Palearctic region (Japan) and in the Chinese transition zone (biogeographic regionalization sensu Morrone 2015). The distribution of these subtribes overlaps in the Chinese transition zone, with one species of *Scolytocis* and four of *Syncosmetus* (Lopes-Andrade and Grebennikov 2015), and in Japan, with one species of *Xylographella* and two of *Syncosmetus* (Lopes-Andrade 2008). Up to date there is only one species of Xylographhellini from the Australian region, *Scolytocis novaezelandiae* Lopes-Andrade, 2008. Whereas the Oriental Xylographhellini com-prise *Xylographella speciosa* Lopes-Andrade, 2008 and six species of *Scolytocis*: *Scolytocis malayanus* Lopes-Andrade, 2008, *Scolytocis philippinensis* Lopes-Andrade, 2008, *Scolytocis samoensis* Blair, 1928, *Scolytocis thayerae* Lopes-Andrade, 2008, *Scolytocis werneri* Lopes-Andrade, 2008, and *Scolytocis zimmermani* Lopes-Andrade, 2008 (Lopes-Andrade 2008). *Scolytocis* and *Xylographella* were recently redescribed and reported from Australia (Lawrence 2016).

Our objective is to describe three new species of Xylographhellini: *Scolytocis australimontensis* sp. n. and *Xylographella frithae* sp. n. from the Australian region; and *Scolytocis insularis* sp. n. from the Oriental region. Moreover, *Sc. samoensis* is recorded for the first time from Guam (Mariana Islands) and redescribed. Identification keys to Australian and Oriental *Scolytocis* and to all known species of *Xylographella* are also provided.

MATERIAL AND METHODS

Generic limits follow Lopes-Andrade (2008) and Lawrence (2016). The number of available specimens was low, but some were dissected in attempts to find males. Holotypes were not dissected, so their sexes are undetermined. Among the paratypes two males of *Xylographella frithae* sp. n. were found, one from Paluma and the other from Hugh Nelson Range, Queensland. The tegmen shown in Fig. 22 is of a paratype from the type locality (Paluma). The sternite VIII and aedeagus shown in Figs 21 and 23, respectively, are of a paratype from Hugh Nelson. The aedeagus extracted from a male paratype from the type locality was a bit damaged during dissection, but it was carefully compared to the aedeagus extracted from the male from Paluma. We were unsuccessful in finding males in good condition from the two new species of *Scolytocis*. We recognized two males *Sc. samoensis* from Guam, but their genitalia were very membranous and deformed by agglomeration of nematodes. The following female paratypes were dissected: three *Sc. australimontensis* sp. n. (from Mount Haig, Hugh Nelson Range and Mossman Bluff Track, Queensland); one *Sc. insularis* sp. n. from Kolonia (Pohnpei Island, formerly known as Ponape Island); two *Sc. samoensis* from Guam; and one *X. frithae* sp. n. from Mount Lewis (Queensland). We provide only the ratio of gula width to head width, which seems to be the same for both sexes in species of *Scolytocis* and *Xylographella* (Lawrence 2016).

The terminology for the external morphology of adult ciids follows Lawrence et al. (2011), Lawrence (2016) and Lopes-Andrade and Lawrence (2005, 2011). The following abbreviations are used for measurements (in mm) and ratios: BW (basal width of scutellar shield), CL (length of antennal club measured from base of the eighth to apex of the tenth antennomere in *Xylographella*; from base of the seventh to apex of the ninth antennomere in *Scolytocis*), EL (elytral length along the midline), EW (greatest width of elytra together), FL (length of antennal funicle measured from base of the third to apex of the seventh antennomere in *Xylographella*, or to the apex of the sixth antennomere in *Scolytocis*), GD (greatest depth of body measured in lateral view), GW (greatest diameter of eye), PL (pronotal length along midline), PW (greatest pronotal width), TL (total length counted as EL + PL, i.e. excluding head). The GD/EW and TL/EW ratios indicate degree of body convexity and elongation, respectively.

Scolytocis danae Lopes-Andrade & Grebennikov, 2015, is included in the key to Oriental *Scolytocis*. Although it is known only from the Chinese transition zone, it is morphologically closely related to the oriental *Sc. philippinensis*, *Sc. thayerae* and *Sc. zimmermani* (Lopes-Andrade and Grebennikov 2015). Only a single specimen of *Sc. danae* is known; the lowest limit of its TL provided in the key was estimated as being 2.10 mm considering that the highest TL in a species of *Scolytocis* is about 15% more than the lowest TL (see measurements in Lopes-Andrade 2008).

Transcription of labels, dissection, photography and measurement of specimens follow the methods provided by Araujo and Lopes-Andrade (2016). The distribution map (Fig. 24) was created in the freeware QGIS 2.14.2. The examined specimens were deposited in the following collections: ANIC – Australian National Insect Collection, CSIRO Entomology (Canberra,

Australia); CELC – Coleção Entomológica do Laboratório de Sistemática e Biologia de Coleoptera, Universidade Federal de Viçosa (Viçosa, Minas Gerais, Brazil); QMBA – Queensland Museum (Brisbane, Australia).

TAXONOMY

Key to Australian and Oriental species of *Scolytocis* Blair

1 Metatibiae with outer edge straight (Fig. 5) to barely rounded (Fig. 10), usually with a clear distinction between outer and apical edges; spines of outer edge separated by one spine-width or more (Figs 5, 10, 15). If distinction of outer and apical edges of metatibiae is not clear (Fig. 10), then TL is less than 1.16 mm 5

1' Metatibiae with outer edge broadly rounded, without a distinction between outer and apical edges; spines of outer edge very close to each other at apical half and TL at least 1.26 mm 2

2 Apical antennomere of club longer than preceding two antennomeres together 4

2' Apical antennomere of club shorter than preceding two antennomeres together 3

3 Prosternum with longitudinal carina in front of coxae conspicuous. TL less than 1.40 mm. Known from the Philippines.. *Scolytocis philippinensis* Lopes-Andrade, 2008

3' Prosternum devoid of a longitudinal carina in front of coxae. TL more than 2.10 mm. Known from southeast China ....
.......*Scolytocis danae* Lopes-Andrade & Grebennikov, 2015

4 Prosternum biconcave. TL more than 1.70 mm. Known from the Philippines ....
.... *Scolytocis thayerae* Lopes-Andrade, 2008

4' Prosternum concave. TL less than 1.70 mm. Known from Fiji .... *Scolytocis zimmermani* Lopes-Andrade, 2008

5 Pronotum with a rugose border along the posterior edge ..8

5' Pronotum lacking a rugose border along the posterior edge, the surface similar to that of pronotal disc 6

6 CL/FL at least 1.80. Known from Malaysia ....
.... *Scolytocis malayanus* Lopes-Andrade, 2008

6' CL/FL 1.60 or less 7

7 Metatibiae with a clear distinction between outer and apical edges (Fig. 15); outer edge straight and bearing few spines (usually three; Fig. 15). Known from Samoa and Guam .... *Scolytocis samoensis* Blair, 1928

7' Metatibiae without a clear distinction between outer and apical edges (Fig. 10); outer edge slightly rounded and bearing much more than three spines (Fig. 10). Known from the Pohnpei Island (Micronesia) ....
....*Scolytocis insularis* sp. n.

8   Pronotum with interspaces of punctures smooth. Known
     from Australia................ *Scolytocis australimontensis* sp. n.

8' Pronotum with interspaces of punctures microreticulate... 9

9   Posterior pronotal edge with a narrow rugose border.
     Metaventrite bearing a conspicuous and long discrimen.
     Known from the Philippines..
     .... *Scolytocis werneri* Lopes-Andrade, 2008

9' Posterior pronotal edge with a broad rugose border. Dis-
     crimen short, not reaching the middle of metaventrite.
     Known from Northern New Zealand...............................
     .... *Scolytocis novaezelandiae* Lopes-Andrade, 2008

Scolytocis australimontensis sp. n.

http://zoobank.org/F5B169D2-9A4D-499A-85DB-F56BC5E16167
Figs 1–5, 24

Scolytocis sp. in Lawrence (2016: 198).

Type locality. Mount Haig, state of Queensland, northeastern Australia (17°06'S, 145°29'E).

Diagnosis. *Scolytocis australimontensis* sp. n. can be distinguished from other Australian and Oriental species of *Scolytocis*
by the combination of the following features: pronotum with a
conspicuous rugose border along the posterior edge and smooth
interspaces of punctures; metatibiae with a clear distinction between the outer and apical edges, the outer edge being straight
and bearing at least five spines separated from each other by
more than a spine-width. *Scolytocis novazelandiae* has a similar
posterior pronotal border, but interspaces of punctures are
microreticulate. *Scolytocis werneri* has a comparatively narrower
posterior pronotal border and interspaces of punctures are
coarsely reticulate.

Description, holotype (Figs 1–5). Adult fully pigmented.
Measurements in mm: TL 1.15, PL 0.38, PW 0.50, EL 0.78, EW
0.58, GD 0.50. Ratios: PL/PW 0.75, EL/EW 1.35, EL/PL 2.07, GD/
EW 0.87, TL/EW 2.00. Body elongate, convex; dorsum and venter
dark reddish-brown; antennae, palpi and tarsi yellowish-brown;
dorsal vestiture of minute setae, smaller than a puncture-width
and barely discernible even in high magnification (150×), except for the posteriormost portion of elytra with conspicuous
setae (easily seen in lateral view); venter subglabrous. Head with
anteriormost portion visible from above; dorsum with shallow,
coarse punctures, separated from each other by a puncture-width
or less and with smooth interspaces. Antennae bearing nine
antennomeres, as follows (in mm, left antenna measured): 0.06,
0.04, 0.03, 0.02, 0.01, 0.01, 0.03, 0.03, 0.06 (FL 0.07, CL 0.12,
CL/FL 1.71). Eyes finely facetted, each bearing about 70 ommatidia; GW 0.11. Gula 0.52 times as wide as head. Pronotum with
shallow, single punctation; punctures irregular, separated from
each other by a puncture-width or less and with smooth interspaces; anterior edge broadly rounded; lateral edges smooth, not
explanate and not visible when seen from above; posterior edge

with a rugose border along it. Scutellar shield triangular, bearing
fine punctures; BW 0.10. Elytra with shallow, dual punctation;
large punctures coarse, seriate, about twice as large as small
punctures; small punctures sparsely and irregularly distributed;
interspaces of punctures, smooth; elytral apex truncate; apical
declivity concave with conspicuous cuticular globules (Fig. 4, arrows). Hind wings developed, apparently functional. Hypomera
with coarse, shallow punctation; each puncture bearing a fine
decumbent seta; interspaces microreticulate. Prosternum in
front of coxae concave; interspaces microreticulate. Prosternal
process laminate, as long as prosternum at midline, apex acute.
Pro-, meso- and metatibiae (Fig. 5, left metatibia of a paratype)
with similar shape and length, approximately three times
as long as broad; tibiae with distinct apical and outer lateral
edges; outer apical angle rounded; outer edge of tibiae straight
and with about five spines separated from each other by more
than a spine-width; apical edge with about 10 spines very close
to each other. Metaventrite with coarse, shallow punctures;
interspaces microreticulate; discrimen about half the length of
metaventrite at midline. Abdominal ventrites with coarse, small
punctures, separated from each other by a puncture-width or
less; interspaces microreticulate; length of ventrites (in mm,
from base to apex at the longitudinal midline) as follows: 0.15,
0.05, 0.05, 0.05, 0.10.

Measurements (in mm) and ratios (n = 7, including the
holotype): TL 1.10–1.33 (1.16 ± 0.07), PL 0.38–0.43 (0.39 ±
0.02), PW 0.45–0.55 (0.48 ± 0.04), EL 0.70–0.90 (0.77 ± 0.06),
EW 0.50–0.63 (0.55 ± 0.04), GD 0.45–0.55 (0.49 ± 0.03), PL/PW
0.75–0.89 (0.82 ± 0.05), EL/EW 1.33–1.55 (1.41 ± 0.08), EL/PL
1.75–2.12 (1.96 ± 0.14), GD/EW 0.86–0.95 (0.88 ± 0.03), TL/EW
2.00–2.30 (2.13 ± 0.10).

Material examined. Australia: holotype (ANIC) labeled
"17.06S 145.29E QLD, Mt. Haig 1150 m GS1, 1 Dec. 1994 – 3
Jan. 1995, P. Zborowski, FI Trap ANIC [printed]*Scolytocis australimontensis* Souza-Gonçalves & Lopes-Andrade HOLOTYPUS
[printed on red paper]". Paratypes: 8 specimens (3 females and
5 with gender not determined) as follows: one female (ANIC,
dissected) and 2 specimens (CELC), same locality data as holotype; one specimen (CELC) "17.06S 145.37E QLD, Mt. Edith
GS2, 1050 m, 3 Jan. – 4 Feb. 1995, P. Zborowski, FI Trap ANIC
[printed]"; one female (CELC, dissected) "17.27S 145.29E QLD,
Hugh Nelson Rg. GS3 1150 m, 1 Dec. 1994 – 3 Jan. 1995, P.
Zborowski, FI Trap ANIC [printed]"; one specimen (ANIC) "Mt.
Lewis, 800 m, QLD, 26 Dec. 1986, H. & A. Howden, flight inter-
cept trap [printed]"; one specimen (QMBA) "Mt Bartle Frere, N.
Qld. Sth. Peak Summit, 1620 m, 6–8 Nov. 1981, EARTHWATCH/
QLD MUSEUM, *Pyrethrum* knockdown [printed]\\QUEENSLAND
MUSEUM LOAN DATE: Dec. 2001 No. LE 01.28 [printed on green
paper]\\A.N.I.C. COLEOPTERA Voucher No. 83-0880 [printed on
green paper]"; one female (QMBA) "Mossman Bluff Track, 5–10
Km W. Mossman N. Qld, 20 Dec 1989 – 15 Jan 1990, Monteith,
Thompson & ANZSES Site 7,1000 m, flt. intercept [printed]\\
QUEENSLAND MUSUEM LOAN DATE: Dec. 2001 No. LE 01.36

Figures 1–5. Holotype of *Scolytocis australimontensis* sp. n. from Queensland, Australia: (1) dorsal view; (2) lateral view; (3) ventral view; (4) apical declivity of elytra; (5) left metatibia. Scale bars: 0.5 mm (1–3); 0.2 mm (4); 0.1 mm (5).

[printed on green paper]". All paratypes are additionally labeled \ *Scolytocis australimontensis* Souza-Gonçalves & Lopes-Andrade PARATYPUS [printed on yellow paper].

Etymology. The species name derives from the Latin adjectives "australis", which means "of the South" and refers to Australia, and "montensis", which means "of or belonging to mountains", both in the genitive singular. The name is a reference to the Australian mountains where most specimens were collected.

Remarks. This new species was collected only in localities above 800 m, three of them at Australian mountains (Mount Haig, Mount Edith and Mount Bartle Frere) (Fig. 24). This species may be included in the *Sc. werneri* species-group, in which the species have a rugose border along the posterior pronotal edge.

Scolytocis insularis sp. n.

http://zoobank.org/A16E8CB3-2403-495E-AD46-34B12386ED83
Figs 6–10, 24

Type locality. Kolonia, Pohnpei Island, state of Pohnpei, Federated States of Micronesia (6°57′N, 158°12′E).

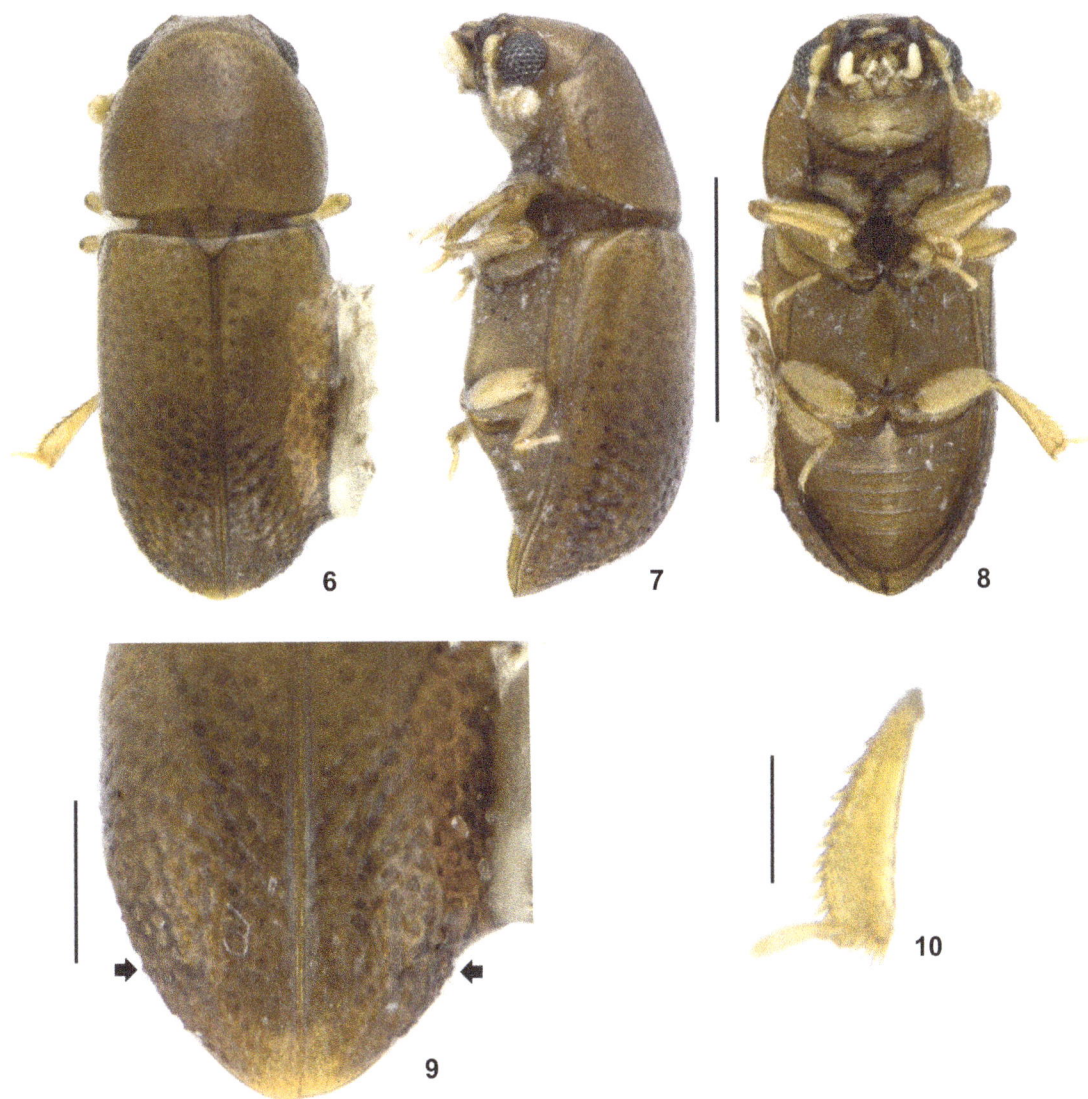

Figures 6–10. Holotype of *Scolytocis insularis* sp. n. from Pohnpei Island, Micronesia: (6) dorsal view; (7) lateral view; (8) ventral view; (9) apical declivity of elytra, (10) left metatibia. Scale bars: 0.5 mm (6-8); 0.2 mm (9); 0.1 mm (10).

Diagnosis. *Scolytocis insularis* sp. n. can be distinguished from other Australian and Oriental species of *Scolytocis* by combination of the following features: pronotum with interspaces of punctures microstriated and posterior edge devoid of a rugose border; metatibiae without a clear distinction between outer and apical edges, the outer edge being slightly rounded.

Description, holotype (Figs 6–10). Adult apparently not fully pigmented but in good condition, except for lacking the right antenna and one tarsus. Measurements in mm: TL 1.08, PL 0.35, PW 0.48, EL 0.73, EW 0.48, GD 0.48. Ratios: PL/PW 0.74, EL/EW 1.53, EL/PL 2.07, GD/EW 1.00, TL/EW 2.26. Body elongate, convex; dorsum and venter yellowish brown; anten-

nae, palpi and tarsi yellowish; dorsal vestiture of minute setae, smaller than a puncture-width and barely discernible even in high magnification (150×), except for the posteriormost portion of elytra with conspicuous setae (easily seen in lateral view); venter subglabrous. Head with anteriormost portion visible from above; dorsum with shallow, coarse, fine punctures, separated from each other by a puncture-width or less and with microreticulate interspaces. Antennae bearing nine antennomeres, as follows (in mm, left antenna measured): 0.06, 0.03, 0.04, 0.01, 0.01, 0.01, 0.03, 0.03, 0.05 (FL 0.07, CL 0.11, CL/FL 1.57). Eyes finely facetted, each bearing about 70 ommatidia; GW 0.11. Gula 0.47 times as wide as head. Pronotum with shallow, single punctation; punctures irregular, fine, separated from each other by a distance of one to two puncture-widths on disc and one puncture-width close to the laterals; interspaces transversely microstriated on disc, diagonally microstriated near lateral edges; anterior edge broadly rounded; lateral edges smooth, not explanate and not visible when seen from above; posterior edge without a rugose border along it. Scutellar shield triangular, bearing small punctures; BW 0.12. Elytra with confuse, shallow punctation; punctures coarsely and irregularly distributed, with somewhat rugose interspaces; elytral apex truncate; apical declivity (posterior one-fourth of elytra) with conspicuous cuticular globules (Fig. 9, arrows). Hind wings developed, apparently functional. Hypomera with coarse, shallow punctation; each puncture bearing a fine decumbent seta; interspaces transversely microstriated. Prosternum in front of coxae biconcave; interspaces transversely microstriated. Prosternal process laminate, as long as prosternum at midline, apex acute. Pro-, meso- and metatibiae (Fig. 10, left metatibia) with similar shape and length, approximately three times as long as broad, expanded from base to basal two-thirds; tibiae with outer edge slightly rounded, devoid of a clear distinction of outer and apical edges; outer edge with about 15 spines, a bit sparser near tibial base and getting closer until apex. Metaventrite with coarse, small punctures; interspaces transversely microstriated; discrimen about two-fifths the length of metaventrite at midline. Abdominal ventrites with coarse, small punctures, separated from each other by a puncture-width or less; interspaces transversely microstriated; length of ventrites (in mm, from base to apex at the longitudinal midline) as follows: 0.12, 0.05, 0.05, 0.05, 0.09.

Measurements (in mm) and ratios (n = 2, including the holotype): TL 1.05–1.15, PL 0.35–0.38, PW 0.48–0.48, EL 0.73–0.78, EW 0.48–0.53, GD 0.48–0.49, PL/PW 0.74–0.79, EL/EW 1.48–1.53, EL/PL 2.07, GD/EW 0.95–1.00, TL/EW 2.19–2.26.

Material examined. Federated States of Micronesia: holotype (ANIC) labeled "PONAPE ISLAND: Colonia, iii.1998, H. S. Dybas, FMHN [printed]\Scolytocis insularis Souza-Gonçalves & Lopes-Andrade HOLOTYPUS [printed on red paper]". Paratype: one female (CELC, dissected), same locality data as holotype and additionally labeled\Scolytocis insularis Souza-Gonçalves & Lopes-Andrade PARATYPUS [printed on yellow paper].

Etymology. The species name is a Latin adjective in the genitive singular and means relative or belonging to an island, in reference to the insular distribution of this species.

Remarks. This new species is known only from the type locality, Kolonia, a coastal town and capital of the state of Pohnpei in the Federated States of Micronesia (Fig. 24). The species does not fit in any previously proposed species-group of Scolytocis. Here, we propose the Sc. insularis species-group to encompass this single species, the group defined by the combination of the following features: biconcave prosternum; outer edge of metatibiae slightly rounded, without a clear distinction between outer and apical edges. The prosternum is also biconcave in species of the Scolytocis danielssoni and the Scolytocis fritzplaumanni

species-groups. However, in the Sc. danielssoni species-group the species are comparatively larger and have a rugose border along the posterior pronotal edge. In the Sc. fritzplaumanni group the species are also larger than Sc. insularis and bear a smooth border along the posterior pronotal edge. Metatibiae with slightly rounded outer edge are also seen in species of the Scolytocis bouchardi and the Scolytocis lawrencei species-group, but they have triconcave and tumid prosternum, respectively, and are exclusively neotropical.

Scolytocis samoensis Blair, 1928
Figs 11–15, 24

Scolytocis samoensis: Blair 1928: 95–96 (description, type species of Scolytocis Blair, 1928, by original designation); Lopes-Andrade 2008: 14, 36 (inclusion in the Sc. lawrencei species-group, taxonomic notes).

Type locality. Pago Pago, Tutuila, Samoa, Polynesia (14°16′S, 170°42′W).

Diagnosis. Scolytocis samoensis can be distinguished from other Australian and Oriental Scolytocis by the following combination of features: pronotum devoid of a rugose border along the posterior edge; metatibiae with a clear distinction between outer and apical edges, the outer edge being straight and usually bearing three well-separated spines (Fig. 15); elytral punctation, seriate.

Redescription based in a specimen from Guam (Figs 11–15). Adult apparently not fully pigmented but in good condition. Measurements in mm: TL 1.03, PL 0.40, PW 0.45, EL 0.63, EW 0.50, GD 0.48. Ratios: PL/PW 0.89, EL/EW 1.25, EL/PL 1.56, GD/EW 0.95, TL/EW 2.05. Body elongate, convex; dorsum and venter yellowish brown; antennae, palpi and tarsi yellowish; dorsal vestiture of minute setae, smaller than a puncture-width and barely discernible even in high magnifications (150×), except for the posteriormost portion of elytra with conspicuous setae (easily seen in lateral view); venter subglabrous. Head with anteriormost portion visible from above; dorsal surface with shallow punctures, separated from each other by a puncture-width or less and with microreticulate interspaces. Antennae bearing nine antennomeres, as follows (in mm, left antenna measured): 0.06, 0.03, 0.02, 0.01, 0.01, 0.01, 0.02, 0.03, 0.03 (FL 0.05, CL 0.08, CL/FL 1.60). Eyes finely facetted, each bearing about 60 ommatidia; GW 0.10. Gula 0.50 as wide as head. Pronotum with shallow, single punctation; punctures irregular, separated from each other by a puncture-width or less; interspaces microreticulate; anterior edge broadly rounded; lateral edges smooth, not explanate and not visible when seen from above; posterior edge without a rugose border along it. Scutellar shield triangular, bearing small punctures; BW 0.08. Elytra with shallow, dual punctation; large punctures coarse, seriate, about twice as large as small punctures; small punctures sparsely and irregularly distributed; interspaces of punctures, smooth; elytral apex truncate; apical declivity

Taxonomy of Xylographellini (Coleoptera: Ciidae) from the Australian and Oriental regions with descriptions... 103

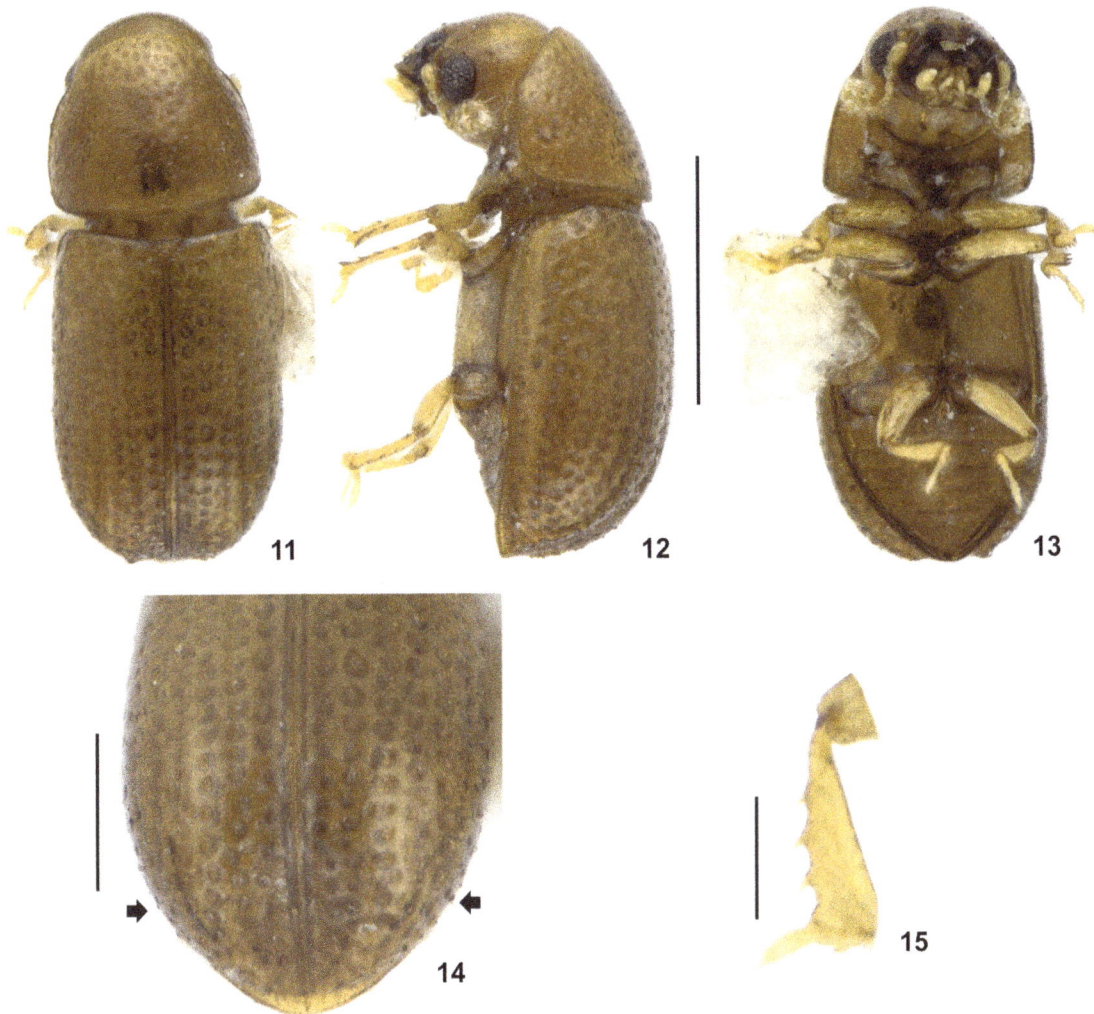

Figures 11–15. *Scolytocis samoensis* from Guam, Micronesia: (11) dorsal view; (12) lateral view; (13) ventral view; (14) apical declivity of elytra; (15) left metatibia. Scale bars: 0.5 mm (11–13); 0.2 mm (14); 0.1 mm (15).

concave with conspicuous cuticular globules (Fig. 14, arrows). Hind wings developed, apparently functional. Hypomera with coarse, shallow punctation; each puncture bearing a fine decumbent seta; interspaces microreticulate. Prosternum in front of coxae biconcave; interspaces microreticulate. Porsternal process laminate, as long as prosternum at midline, apex acute. Pro-, meso- and metatibiae (Fig. 15, left metatibia) with similar shape and length, approximately three times as long as broad; tibiae with a clear distinction between outer and apical edges; outer apical angle somewhat perpendicular; outer edge straight and bearing three well-separated spines; apical edge with about 10 spines very close to each other. Metaventrite with coarse, small punctures; interspaces, microreticulate; discrimen as long as metaventrite at midline. Abdominal ventrites with coarse, small punctures, separated from each other by a puncture-width or

less; interspaces, microreticulate; length of ventrites (in mm, from base to apex at the longitudinal midline) as follows: 0.10, 0.04, 0.04, 0.04, 0.10.

Measurements (in mm) and ratios (n = 15): TL 1.00–1.23 (1.11 ± 0.07), PL 0.38–0.43 (0.40 ± 0.02), PW 0.43–0.53 (0.48 ± 0.03), EL 0.63–0.80 (0.71 ± 0.06), EW 0.48–0.55 (0.53 ± 0.03), GD 0.45–0.53 (0.50 ± 0.03), PL/PW 0.75–0.89 (0.83 ± 0.04), EL/EW 1.25–1.45 (1.34 ± 0.07), EL/PL 1.56–1.94 (1.79 ± 0.12), GD/EW 0.82–1.16 (0.95 ± 0.08), TL/EW 1.95–2.23 (2.10 ± 0.07).

Material examined. Guam: 15 specimens (8 ANIC; 7 CELC, including 2 dissected females) labeled "MARIANAS: Guam, Ritidian Point, 29.v.1945, #2086 FMNH in polypore, H. S. Dybas".

Remarks. *Scolytocis samoensis* is the type species of the genus and was described based on a single specimen from Samoa, possibly a female, deposited in the Bernice Pauahi Bishop Museum (Hawaii, USA). Blair (1928) mentioned the poor condition of the holotype. This species was tentatively included in the *Sc. lawrencei* species-group by Lopes-Andrade (2008) and here we keep the same opinion. The specimens we examined from Guam (Fig. 24) fit the original description by Blair (1928).

Key to species of *Xylographella* Miyatake

1 Pronotum with interspaces of punctures microreticulate. Elytra with apical declivity (posterior one-third of elytra) smooth, lacking raised keels. Known from Japan ....*Xylographella punctata* Miyatake, 1985
– Pronotum with interspaces of punctures smooth. Elytra with apical declivity bearing raised keels 2
2 TL > 2.05 mm. Elytra with apical declivity bearing twelve raised keels (six in each elytron). Known from the Philippines ....*Xylographella speciosa* Lopes-Andrade, 2008
– TL < 1.90 mm. Elytra with apical declivity bearing six raised keels (three in each elytron). Known from Australia ....*Xylographella frithae* sp. n.

Xylographella frithae sp. n.

http://zoobank.org/3500CFF5-B5D1-41F6-8520-8E8C173E0BDB
Figs 16–24

Xylographella sp. in Lawrence (2016: 198).

Type locality. Paluma, state of Queensland, northeastern Australia (18°56′S, 146°10′E).

Diagnosis. *Xylographella frithae* sp. n. differs from *X. punctata* in possessing longitudinal raised keels at elytral declivity and in the smooth interspaces of pronotal punctures. It is closely related to *X. speciosa*, but differs in possessing six longitudinal raised keels at elytral declivity, rather than twelve, and in being comparatively smaller (TL less than 1.90 mm).

Description, holotype (Figs 16–20). Adult fully pigmented. Measurements in mm: TL 1.65, PL 0.60, PW 0.75, EL 1.05, EW 0.80, GD 0.73. Ratios: PL/PW 0.80, EL/EW 1.31, EL/PL 1.75, GD/EW 0.91, TL/EW 2.06. Body elongate, convex; dorsum and venter dark reddish brown; antennae, palpi and tarsi a bit lighter; dorsal vestiture of minute setae, smaller than a puncture-width and barely discernible even in high magnification (150×), except for the posteriormost portion of elytra with conspicuous setae (easily seen in lateral view); venter subglabrous. Head concealed by pronotum and not visible from above; dorsum with shallow punctures, separated from each other by a puncture-width or less and with microreticulate interspaces. Antennae bearing 10 antenommeres, as follows (in mm; left antenna measured): 0.09, 0.04, 0.06, 0.03, 0.02, 0.02, 0.02, 0.04, 0.04, 0.06 (FL 0.15, CL 0.14, CL/FL 0.93). Eyes finely facetted, each bearing about 80 ommatidia; GW 0.15. Gula 0.42 times as wide as head. Pronotum with moderately deep, single punctation; punctures irregular,

separated from each other by a puncture-width or less and with smooth interspaces; anterior edge broadly rounded; lateral edges finely crenulate, not explanate and not visible when seen from above. Scutellar shield triangular, bearing a few punctures near lateral edges; BW 0.11. Elytra with coarse, deep, dual punctation; large punctures coarsely and irregularly distributed, deeper than those on pronotum, about twice as large as small punctures; small punctures sparsely and irregularly distributed; interspaces of punctures, smooth; elytral apex truncate; apical declivity (posterior one-third of elytra) bearing six raised keels (three in each elytron) converging to apex (Fig. 19, arrows). Hind wings developed, apparently functional. Hypomera with coarse, shallow punctation; each puncture bearing a fine decumbent seta; interspaces, microreticulate. Prosternum in front of coxae biconcave; interspaces, microreticulate. Prosternal process laminate, as long as prosternum at midline. Protibiae about three times as long as broad and expanded near apex; outer edge with spines extending from apex to almost its base; inner facet with a conspicuous tuft of long bristles along the apical two-fifths of the inner edge. Meso- and metatibiae (Fig. 20, left metatibia of a paratype) about four times as long as broad; outer edge with spines extending from apex to almost its base. Metaventrite with coarse, small punctures; interspaces, microreticulate; discrimen apparently absent. Abdominal ventrites with coarse, large punctures, separated from each other by a puncture width or less; interspaces, microreticulate; length of ventrites (in mm, from base to apex at the longitudinal midline) as follows: 0.22, 0.09, 0.08, 0.08, 0.17. Male terminalia in paratypes (Figs 21–23): sternite VIII (Fig. 21) with posterior margin rounded, bearing long setae medially; anterior portion with spiculum relictum. Tegmen (Fig. 22, tegmen alone; Fig. 23, tegmen with penis) about 4× as long as wide, sides straight and almost parallel; posterior half membranous and bearing a short median emargination at apex; anterior half sclerotized. Penis (Fig. 23, together with tegmen) about twice as long as tegmen; about 10× as long as wide; bearing paired longitudinal baculi more visible at the posterior portion and united near apex, forming a narrow arch posteriorly; apicalmost portion membranous, slightly expanded and bearing several sensilla at apex.

Measurements (in mm) and ratios (n = 17, including the holotype): TL 1.48–1.88 (1.64 ± 0.10), PL 0.50–0.60 (0.54 ± 0.03), PW 0.65–0.80 (0.70 ± 0.04), EL 0.95–1.28 (1.09 ± 0.08), EW 0.68–0.85 (0.76 ± 0.05), GD 0.65–0.85 (0.71 ± 0.05), PL/PW 0.74–0.81 (0.78 ± 0.02), EL/EW 1.31–1.61 (1.44 ± 0.07), EL/PL 1.75–2.29 (2.01 ± 0.16), GD/EW 0.87–1.00 (0.94 ± 0.04), TL/EW 2.03–2.32 (2.17 ± 0.08).

Material examined. Australia: holotype (ANIC) labeled "19.00S 146.12E, Paluma, QLD, 900 m, 11-vii-80, D. W. Frith [printed] \ J. F. Lawrence, Lot 80–56, *Phellinus pectinatus* [printed] \ *Xylographella frithae* Souza-Gonçalves & Lopes-Andrade HOLOTYPUS [printed on red paper]". Paratypes: 18 specimens (2 males and 2 females) as follows: 3 specimens (CELC), same locality data as holotype; 5 specimens (3 ANIC; 2 CELC) "Paluma, QLD, 11

Figures 16–23. Holotype of *Xylographella frithae* sp. n. from Queensland, Australia: (16) dorsal view; (17) lateral view; (18) ventral view; (19) apical declivity of elytra; (20) left metatibia. Dissected male terminalia of paratypes: (21) sternite VIII; (22) tegmen; (23) aedeagus showing tegmen (teg) and penis (pen). Scale bars: 0.5 mm (16–8); 0.2 mm (19); 0.1 mm (20–23).

July 1980, D. Frith [printed]\J. F. Lawrence, Lot 80–56, *Phellinus pectinatus* [printed]"; one male (CELC, dissected) "Paluma, QLD, 11–12 Dec. 78, D. Frith [printed]\J. F. Lawrence, Lot 78–203, *Nigrofomes melanoporus* [printed]"; one specimen (QMBA) "NEQ: 16°31'S x 146°16'E, Mt Lewis Rd (Hut), 14 July 1996, 1200 m, G. B. Monteith, *Pyrethrum*, trees [printed]\QUEENSLAND MUSEUM LOAN DATE: March 2001 No. LE 01.11"; one female (CELC, dissected) "Mt. Lewis, 8 Km NW of Julatten, N. QLD, 8 Jan. – 2 Feb. 1987, R. Storey & H. Howden [printed]"; one specimen (ANIC) "Mt. Lewis, 800 m, QLD, 26 Dec. 1986, H. & A. Howden, flight intercept trap [printed]"; one specimen (CELC) "17.27S

145.29E QLD, Hugh Nelson Rg. GS3 1150 m, 1 Dec. 1994 – 3 Jan. 1995, P. Zborowski, FI Trap JCU (East) [printed]"; one male (ANIC, dissected) "Hugh Nelson Ra., 21 Km S Atherton, N. Qld, 1.xii.1983 – 9.1.1984, Storev & Brown [printed]\MDPI Intercept Trap. Site No. 16 [printed]\On loan from: Dept. Prim. Industries Mareeba, Qld. Aust. [printed on green paper]"; one specimen (QMBA) "Mossman Bluff Track, 5–10 Km W. Mossman N. Qld, 20 Dec 1989 – 15 Jan 1990, Monteith, Thompson & ANZSES Site 6,860 m, flt. intercept [printed]\QUEENSLAND MUSEUM LOAN DATE: Dec. 2001 No. LE 01.29 [printed on green paper]"; 3 specimens (one CELC, dissected female; 2 ANIC) "Mossman

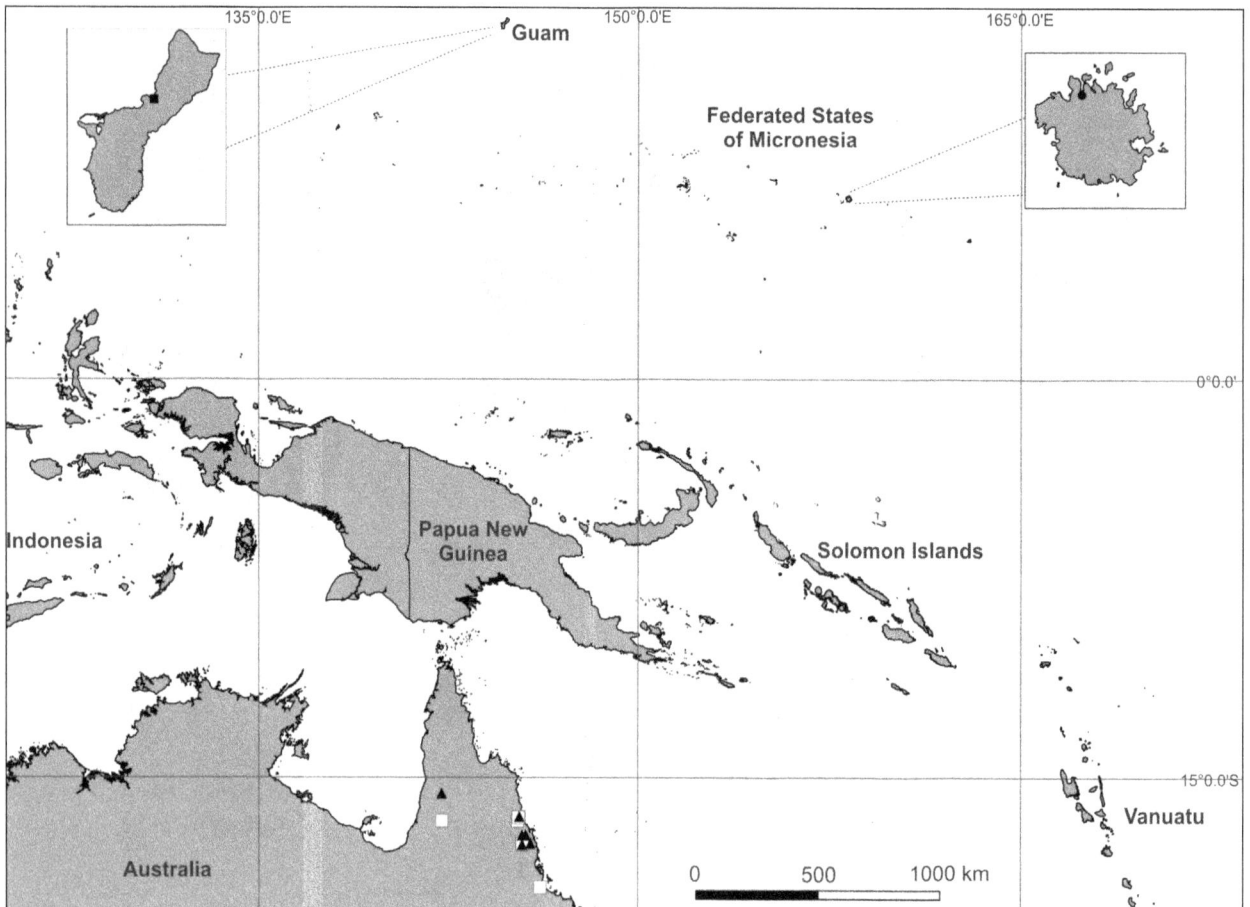

Figure 24. Distribution of the new species of Xylographellini in the Australian and Oriental regions, and new record for *Scolytocis samoensis* Blair from Guam: *Sc. australimontensis* sp. n. (black triangle); *Sc. insularis* sp. n. (black circle); *Sc. samoensis* (black square); *X. frithae* sp. n. (white square).

Gorge, NP QLD, 6 Km SW of Mossman 50 m, 11 July 1982, S. & J. Peck, SBP6 [printed]\J. F. Lawrence, lot 82–28, *Nigroporus* [printed]". All paratypes are additionally labeled "*Xylographella frithae* Souza-Gonçalves & Lopes-Andrade PARATYPUS [printed on yellow paper]".

Host fungi. *Phylloporia pectinata* (Klotzsch) Ryvarden (Hymenochaetaceae), two records; *Nigrofomes melanoporus* (Mont.) Murrill (Polyporaceae), one record; *Nigroporus* Murrill (Polyporaceae), one record.

Etymology. The new species is named in honor of the ornithologist Dawn Whyatt Frith, who collected all specimens from the type locality. The species name is Latinized from "Frith" using the feminine suffix in the genitive singular (*-ae*).

Remarks. This is the first described species of *Xylographella* from the Australian region. The genus encompasses only two other species: *X. punctata* Miyatake, 1985 and *X. speciosa* Lopes-Andrade, 2008, the former from the Japanese islands

of Honshu and Shikoku, and the latter from the provinces of Mindanao and Luzon in the Philippines. Images of *X. frithae* were recently provided by Lawrence (2016).

ACKNOWLEDGEMENTS

We would like to especially thank John Francis Lawrence for giving us the opportunity to work on these Xylographellini and to the staff of ANIC for managing the loan of ciids. We also thank two anonymous reviewers and the associate editor Ângelo P. Pinto for valuable corrections to the text. The senior author I.S.G. thanks the Graduate Program in Ecology (Universidade Federal de Viçosa, Brazil) for the academic support during his master degree. Financial support was provided by Fundação de Amparo à Pesquisa do Estado de Minas Gerais (FAPEMIG; Edital 01/2016 – Demanda Universal, APQ-02675–16), Conselho Nacional de Desenvolvimento Científico e Tecnológico (CNPq;

research grant to the junior author C.L.A., 307116/2015–8) and Coordenação de Aperfeiçoamento de Pessoal de Nível Superior (CAPES; master degree grant to I.S.G.).

REFERENCES

Araujo LS, Lopes-Andrade C (2016) A new species of *Falsocis* (Coleoptera: Ciidae) from the Atlantic Forest biome with new geographic records and an updated identification key for the species of the genus. Zoologia 33: e20150173. https://doi.org/10.1590/S1984-4689zool-20150173

Blair KG (1928) Heteromera, Bostrichoidea, Malacodermata and Buprestidae. In: Insects of Samoa and Other Samoan Terrestrial Arthropoda. Part 45. Coleoptera. Fasc. 2. British Museum (Natural History), London, 67–109.

Lawrence JF (2016) The Australian Ciidae (Coleoptera: Tenebrionoidea): A Preliminary Revision. Zootaxa 4198: 1–208. https://doi.org/10.11646/zootaxa.4198.1.1

Lawrence JF, Ślipiński A, Seago AE, Thayer MK, Newton AF, Marvaldi AE (2011) Phylogeny of the Coleoptera based on morphological characters of adults and larvae. Annales Zoologici 61: 1–217. https://doi.org/10.3161/000345411X576725

Lopes-Andrade C (2008) An essay on the tribe Xylographellini (Coleoptera: Tenebrionoidea: Ciidae). Zootaxa 1832: 1–110.

Lopes-Andrade C, Grebennikov VV (2015) First record and five new species of Xylographellini (Coleoptera: Ciidae) from China, with online DNA barcode library of the family. Zootaxa 4006: 463–480. https://doi.org/10.11646/zootaxa.4006.3.3

Lopes-Andrade C, Lawrence JF (2005) *Phellinocis*, a new genus of Neotropical Ciidae (Coleoptera: Tenebrionoidea). Zootaxa 1034: 43–60.

Lopes-Andrade C, Lawrence JF (2011) Synopsis of *Falsocis* Pic (Coleoptera: Ciidae), new species, new records and an identification key. ZooKeys 145: 59–78. https://doi.org/10.3897/zookeys.145.1895

Miyatake M (1985) Ciidae. In: Kurasawa Y, Hisamatsu S, Sasaji H (Eds) The Coleoptera of Japan in Color, 3. Hoikusha, Osaka, 278–285.

Morrone JJ (2015) Biogeographical regionalization of the world: a reappraisal. Australian Systematic Botany 28: 81–90. https://doi.org/10.1071/SB14042

Author Contributions: ISG and CLA participated equally in the preparation of this article.

Competing Interests: The authors have declared that no competing interests exist.

12

Description of ten additional ossicles in the foregut of the freshwater crabs *Sylviocarcinus pictus* and *Valdivia serrata* (Decapoda: Trichodactylidae)

Renata C. Lima-Gomes[1], Jô de Farias Lima[2], Célio Magalhães[3]

[1]*Programa de Pós-Graduação em Biologia de Água Doce e Pesca Interior, Instituto Nacional de Pesquisas da Amazônia. Caixa Postal 2223, 69080-971 Manaus, AM, Brazil.*
[2]*Empresa Brasileira de Pesquisa Agropecuária – Embrapa Amapá. Rodovia Juscelino Kubitschek, km 5, 2600, Caixa Postal 10, 68906-970 Macapá, AP, Brazil.*
[3]*Coordenação de Biodiversidade, Instituto Nacional de Pesquisas da Amazônia. Caixa Postal 2223, 69080-971 Manaus, AM, Brazil.*
Corresponding author: Renata C. Lima-Gomes (renatacslima@yahoo.com.br)

http://zoobank.org/2017D464-5A2D-4524-8D86-5052C7AE4907

ABSTRACT. The morphology of stomach ossicles of decapod crustaceans provides valuable information on their phylogeny and biology. We herein described ten new ossicles in the foreguts of two trichodactylid crabs, *Sylviocarcinus pictus* (H. Milne-Edwards, 1853) and *Valdivia serrata* White, 1847, in addition to previously described 38 ossicles, which are also recognized and listed. Five specimens each of *S. pictus* and *V. serrata* were selected for morphological analysis of gastric ossicles. The stomachs were obtained after removing the carapace, and they were fixed in 10% formalin for 24 hours. After this procedure, the stomachs were immersed in a solution of 10% Potassium Hydroxide (KOH) and heated to 100 °C during 60 minutes for tissue maceration. At this point, the clean skeletons were colored by adding 1% Alizarin Red to the KOH solution in order to facilitate visualization of the internal structures such as the setae and ossicles. The ten new ossicles are: dorsomedial cardiac plate; dorsolateral cardiac plate; suprapectineal lateral ossicle; inferior cardiac valve; lateral mesopyloric ossicle; ampullary roof-medium portion ossicle; process of the ampullary roof-upper portion; lateral-inferior post-ampullary plate; pleuro-pyloric valve's ossicle; and lateral pleuro-pyloric plate. Some ossicles are thin plates that together with the main ossicles assist in the structure and support of the stomach, which are similar in the two species studied herein. The current knowledge on gastric ossicles will be useful in establishing taxonomic characters, which can evaluate phylogenetic relationships among brachyuran crabs.

KEY WORDS. Amazon, anatomy, foregut, morphology, Neotropical, stomach, Valdiviini.

INTRODUCTION

The stomach of decapod crustaceans is composed of a muscular and nervous complex called gastric mill (Meiss and Norman 1977), where a system of striated muscles performs movements of skeletal elements that work together to break and grind large particles of food in the cardiac chamber. The main skeletal elements consist of the following ossicles: mesocardiac, pterocardiac, pyloric, exopyloric, zygocardiac (that supports the lateral teeth), propyloric and urocardiac (that supports the medial tooth) (Factor 1989). In addition to the support of the gastric skeleton, the ossicles assist in crushing and filtration activities during the feeding process. Many studies have been undertaken to understand how this complex operates in different decapods since the gastric skeleton can provide valuable information on their phylogeny and biology, especially their feeding habits (Felgenhauer and Abele 1983, 1985, Brösing et al. 2002, Abrunhosa et al. 2003, Brösing et al. 2006, Brösing 2010, Alves et al. 2010). The morphology of stomach ossicles can also be an important source for taxonomic characters and is potentially useful for studying phylogenetic relationships in different groups of decapods (Sakai 2005, Sakai et al. 2006, Naderloo et al. 2010, Brösing and Türkay 2011).

Studies on the gastric skeleton of Amazonian decapods are very recent, and only a few have been developed for brackish and freshwater decapods. Alves et al. (2010) published the only study for the Trichodactylidae. They described 38 ossicles for *Valdivia serrata* White, 1847 and *Sylviocarcinus pictus* (H. Milne Edwards, 1853) (Valdiviini) and 37 ossicles for *Dilocarcinus septemdentatus* (Herbst, 1783) (Dilocarcinini). A total of 48 ossicles have been recognized in Gecarcinidae members (Jô de Farias Lima, unpublished data). In the present work, we describe ten additional ossicles found in the foreguts of *S. pictus* and *V. serrata*, as this complementary information may be useful for studies on the trophic ecology and phylogenic relationships of the trichodactylid crabs.

MATERIAL AND METHODS

Uncatalogued specimens of *S. pictus* and *V. serrata* were obtained from the crustacean collection of the Instituto Nacional de Pesquisas da Amazônia, Manaus, Brazil. The stomachs of five specimens of each species were analyzed: four males and a female of *S. pictus*, and two males and three females of *V. serrata*. The stomachs of crabs do not differ in regard to gender and so did not matter the number of each gender. The stomachs were obtained after carapace removal and fixed in 10% formalin for 24 hours. For tissue maceration, the stomachs were cooked for 60 minutes in 10% Potassium Hydroxide (KOH) solution and heated to 100 °C (Mocquard 1883, Brösing et al. 2002). The cleaned skeleton was then colored by adding 1% Alizarin Red to the KOH solution to facilitate visualization of the internal structures such as the setae and ossicles (Brösing et al. 2002).

The nomenclature and abbreviations used in the morphological description of the gastric skeleton followed Alves et al. (2010); the degree of calcification of the ossicles was described as: (I) mild calcification (membranous aspect); (II) moderate calcification (cartilaginous aspect); (III) strong calcification (opaque aspect); (IV) free ossicle (when connected to adjacent ossicle by thin and pliable membrane); (V) partially fused ossicles (when connected by rigid membrane cartilaginous aspect or clearly incomplete fusion); and (VI) fused ossicles (indistinct separation) (Jô de Farias Lima, unpublished data). The roman numerals in the table and figures refer to ossicles described by Alves et al. (2010) as well as the ossicles presented herein. The complete list of the names and abbreviations of all the described ossicles is in Table 1.

RESULTS

All the 38 ossicles previously described by Alves et al. (2010) were recognized and listed in Table 1, in addition to the ten new ossicles indentified in the present study. Some are thin plates that together with the main ossicles assist in the structure and support of the stomach.

The stomach ossicles are grouped according to the following regions of the gastric skeleton: 10 ossicles of the gastric mill; 11 ossicles of the lateral supporting cardiac region; 4 ossicles of

the cardio-pyloric valve; 6 ossicles supporting the dorsal pyloric stomach; 9 ossicles supporting the ventral pylorus and bulb; 3 ossicles supporting the supra-ampullary region; and 5 ossicles supporting the lateral pylorus region (Table 1). In total, the gastric skeleton consists of 48 ossicles, which are similar in the two species studied herein (Figs 1–8).

The ten new ossicles are described below and illustrated in Figs 1–9.

Ossicles of the gastric mill: Dorsomedial cardiac plate (VIIa) (Figs 3, 4, 7–9) – paired, compressed, devoid of calcification, located close to the posterior portion of the urocardiac, anterior region curved laterally.

Dorsolateral cardiac plate (VIIb) (Figs 3, 4, 7, 8 and 10) – paired, compressed, mildly calcified, oblong, dorsally located in the cardiac stomach next to the urocardiac ossicle.

Lateral supporting cardiac ossicles: Suprapectineal lateral ossicle (VIIIa) (Figs 3, 4, 7, 8 and 11) – paired, compressed, mildly calcified, located laterally in the cardiac stomach, between the pectineal and subdentate ossicles. Inferior cardiac valve (XIIa) (Figs 3, 4, 7, 8 and 16) – paired, moderately calcified, tapering, located in the bottom lateral portion of the cardiac stomach, loosely connected to the posterior cardiac plate and to the post-pectineal keel ossicle. Mesial portion elongated, covered with tufts of long setae.

Supporting ossicles of the dorsal pyloric chamber: Lateral mesopyloric ossicle (XIXa) (Figs 3, 4, 7, 8 and 12) – paired, moderately calcified, compressed dorsoventrally, slightly concave dorsally, moderately sinuous, located laterally in the upper portion of the pyloric chamber, next to the anterior mesopyloric ossicle.

Supporting ossicles of the ventral pylorus and ampullae: Ampullary roof-mediun portion ossicle (XXVa) (Figs 3, 4, 7, 8 and 17) – paired, mildly calcified, flat, elipsoid, located between the upper and lower portions of the ampullary roof. Process of the ampullary roof-upper portion (XXVIa) (Figs 3, 4, 7, 8 and 17) – unpaired, heavily calcified consisting of two subtriangular portions obliquely positioned in relation to each other, and located laterally in the pyloric chamber, connected to the following ossicles: anterior pleuro-pyloric (strong connection), pleuro-pyloric valve (partially fused), and the upper ampullary roof (partially fused).

Lateral-inferior post-ampullary plate (XXVIIa) (Figs 3, 4, 7, 8 and 14) – paired, compressed, mildly calcified, subtriangular, located immediately adjacent to the inferior ampullary ossicle.

Supporting ossicles of the lateral pylorus: Pleuro-pyloric valve's ossicle (XXXIa) (Figs 3, 4, 7, 8 and 13) – paired, anterior portion heavily calcified (bar-shaped) and greatly reduced; posterior portion expanded as mildly calcified, translucent membrane, located laterally in the pyloric stomach, just adjacent to the anterior pleuro-pyloric ossicle. Lateral pleuro-pyloric plate (XXXIIa) (Figs 3, 4, 7, 8 and 15) – paired, compressed, mildly calcified, located on the lateral portion of the stomach pyloric, above the median pleuro-pyloric ossicle.

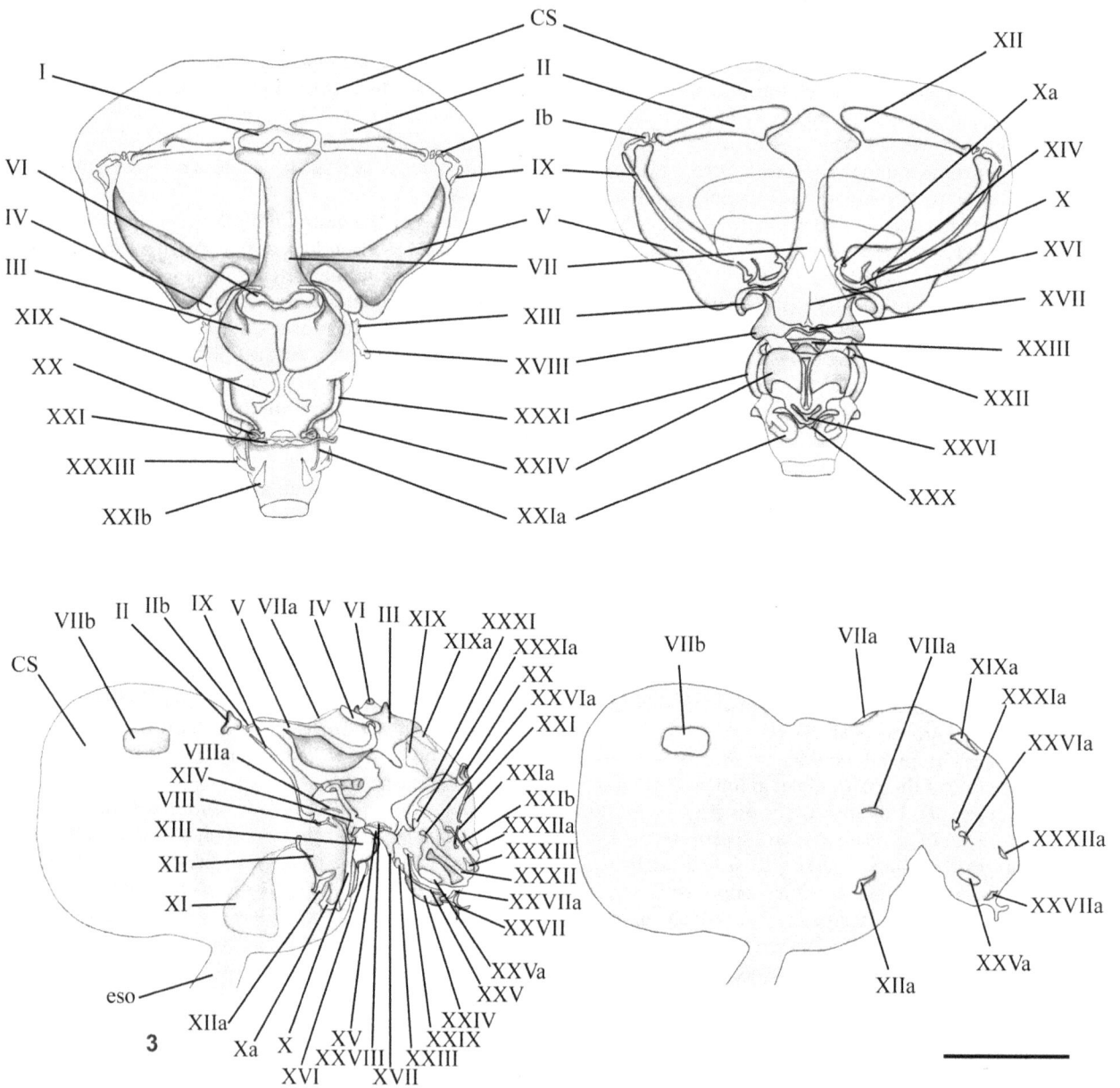

Figures 1–4. Foregut of *Sylviocarcinus pictus*: (1) dorsal view; (2) ventral view; (3) lateral view; (4) lateral view with the additional ossicles found in the present study. (CS) Cardiac sac, (eso) esophagus. Scale bar = 5 mm.

DISCUSSION

We recognized 48 ossicles in the gastric skeletons of *S. pictus* and *V. serrata* whereas Alves et al. (2010) recognized 38 ossicles for the same species (Table 1). This discrepancy may be due to methodological issues or interpretation, as discussed below.

The 'dorsomedial cardiac plate' and the 'dorsolateral cardiac plate', the 'lateral inferior post-ampullary plate' and

the 'lateral pleuro-pyloric plate' ossicles were not mentioned by Alves et al. (2010). These structures are very thin due to their low degree of calcification, which makes their visualization and identification somewhat difficult, especially if the cooking time is exceeded, the KOH concentration is higher than the suggested method, or the stomach is not properly colored during the preparation process. The 'lateral supra-pectineal ossicle' was also not described by Alves et al. (2010); this ossicle probably

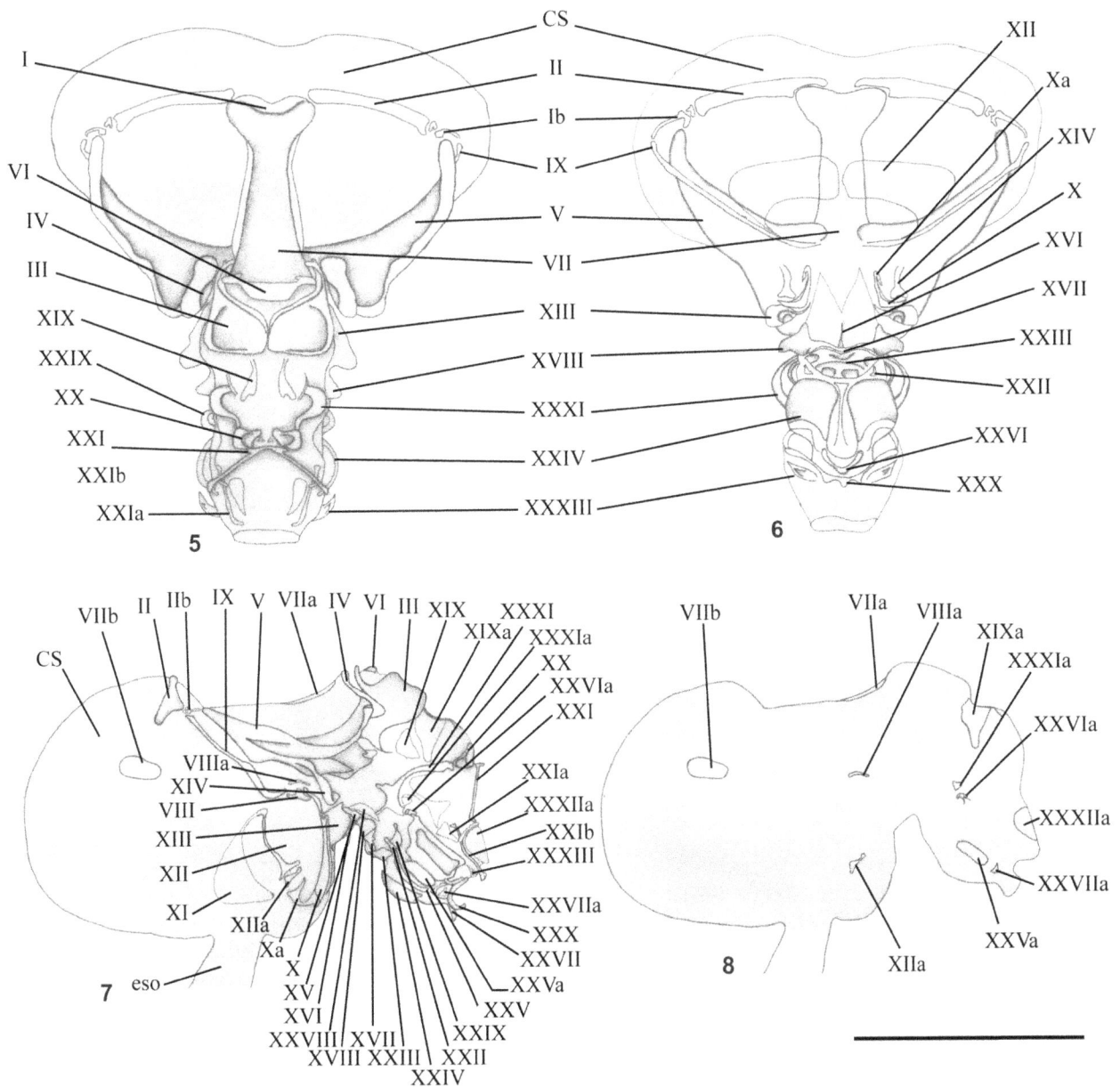

Figures 5–8. Foregut of *Valdivia serrata*: (5) dorsal view; (6) ventral view (7) lateral view; (8) lateral view with the additional ossicles found in the present study. (CS) cardiac sac, (eso) esophagus. Scale bar = 5 mm.

went unnoticed because it is mildly calcified and translucent. In addition, some ossicles are minuscule, which could have led to them being interpreted as a single entity, rather than individualized structures. This might have been the case for the 'ampullary roof-lower portion ossicle', 'ampullary roof-median portion ossicle' and 'lateral supra-pectineal ossicle'.

The 'lateral mesopyloric ossicle' was first recognized by Brösing et al. (2002) who proposed it as a new ossicle to the

foregut ossicle-system of *Dromia wilsoni*, *Dromia personata* and *Lauridormis intermedia*. However, Brösing (2010) reported the absence of this ossicle in *Ocypode gaudichaudi* (H. Milne-Edwards & Lucas, 1843) and *O. cursor* (Linnaeus, 1758). This ossicle was recognized as a poorly calcified structure in *Ocypode quadrata* (Fabricius, 1787) and treated as new (Jô de Farias Lima, unpublished data). Thus, the 'lateral mesopyloric' is most probably present in *O. gaudichaudi* and *O. cursor* as well. Perhaps its low degree

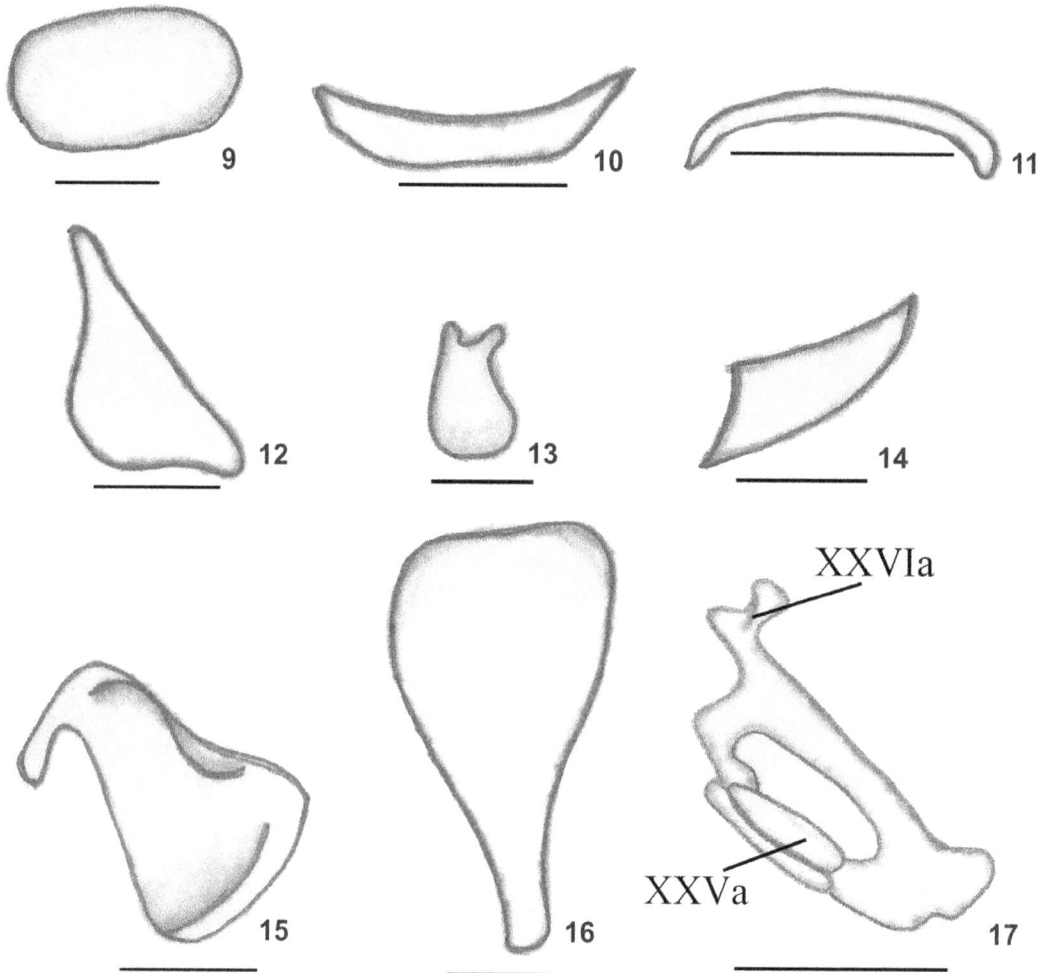

Figures 9–17. The ten additional ossicles of the foregut of *Sylviocarcinus pictus* and *Valdivia serrata*, found in the present study and not previously recognized by Alves et al. (2010): (9) dorsolateral cardiac plate (VIIb); (10) dorsomedial cardiac plate (VIIa); (11) suprapectineal lateral ossicle (VIIIa); (12) lateral mesopyloric ossicle (XIXa); (13) pleuro-pyloric valve's ossicle (XXXIa); (14) lateral-inferior post-ampullary plate (XXVIIa); (15) lateral pleuro-pyloric plate (XXXIIa); (16) inferior cardiac valve (XIIa); (17) process of the ampullary roof-upper portion (XXVIa), ampullary roof-mediun portion ossicle (XXVa) Scale bars: 9-15, 17 = 2 mm, 16 = 1 mm.

of calcification precluded it from being recognized by Brösing (2010). A similar situation could also have occurred during the study of Alves et al. (2010) in which poorly calcified ossicles might have gone unnoticed and hence were not mentioned.

The 'inferior cardiac valve' looks like a very small ossicle from an external perspective coming from the outer side of the stomach, but it is quite distinct and easily recognizable from a point of view coming from the inner surface of the stomach. However, if a disruption of the heart sac occurs during dissection of the stomach, this structure can be lost.

The 'ossicles of the ampullary roof-mediun portion ossicle', the 'ampullary roof-upper portion ossicle', and the 'pleu-ro-pyloric valve's ossicle', may have been overlooked because they are fused to each other, appearing to be a single piece. The 'ampullary roof-mediun portion ossicle' may have been confused or even not noticed because they are very small and are situated between the 'ampullary roof-upper portion ossicle' and 'ampullary roof-lower portion ossicle'.

A careful recognition of these stomach ossicles and enhanced understanding of the gastric mill complex can provide useful information to establish relevant characters which in turn will be helpful for tracing affinities and evaluating phylogenetic relationships not only in trichodactylid crabs but also among other taxa of Eubrachyura.

Table 1. Nomenclatures and abbreviations used in morphological descriptions ossicles of the stomachs of *Sylviocarcinus pictus* and *Valdivia serrata*, and the comparison between the study of Alves et al. (2010) and the present paper.

Name of the ossicles	Number assigned to the ossicles	Alves et al. (2010)	Present paper
Ossicles of the gastric mill			
Mesocardiac ossicle	(I)	X	X
Pterocardiac ossicle	(II)	X	X
Post-pterocardiac ossicle	(IIb)	X	X
Pyloric ossicle	(III)	X	X
Exopyloric ossicle	(IV)	X	X
Zygocardiac ossicle	(V)	X	X
Propyloric ossicle	(VI)	X	X
Urocardiac ossicle	(VII)	X	X
Pectineal ossicle	(VIII)	X	X
Dorsomedian cardiac plate	(VIIa)	–	X
Dorsolateral cardiac plate	(VIIb)	–	X
Lateral supporting cardiac ossicles			
Suprapectineal lateral ossicle	(VIIIa)	–	X
Prepectineal ossicle	(IX)	X	X
Postpectineal ossicle	(X)	X	X
Quill of the postpectineal ossicle	(Xa)	–	X
Anterior lateral cardiac plate	(XI)	X	X
Posterior lateral cardiac plate	(XII)	X	X
Inferior cardiac valve	(XIIa)	–	X
Inferior lateral cardiac	(XIII)	X	X
Subdentate	(XIV)	X	X
Lateral cardiac-pyloric ossicle	(XV)	X	X
Ossicles of the cardio-pyloric valve			
Anterior ossicle of the cardio-pyloric valve	(XVI)	X	X
Posterior lateral cardiac plate	(XVII)	X	X
Lateral ossicle of the cardio-pyloric valve	(XVIII)	X	X
Supporting ossicles of the dorsal pyloric chamber			
Anterior mesopyloric ossicle	(XIX)	X	X
Lateral mesopyloric ossicle	(XIXa)	–	X
Posterior mesopyloric ossicle	(XX)	X	X
Uropyloric ossicle	(XXI)	X	X
Infra-uropyloric fragment	(XXIa)	X	X
Posterior uropyloric ossicle	(XXIb)	X	X
Supporting ossicles of the ventral pylorus and ampullae			
Preampullary ossicle	(XXII)	X	X
Anterior inferior pyloric ossicle	(XXIII)	X	X
Inferior ampullary ossicle	(XXIV)	X	X
Ampullary roof ossicle, lower portion	(XXV)	X	X
Ampullary roof-mediun portion ossicle	(XXVa)	–	X
Ampullary roof ossicle, upper portion	(XXVI)	X	X
Process of the ampullary roof-upper portion	(XXVIa)	–	X
Posterior inferior pyloric ossicle	(XXVII)	X	X
Lateral-inferior post-ampullary plate	(XXVIIa)	–	X
Supporting ossicles of the supra-ampullary			
Anterior supra-ampullary ossicle	(XXVIII)	X	X
Middle supra-ampullary ossicle	(XXIX)	X	X
Posterior supra-ampullary ossicle	(XXX)	X	X
Supporting ossicles of the lateral pylorus			
Anterior pleuropyloric ossicle	(XXXI)	X	X
Pleuro-pyloric valve's ossicle	(XXXIa)	–	X
Middle pleuropyloric ossicle	(XXXII)	X	X
Lateral pleuro-pyloric plate	(XXXIIa)	–	X
Posterior pleuropyloric ossicle	(XXXIII)	X	X
Cardiac-pyloric valve (v.c.p.)	(v.c.p.)	X	X

ACKNOWLEDGMENTS

The authors are grateful to Michael Türkay (*in memoriam*) for his constructive comments to an early version of the manuscript; and to Felipe B. R. Gomes for his comments and preliminary English revision, and to Colleen L, Flannagan for the final English revision; and to Leandro M. Sousa for the figures edition. RCLG thanks the team of the Laboratory of Limnology of the Instituto Nacional de Pesquisas da Amazônia, particularly Josedec Faria Monteiro, for their help in laboratory procedures, and the Conselho Nacional de Desenvolvimento Científico e Tecnológico – CNPq for a Master's Degree scholarship (process 134785/2011-8). CM also thanks CNPq for an ongoing research grant (process 304736/2015-5).

REFERENCES

Alves SM, Abrunhosa FA, Lima JF (2010) Foregut morphology of Pseudothelphusidae and Trichodactylidae (Decapoda: Brachyura) from northeastern Pará. Zoologia 27: 228–244. https://doi.org/10.1590/S1984-46702010000200011

Brösing A, Richter S, Scholtz G (2002) The foregut ossicle-system of *Dromia wilsoni*, *Dromia personata* and *Lauridormis intermedia* (Decapoda, Brachyura, Dromiidae), studied with a new staining method. Arthropod Structure and Development 30: 329–338. https://doi.org/10.1016/S1467-8039(02)00009-9

Brösing A (2010) Recent developments on the morphology of the brachyuran foregut ossicles and gastric teeth. Zootaxa 2510: 1–44.

Brösing A, Richter S, Scholtz G (2006) Phylogenetic analysis of the Brachyura (Crustacea, Decapoda) based on characters of the foregut with establishment of a new taxon. Journal of Zoological Systematics and Evolutionary Research 45: 20–32. https://doi.org/10.1111/j.1439-0469.2006.00367.x

Brösing A, Türkay M (2011) Gastric teeth of some Thoracotreme crabs and their contribution to the Brachyuran Phylogeny. Journal of Morphology 272: 1109–1115. https://doi.org/10.1002/jmor.10967

Factor JR (1989) Development of the feeding apparatus in deca-
pod crustaceans. In: Felgenhauer BE, Watling L, Thistle AB (Eds)
Functional Morphology of feeding and grooming in Crustacea.
Rotterdam, A.A. Balkema, Crustacean Issues 6, 185–203.

Felgenhauer BE, Abele LG (1983) Phylogenetic relationships
among shrimp-like decapods (Penaeoidea, Caridea, Stenopo-
didea). In: Schram FR (Ed.) Crustacean phylogeny. Rotterdam,
A. A. Balkema, Crustacean issues 1, 291–311.

Felgenhauer BE, Abele LG (1985) Feeding structures of two atyid
shrimps, with comments on caridean phylogeny. Journal of Crustacean Biology 5: 397–419. https://doi.org/10.2307/1547911

Meiss DE, Norman RS (1977) A comparative study of the stomato-
gastric system of several decapod Crustacea, I. Skeleton. Journal of Morphology 152: 21–54. https://doi.org/10.1002/
jmor.1051520103

Mocquard F (1883) Recherches anatomiques sur L'estomac des
Crustacés podophthalmaires. Annales des Sciences Naturelles,
Série Zoologie, 16: 1–311.

Naderloo R, Türkay M, Chen HL (2010) Taxonomic revision of the
wide-front fiddler crabs of the *Uca lactea* group (Crustacea:
Decapoda: Brachyura: Ocypodidae) in the Indo-West Pacific.
Zootaxa 2500: 1–38.

Sakai K (2005) The diphyletic nature of the infraorder Thalassini-
dea (Decapoda, Pleocyemata) as derived from the morphology of the gastric mill. Crustaceana 77: 1117–112. https://doi.
org/10.1163/1568540042900268

Sakai K, Türkay M, Yang SL (2006) Revision of the *Helice/Chasmag-
nathus* complex (Crustacea: Decapoda: Brachyura). Abhandlungen Senckenbergische Naturforschende Gesellschaft 565: 1–76.

Author Contributions: RCL-G collected the data; CM identified
the specimens; RCL-G made the analysis and illustrations; RCL-G,
CM and JFL wrote the text.

Competing Interests: The authors have declared that no competing
interests exist.

13

Anatomical and histological study of the liver and pancreas of two closely related mountain newts *Neurergus microspilotus* and *N. kaiseri* (Amphibia: Caudata: Salamandridae)

Somaye Vaissi[1], Paria Parto[1], Mozafar Sharifi[1]

[1]*Department of Biology, Faculty of Science, Razi University. Baghabrisham 6714967346, Kermanshah, Iran.*
Corresponding author: Mozafar Sharifi (sharifimozafar2012@gmail.com)

http://zoobank.org/72F19481-A9DB-4856-B505-B8136D3ED6B0

ABSTRACT. Anatomical and histological examinations were conducted on the digestive glands of two closely related mountain newts, *Neurergus microspilotus* (Nesterov, 1916) and *Neurergus kaiseri* Schmidt, 1952. In *N. microspilotus* and *N. kaiseri* the major digestive glands comprise a very large liver and a small pancreas. In both species the liver has two distinct lobes, right and left. Histologically, the parenchyma of the liver of both species is contained within a thin capsule of fibroconnective tissue. Glycogen deposits and fat storage often dissolve during the routine histological process and produce considerable histological variability. Sinusoids are lined with endothelial cells forming a very thin epithelial sheet, with discontinuous basement membrane. Bile ducts also occur within the parenchyma of the liver. The ducts are lined by simple cuboidal epithelium. The gall bladder is a storage depot for bile. Its mucosa is thrown into numerous folds. The epithelial lining of the tunica muscularis is arranged circularly. There is a lot of pigmentation in the hepatic parenchyma. The pancreas in *N. microspilotus* and *N. kaiseri* is roughly triangular in shape, and lies rather to the dorsal side of the duodenum, between it and the stomach. The exocrine portion of the pancreas consists of clusters of pyramidal cells, which are mostly organized in acini. In both species the cells have a dark basophilic cytoplasm, distinct basal nuclei, and many large eosinophilic zymogen granules containing enzymes responsible for the digestion of proteins, carbohydrates, fats and nucleotides.

KEY WORDS. Digestive glands, light microscopy, Hematoxylin-Eosin, Periodic acid–Schiff (PAS).

INTRODUCTION

The digestive system of vertebrates demonstrates various structural and functional adaptations to their diverse feeding habits. The digestive tract also represents a functional link between foraging activity and energy conservation through energy allocation for various activities (Secor 2005, Romão et al. 2011). Over the last decades, field observations and experimental laboratory studies have shown that the anatomy and physiology of the digestive tract of many species are flexible, and can change in response to variation in environmental conditions (McWilliams and Karasov 2001). A variety of glands are present within the digestive tract. The liver and pancreas are major secretory structures that lie across the stomach and duodendum and are derived from the embryonic gut. The liver is the largest of the digestive glands, serving as a nutrient storage organ and producer of bile (Vitt and Caldwell 2009). The bile drains from the liver into the gallbladder and then moves via the bile duct into the duodenum, where it assists in the breakdown of food. The amphibian liver is located posterior and ventral to the heart, and the gross anatomy of the former varies depending on the taxonomic group, but generally conforms to the body shape of the amphibian. Anurans have a bilobate liver, while caudate have a slightly elongated and emarginated liver, and in the caecilians it is slightly emarginated and is very elongated. The gall bladder is intimately associated with the liver in many groups of vertebrates, with a bile duct connecting it to the duodenum. The pancreas is a smaller, diffuse gland. It secretes digestive fluids into the duodenum and also its endocrine portion produces insulin (Vitt and Caldwell 2009).

In Iran, the genus *Neurergus* has a relatively wide geographic distribution, ranging from the southern Zagros Mountains to the mid-Zagros range, and extending into Iraq and southern Turkey (Baloutch and Kami 1995). Afroosheh et al. (2016) demonstrated that *Neurergus microspilotus* (Nesterov, 1916) occurs in 42 highland streams in the mid Zagros

Mountains, at elevations ranging between 630–2057 m.a.s.l. *N. microspilotus* is listed as a Critically Endangered species by the International Union for Conservation of Nature (IUCN) (Sharifi et al. 2009, IUCN 2011). *Neurergus kaiseri* Schmidt, 1952 is endemic to first order streams at elevations ranging between 800 and 1500 m a.s.l., and occurs in 36 highland streams (Mobaraki et al. 2014). *N. kaiseri* has also been evaluated as being vulnerable species by IUCN criteria (IUCN 2016). This species has also been amended to the Appendix I of the Convention to the International Trade to Endangered Species (CITES). *N. microspilotus* is slightly larger than *N. kaiseri* and can be found in different climatic regions. Although both species of *Neurergus* occur in highlands' first order streams, the macro-ecology of these two areas (mid-Zagros and southern Zagros) are distinctively different. In the southern Zagros Range, where *N. kaiser*, occurs, the climate is warm without freezing temperatures in the winter, while in western Zagros, where *N. micropilotus* occurs, the climate is cold with pronounced seasonal variations, including a prolonged winter freezing. In both areas the mountain newts are top predators of the diverse benthic macro-invertebrates (Sharifi and Assadian 2004).

The main objective of this study is to describe the digestive gland (including the liver and the pancreas) of two critically endangered mountain newts. We compare and contrast the specific similarities and differences in the anatomy and histology of these two digestive organs.

MATERIAL AND METHODS

Several newts of *N. microspilotus* and *N. kaiseri* were collected from Kavat Stream (34°53N, 46°31E) in the mid-Zagros in western Iran, and Bozorgab Stream in the southern Zagros Mountains (32°56N. 48°28E) in spring 2012 (April to May), respectively, and were kept captive at a breeding facility (CBF) in the Razi University (Sharifi and Vaissi 2014). Permits for collections for the scientific study of *N. microspilotus* were issued by the Regional Office of Environment in Kermanshah Province; and for *N. kaiseri*, from the equivalent office in Khoramabad Province. The newts were maintained in a 75 × 45 ×35 cm glass aquarium, supplied with local water and were fed earthworms or blood worms. Two females and two males of *N. microspilotus* and *N. Kaiseri*, which died at the CBF, were subjected to the present histological study. All animals were in resting condition and each with a body length of about 173.91 ± 17.75 mm for *N. kaiseri* and 192.35 ± 10.20 mm for *N. microspilotus*. The body length was measured as the distance from the tip of the snout to the posterior border of the cloacal opening. The body was divided into five parts. The specimens were fixed in 10% formaldehyde and dehydrated in a series of ethanol treatments, starting from the 70% storing solution, then were cleared in xylene, embedded in paraffin, and serially sectioned at 7 µm with a rotary microtome. The sections were stained with Hematoxylin-Eosin for general morphology and PAS for identifying carbohydrates according

to the protocol of Luna (1968). Sections were observed with an Olympus microscope (Leica Galen III) and were photographed with a digital camera (Leica with Dinocapture 2) mounted to a microscope.

RESULTS

The livers in *N. microspilotus* and *N. kaiseri* are similar and have two distinct lobes, right and left. The left lobe is longer than the right, with a sharp distal end, while the distal end of the right lobe is attached to a spine-shaped accessory process on its medial surface. In both species, it lies ventral to the stomach, and, when fresh, is dark red in color. A thin layer of serous membrane with scattered melanin pigment covered the liver. In the two species the liver is an elongate organ with its anterior end attached to the transverse septum, and extending at least as far posteriorly as the duodenum. In every case the major part of the liver lies on the right side of the body cavity, leaving room for the stomach on the left, and the liver completely suspended by mesenteries. There is a gall bladder lying just dorsal to the right lobe of the liver. The main fissure of the liver is long but does not penetrate deep into the liver on its ventral surface. Therefore, the lobs are less evident on this surface than in dorsal section (Figs 1–4).

Analogous in histology, the parenchyma of the liver in *N. microspilotus* and *N. kaiseri* is contained within a thin capsule of fibroconnective tissue. Thin septa originate from the capsule and divide the liver into incomplete lobules. Hematopoitic tissue is located in the subcapsular region, in multiple layers. The parenchyma itself is primarily composed of polyhedral hepatocytes, typically with central nuclei. Fat storage often dissolved during the routine histological process, and glycogen mass, look like scattered red dots in the cytoplasm and produce considerable histological variability. The histology of the liver of newts differs from that of mammals in that there is a pronounced tendency for the disposition of the hepatocytes in lobules, and the typical portal triads of the mammalian liver are rarely seen. Sinusoids are lined with endothelial cells forming a very thin cytoplasmic sheet. The nuclei of these cells are elongated and protrude into the sinusoidal lumen. The endothelium is fenestrated by small pores. Melanomacrophages can be seen on the sinusoidal wall and also on the hematopoitic component of the liver, and they have melanosyntethic activity. Bile ducts also occur within the parenchyma of the liver. Originating between adjacent hepatocytes, bile canaliculi anastomose to produce the canal of Herring, which has a larger diameter. The ducts are lined by simple cuboidal epithelium (Figs 5–8). Hepatocyte nuclei were round with blue-violet color. The bile drains into the duodenum by the common bile duct. Smaller ducts within the liver are lined with a single layer of cuboidal epithelial cells. The gall bladder is a storage depot for bile. Its mucosa is thrown into numerous folds. The epithelial lining of the bladder is simple columnar and the tunica muscularis is arranged circularly (Figs 9–10).

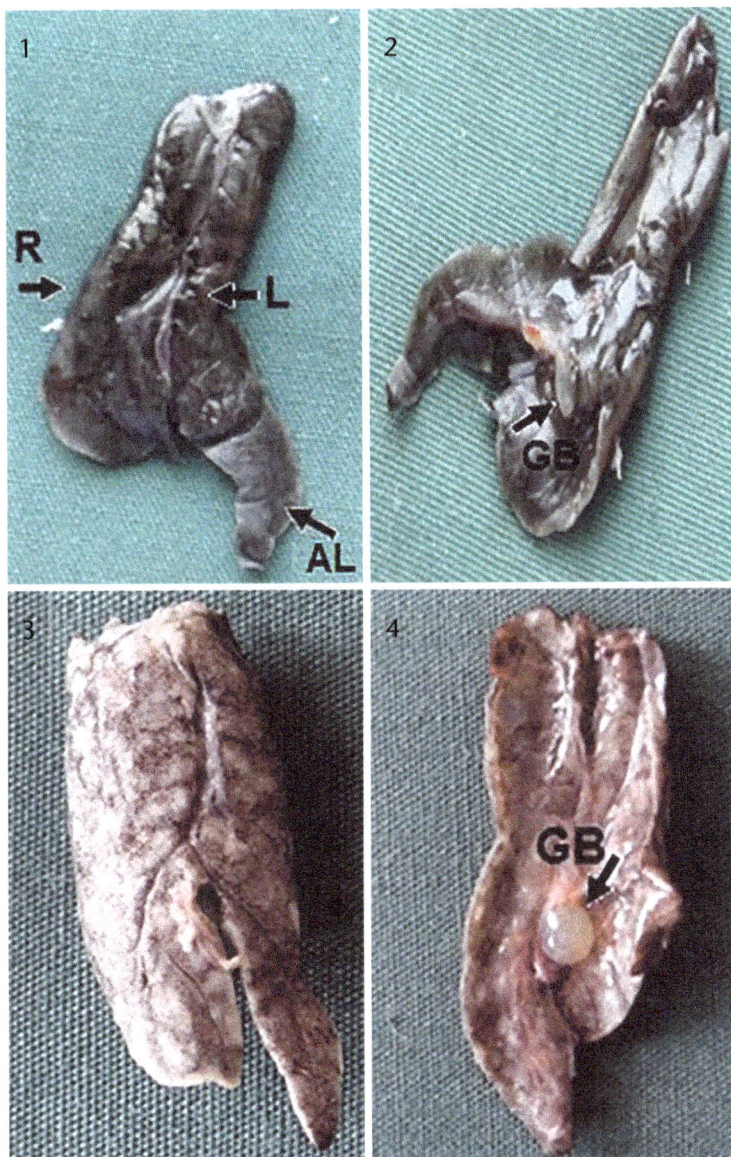

Figures 1–4. The liver of *N. microspilotus* (1, 2) and *N. kaiseri* (3, 4). (1, 3) Dorsal surface; (2–4) ventral surface. (R) Right, (L) left, (AL) accessory lobe, (GB) gall bladder.

The pancreas in *N. microspilotus* and *N. kaiseri* are similar in appearance and both are roughly oblong glands that lie posterior to the greater curvature of the stomach, and are connected to the duodenum (Figs 11–14). The pancreas is made up of small clusters of glandular epithelial cells. About 1% of the cells are organized into clusters called pancreatic islets (islets of Langerhans). They form the endocrine portion. The remaining 99% of the cells are arranged in clusters called acini and constitute the exocrine portion. The exocrine portion of the pancreas consists of clusters of pyramidal cells mostly organized in acini. The cells have a dark basophilic cytoplasm, distinct basal nuclei, and many large apical eosinophilic zymogen granules containing enzymes responsible for the digestion of proteins, carbohydrates, fats and nucleotides, which is called pancreatic juice. Enzymes are delivered into to the duodenum via the pancreatic ductules, which coalesce to form the main pancreatic duct. This latter opens, distinctly or after rejoining the common bile duct, into the duodenum. The pancreatic ductules and the main pancreatic duct are lined with cuboidal to columnar epithelium, respectively (Figs 11–14).

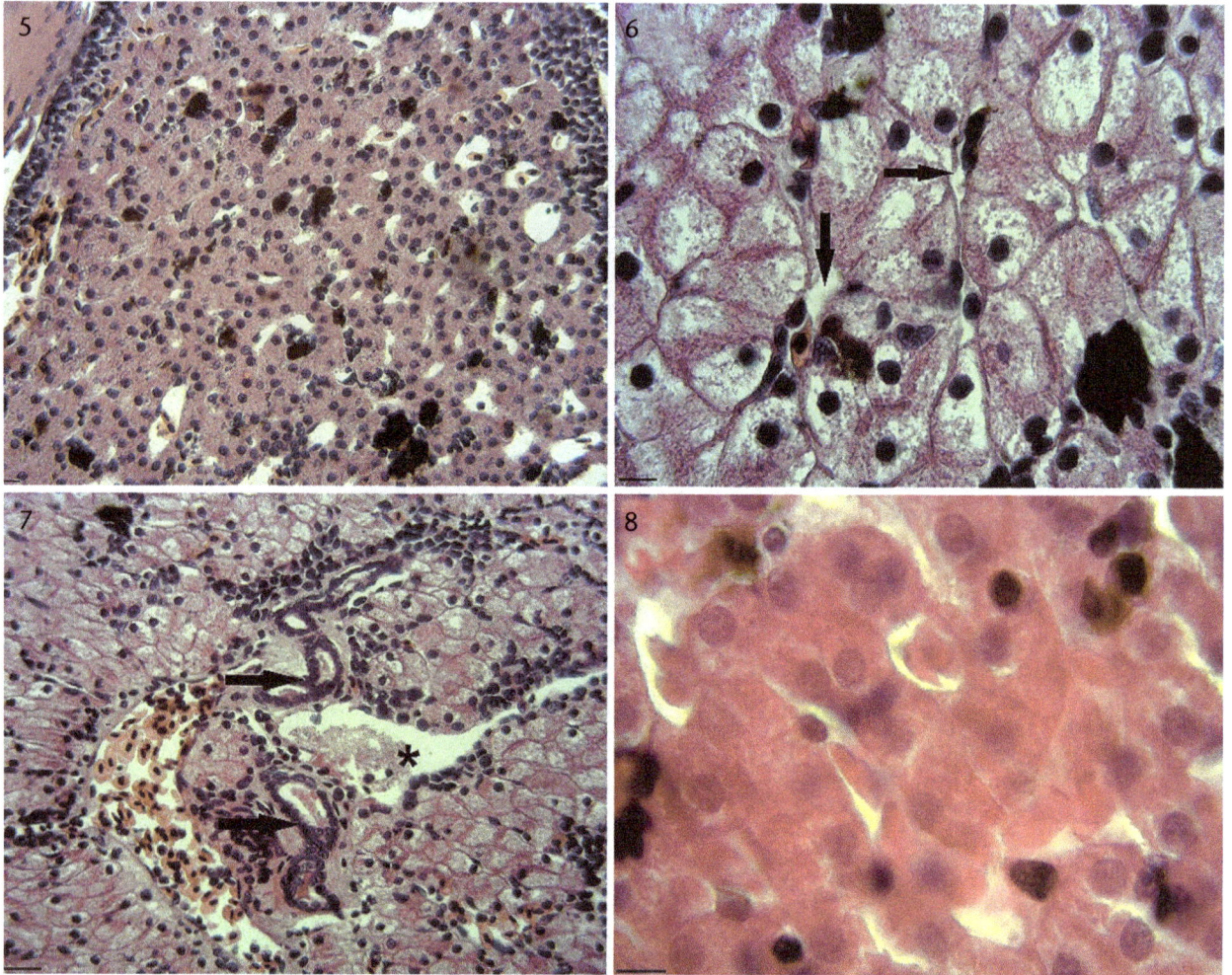

Figures 5–8. Liver of *N. microspilotus*. (5) The liver tissue demonstrates the sponge-like appearance of the parenchyma, which is composed of polyhedral hepatocyts. Numerous dark brown spots are small melanomacrophage centers (H&E, ×1000). (6) Cords of hepatocyte separated by sinusoids (arrows) containing erythrocyte. Hepatocytes are large cells with central nuclei (H&E, ×4000). (7) Central vein (*) and intrahepatic ducts (arrow) are seen in this picture (H&E, ×2500). (8) Liver parenchyma (PAS, ×1000). One of the liver's most metabolic functions is storage of glycogen. At this high magnification, one can see that the hepatocytes are strongly stained in magenta by the PAS method; this reaction reveals the presence of red granules including glycogen.

DISCUSSION

In most amphibian species, the liver is divided into right and left lobes (Grafflin 1966). However, the Taiwanese frog, *Hoplobatrachus regulosus* (Wiegmann, 1834), and Chinese Fire-bellied Newt, *Cynops orientalis* (David, 1873), have three and five lobes, respectively (Chen et al. 2003, Xie et al. 2011). In *N. microspilotus* and *N. kaiseri*, as well as in *Salamandrina* Fitzinger, 1826 (Francis 1934, Wonderly 1936), the liver is large, and only very slightly lobed. *Neurergus* is phylogenetically closer to the *C. orientalis*, they are both in the subfamily Pleurodelinae, so it is unusual that the liver of *Neurergus* resembles that of Salamandra,

which is in another subfamily, Salamandrinae. This should be discussed. However, the hepatic structure normally varies in direct relationship to gender, age, available food (especially with regard to glycogen and fat content), or temperature, and with endocrine influences strongly connected to the environmentally regulated breeding conditions.

The microscopic analysis in *N. microspilotus* and *N. kaiseri* revealed that the liver in these species is covered by a thin layer of connective tissue, forming the hepatic capsule, which according to Schaffner (1998), is common to all vertebrates. According to Ross et al. (2003), this capsule contributes to the division of the parenchyma into structural units, called hepatic

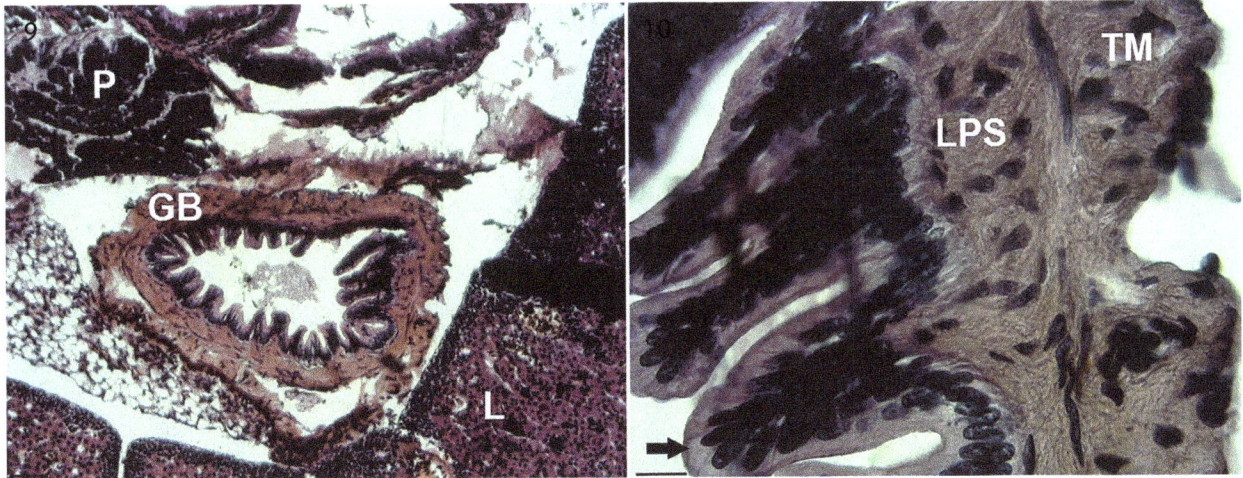

Figures 9–10. (9) Gall bladder of *N. microspilotus* (H&E, ×300). (10) Gall bladder wall consists of a simple columnar epithelium (arrow) supported by underlying fibrovascular lamina propria submucosa (LPS) (H&E, ×2500). The epithelial cells are very tall and possess elongated nuclei basally located. These lining cells consecrate bile. (GB) Gall bladder, (P) Pancreas, (TM) Tunica muscularis.

lobules. These are polygonal in shape and are separated by a thin layer of connective tissue, but the trabecules that have a greater quantity of this tissue allow visualization of the interlobular bile ducts, branches of portal vein and of hepatic artery. The central point of the liver is the hilus, through which the portal vein and the liver artery pass. Haar and Hightower (1976) and Xie et al. (2011) described that fine structural characteristics of hepatocytes in the newt *Notophthalmus viridescens* (Rafinesque, 1820) and *C. orientalis* included abundant lipid and glycogen inclusions. Melanophores with developing melanosomes are situated throughout the hepatic parenchyma. These results are similar to our observation in *N. microspilotus* and *N. kaiseri*.

In the hepatic parenchyma of *N. microspilotus* and *N. kaiseri* a large quantity of melanomacrophage centers, as indicated in the Fig. 5, is present. These, also known as macrophage aggregates, are distinctive groupings of pigment-containing cells called melanomacrophages. They are contained in the tissues of amphibians, reptiles and some fish, normally in the liver (Agius and Roberts 2003). According to Frye (1991), these cells are numerous in amphibians and reptiles, except among snakes, in which they are less plentiful (Hack and Helmy 1964). These cells have various functions, among which the synthesis of melanin, fagocytosis and neutralization of free radicals (Guida et al. 2004). The numbers of hepatic melanomacrophages in the amphibian liver are influenced by seasonal variation in some species, and increase with age and with antigenic stimulation in all species (Sichel et al. 2002). In *N. microspilotus* and *N. kaiseri* there is a gall bladder lying just dorsal to the right lobe of the liver. The gall bladder is a storage depot for bile. Its mucosa is thrown into numerous folds. The epithelial lining of the tunica muscularis is arraigned circularly.

The pancreas contains two distinct populations of cells, the exocrine cells, which secrete enzymes into the digestive tract, and the endocrine cells, which secrete hormones into the bloodstream (Slack 1995). The pancreas arises from the endoderm as a dorsal and a ventral bud, which fuse together to form the single organ. Mammals, birds, reptiles and amphibians have a pancreas with similar histology and mode of development, while in some fish, the islet cells are segregated as Brockmann bodies (Slack 1995). The pancreas in *N. microspilotus* and *N. kaiseri* are roughly triangular in shape, and lie rather to the dorsal side of the duodenum, between it and the stomach. In *N. microspilotus* and *N. kaiseri* the exocrine pancreas is a lobulated, branched, acinar gland. The secretory cells are grouped into acini and are pyramidal in shape, with basal nuclei, regular arrays of rough endoplasmic reticulum, a prominent Golgi complex and numerous secretory (zymogen) granules, containing the digestive enzymes. The lumina of the acini are small and may be terminal or intercalary. At the junction of the acini and ducts are low cuboidal centroacinar cells. The ducts proper are lined with columnar epithelial cells, and in the larger ducts are found small numbers of goblet and brush cells similar to those of the intestine. The acini and smaller ducts are invested with a delicate, loose connective tissue, which becomes more extensive around the larger ducts.

Finally, a number of infectious diseases such as Ranavirosis (Stöhr et al. 2013), Chytridiomycosis (Spitzen-van-der-Sluijs et al. 2011, Bogaerts et al. 2012, Parto et al. 2013, Sharifi et al. 2014), Red leg syndrome (Parto et al. 2014) and Rickettsial inclusions (Vaissi et al. 2017) have been recently reported in specimens belonging to *Neurergus*, both in the wild and in captivity. Internally, diseases commonly affects the liver and pancreas of amphibian (Bollinger et al. 1999, Green 2001, Wright 2006, Parto et al. 2014). The development and refinement of amphibian medicine remains

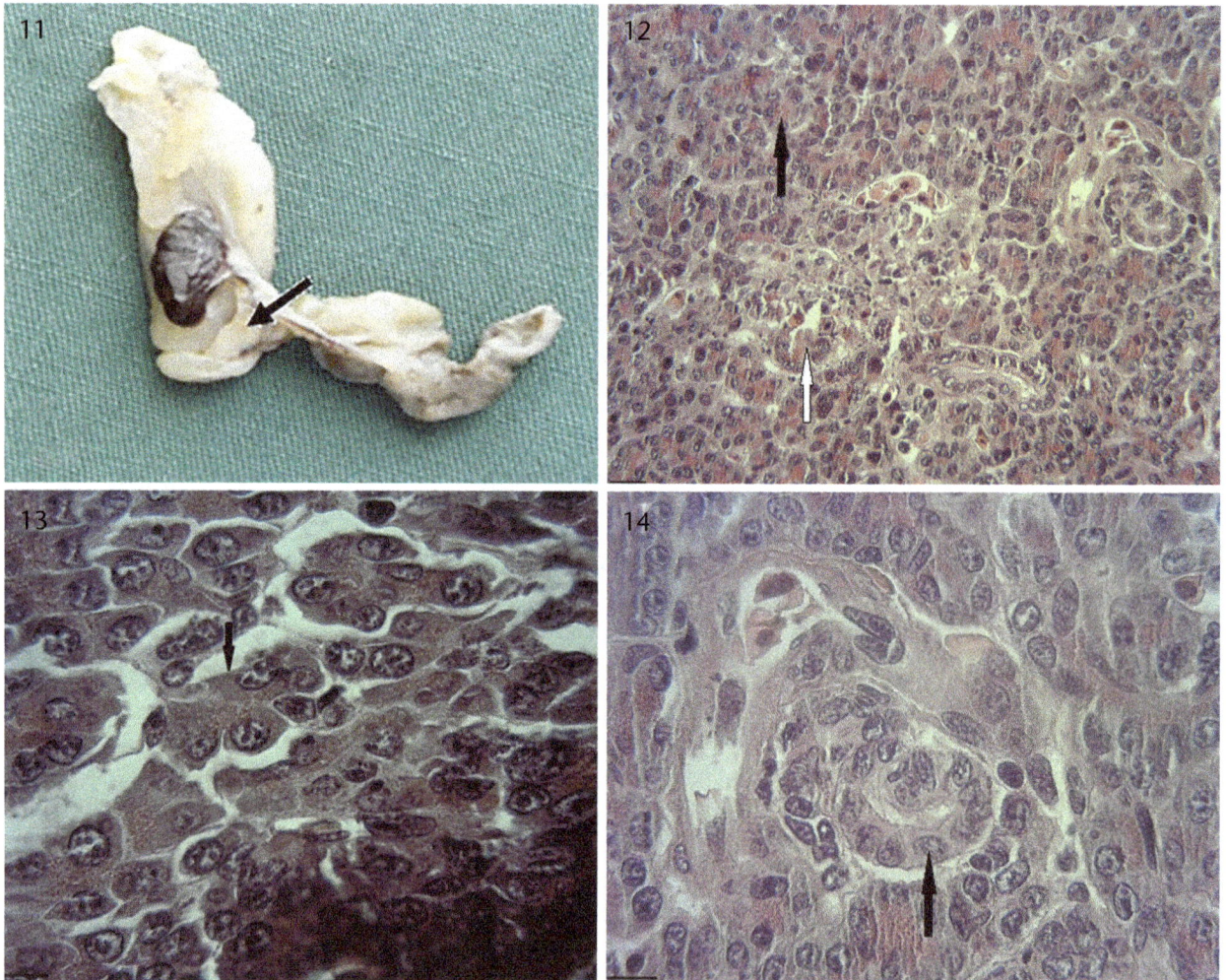

Figures 11–14. Pancreas of *N. kaiseri*. (11) Pancreas is a triangular organ and it's situated in the curvature of duodenum. (12) Its composed of numerous masses of exocrine acini (black arrow) which secret digestion enzyme. Langerhounse Island (white arrow) is also present (H&E, ×1000). (13) Acini (arrow) is enzyme secreting units of exocrine portion of pancreas. Each acinius is an ovoid elliptical cluster of pyramid-shaped secretory cells surrounding the lumen. In the apical portion of the cells these are aggregated bright eosinophilic zymogens granules. The round or flattened cell nuclei are located basally (H&E, ×4000). (14) Pancreatic acini drain into a branched system of variously sized ducts. In this Image the duct (arrow) is surrounded by a simple cuboidal epithelium (H&E, ×2500).

an ongoing practice that reflects the unique life history of these animals and our growing knowledge of amphibian diseases (Densmore and Green 2007). Also, a number of morphological studies that have been conducted might be useful in developing a conservation medicine for the Iranian newts (Sharifi et al. 2013, Parto et al. 2014). The findings of this study demonstrate that the morphological description of the digestive gland of *N. microspilotus* and *N. kaiseri* are very similar and can be extended to other newts. Results obtained in the current study are important for understanding the digestive processes, underpinning physiological, pathological

and phylogenetic studies (Akiyoshi and Inoue 2012), and for the management and conservation, including preventive and therapeutic medicine, of these animals.

REFERENCES

Afroosheh M, Akmali V, Esmaili S, Sharifi M (2016) Distribution and abundance of the endangered yellow spotted mountain newt *Neurergus microspilotus* (caudata: salamandridae) in western Iran. Herpetological Conservation and Biology 11: 52–60.

Agius C, Robert RJ (2003) Melanomacrophage centers and their role in fish pathology. Journal of Fish Diseases 26: 499–509. https://doi.org/10.1046/j.1365-2761.2003.00485.x

Akiyoshi H, Inoue AM (2012) Comparative histological study of hepatic architecture in the three orders amphibian livers. Comparative Hepatology 11: 2. https://doi.org/10.1186/1476-5926-11-2

Baloutch M, Kami HG (1995) Amphibians of Iran. Tehran University Publications 177: 91–99.

Bogaerts S, Janssen H, Macke J, Schultschik G, Ernst K, Maillet F, Bork C, Pasmans F, Wisniewski P (2012) Conservation biology, husbandry, and captive breeding of the endemic Anatolia newt, Neurergus strauchii Steindachner (1887) (Amphibia: Caudata: Salamandridae). Amphibian and Reptile Conservation 6: 9–29.

Bollinger TK, Mao J, Schock D, Brigham RM, Chinchar VG (1999) Pathology, isolation, and preliminary molecular characterization of a novel iridovirus from tiger salamanders in Saskatchewan. Journal of Wildlife Diseases 35: 413–429. https://doi.org/10.7589/0090-3558-35.3.413

Chen XQ, Jiang JP, Lin W (2003) Histology of the digestive gland in Hoplobatrachus rugulosus. Journal of Fujian Normal University 19: 117–20.

Densmore CL, Green DE (2007) Diseases of amphibians. ILAR Journal 48: 235–254. https://doi.org/10.1093/ilar.48.3.235

Francis ETB (1934). The anatomy of the salamander. London, Oxford, The Clarendon Press.

Frye FL (1991) Reptile care: An atlas of diseases and treatments. Neptune, TFH Publications, 324 pp.

Grafflin AL (1966) In vivo studies of hepatic structure and function in the salamander. The Anatomical Record 115: 53–61. https://doi.org/10.1002/ar.1091150105

Green DE (2001) Pathology of amphibia. In: Wright KM, Whitaker BR (Eds) Amphibian Medicine and Captive Husbandry. Malabar, Krieger Publishing Company, 401–485.

Guida G, Zanna P, Gallone A, Argenzio E, Cicero R (2004) Melanogenic response of the Kupffer cells of Rana esculenta L. to melanocyte stimulating hormone. Pigment Cell Research 17: 128–134. https://doi.org/10.1046/j.1600-0749.2003. 00118.x

Haar JL, Hightower JA (1976) A light and electron microscopic investigation of the hepatic parenchyma of the adult newt, Notophthalmus viridescens. The Anatomical Record 185: 313–323. https://doi.org/10.1002/ar.1091850305

Hack MH, Helmy FM (1964) Analysis of melanoprotein from Amphiuma liver and from a human liver melanoma. Proceedings of the Society for Experimental Biology and Medicine 116: 348. https://doi.org/10.3181/00379727-116-29244

IUCN (2011) Red list of threatened species. International Union for Conservation of Nature, version 2011.1 http://www.iucnredlist.org [Accessed: June 2013]

IUCN (2016) Neurergus kaiseri. The IUCN Red List of Threatened Species 2016: e.T59450A49436271. http://dx.doi.org/10.2305/IUCN.UK.2016-3.RLTS.T59450A49436271.en. [Accessed: 14 April 2017]

Luna LG (1968) Manual of histologic staining methods of the Armed Forces Institute of the pathology. New York, McGraw-Hill, 3rd ed.

McWilliams SR, Karasov WH (2001) Phenotypic flexibility in digestive system structure and function in migratory birds and its ecological significance. Comparative biochemistry and physiology. Molecular and Integrative Physiology, Part A, 128: 579–593.

Mobaraki A, Mohsen Amiri M, Alvandi R, Tehrani ME, Kia HZ, Khoshnamvand A, Bali A, Forozanfar E, Browne RK (2014) A conservation reassessment of the Critically Endangered, Lorestan newt Neurergus kaiseri (Schmidt, 1952) in Iran. Amphibian and Reptile Conservation 9: 16–25.

Parto P, Vaissi S, Farasat H, Sharifi M (2013) First report of Chytridiomycosis (Batrachochytrium dendrobatidis) in endangered Neurergus microspilotus in western Iran. Global Veterinaria 11: 547–551.

Parto P, Haghighi ZMS, Vaissi S, Sharifi M (2014) Microbiological and histological examinations in endangered Neurergus kaiseri tissues displaying Red-leg syndrome. Asian Herpetological Research 5: 204–208. https://doi.org/10.3724/SP.J.1245.2014.00204

Romão MF, Santos ALQ, Lima CF, Desimone SS, Silva JMM, Hirano LQ, Viera LG, Pinto JGS (2011) Anatomical and topographical description of the digestive system of Caiman crocodilus (Linnaeus 1758), Melanosuchus niger (Spix, 1825) and Paleosuchus palpebrosus (Cuvier, 1807). Journal of Morphology 29: 94-99. https://doi.org/10.4067/S0717-95022011000100016

Ross M, Kaye G, Pawlina W (2003) Histology: a text and atlas. Baltimore, Lippincott Williams and Wilkens, 4th ed., 864p.

Schaffner F (1988) The liver. In: Gans C (Ed.) Visceral organs. Philadelphia, Saunders, p. 485-531.

Secor SM (2005) Evolutionary and cellular mechanisms regulating intestinal performance of amphibians and reptiles. Integrative and Comparative Biology 45: 282–294. https://doi.org/10.1093/icb/45.2.282

Sharifi M, Assadian S (2004) Distribution and conservation status of Neurergus microspilotus (Caudata: Salamandridae) in western Iran. Asiatic Herpetological Research 10: 224–229.

Sharifi M, Vaissi S (2014) Captive breeding and trial re-introduction of the endangered yellow spotted mountain newt Neurergus microspilotus (Caudata: Salamandridae) in western Iran. Endangered Species Research 23: 159–166. https://doi.org/10.3354/esr00552

Sharifi M, Bafti S, Papenfuss T, Anderson S, Kuzmin S, Rastegar-Pouyani N (2009) Neurergus microspilotus. In: IUCN (Ed.) Red List of Threatened Species. International Union for Conservation of Nature, version 2012.2, available online at: http://www.iucnredlist.org [Accessed: 23 April 2015]

Sharifi M, Farasat H, Vaissi S, Parto P, Siavosh Haghighi ZM (2014) Prevalence of the amphibian pathogen Batrachochytrium dendrobatidis in endangered Neurergus microspilotus (Caudata: Salamandridae) in Kavat stream, western Iran. Global Veterinaria 12: 45–52.

Sharifi M, Naderi B, Hashemi R (2013) Suitability of the photographic identification method as a tool to identify the endangered yellow spotted newt, *Neurergus microspilotus* (Caudata: Salamandridae). Russian Journal of Herpetology 20: 4–264.

Sichel G, Scalia M, Corsaro C (2002) Amphipia Kupffer cells. Microscopy Research and Technique 57: 477–490. https://doi.org/10.1002/jemt.10101

Slack JMW (1995) Developmental biology of the pancreas. Development 121: 1569–1580.

Spitzen-van-der-Sluijs A, Martel A, Wombwell E, Van Rooij P, Zollinger R, Woeltjes T, Rendle M, Haesebrouck F, Pasmans F (2011) Clinically healthy amphibians in captive collections and at pet fairs: A reservoir of *Batrachochytrium dendrobatidis*. Amphibia-Reptilia 32: 419–423. https://doi.org/10.1163/017353711X579830

Stöhr AC, Fleck J, Mutschmann F, Marschang RE (2013) Ranavirus infection in a group of wild-caught Lake Urmia newts *Neurergus crocatus* imported from Iraq into Germany. Diseases of Aquatic Organisms 103: 185–189. https://doi.org/10.3354/dao02556

Vaissi S, Parto P, Haghighi ZMS, Sharifi M (2017) Intraerythrocytic rickettsial inclusions in endangered Kaiser's mountain newt, *Neurergus kaiseri* (Caudata: Salamandridae). Journal of Applied Animal Research 45: 505–507. https://doi.org/10.1080/09712119.2016.1220385

Vitt LJ, Caldwell JP (2009) Herpetology. New York, Elsevier.

Wonderly DE (1936) A comparative study of the cross anatomy of the digestive system of some North American salamanders. Herpetological Society 4: 31–48. https://doi.org/10.2307/1562578

Wright KM (2006) Overview of amphibian medicine, p. 941–971. In: Mader DR (Ed.) Reptile Medicine and Surgery. St. Louis, Saunders, Elsevier, 2nd ed. https://doi.org/10.1016/B0-72-169327-X/50079-1

Xie ZH, Zhong HB, Li HJ, Hou YJ (2011) The structural organization of the liver in the Chinese fire-bellied newt (*Cynops orientalis*). International Journal of Morphology 29: 1317–1320. https://doi.org/10.4067/S0717-95022011000400041

Author Contributions: SV and MSH collected the newts; SV and PP designed the experiments; SV conducted the experiments; SV and PP described the anatomy and histology of the specimens; SV and MSH wrote the paper.

Competing Interests: The authors have declared that no competing interests exist.

14

Feeding behavior by hummingbirds (Aves: Trochilidae) in artificial food patches in an Atlantic Forest remnant in Southeastern Brazil

Lucas L. Lanna[1], Cristiano S. de Azevedo[1], Ricardo M. Claudino[1], Reisla Oliveira[1], Yasmine Antonini[1]

[1]Instituto de Ciências Exatas e Biológicas, Universidade Federal de Ouro Preto. Campus Morro do Cruzeiro, Bauxita, 35400-000 Ouro Preto, MG, Brazil.
Corresponding author: Cristiano S. de Azevedo (cristianoroxette@yahoo.com)

http://zoobank.org/8AE55EE4-4553-46C1-9152-7828A6816F38

ABSTRACT. During flight, hummingbirds achieve the maximum aerobic metabolism rates within vertebrates. To meet such demands, these birds have to take in as much energy as possible, using strategies such as selecting the best food resources and adopting behaviors that allow the greatest energy gains. We tested whether hummingbirds choose sources that have higher sugar concentrations, and investigated their behaviors near and at food resources. The study was conducted at Atlantic forest remnant in Brazil, between June and December 2012. Four patches were provided with artificial feeders, containing sucrose solutions at concentrations of 5%, 15%, 25% and 35% weight/volume. Hummingbird behaviors were recorded using the ad libitum method with continuous recording of behaviors. The following species were observed: the Brazilian ruby *Clytolaema rubricauda* (Boddaert, 1783), Violet-capped woodnymph *Thalurania glaucopis* (Gmelin, 1788), Scale-throated hermit *Phaethornis eurynome* (Lesson, 1832), White-throated hummingbird *Leucochloris albicollis* (Vieillot, 1818), Versicoloured emerald *Amazilia versicolor* (Vieillot, 1818), Glittering-bellied emerald *Chlorostilbon lucidus* (Shaw, 1812) and other *Phaethornis* spp. *C. rubricauda*, *P. eurynome* and *Phaethornis* spp. visited the 35%-sucrose feeders more often, while the *T. glaucopis* visited the 25%-sucrose feeders more often. *L. albicollis* and *A. versicolor* visited more often solutions with sugar concentration of 15%. *C. lucidus* visited all patches equally. Three behavioral strategies were observed: 1) *C. rubricauda* and *T. glaucopis* exhibited interspecific and intraspecific dominance; 2) the remaining species exhibited subordinance to the dominant hummingbirds, and 3) *P. eurynome* and *Phaethornis* spp. adopted a hide-and-wait strategy to the dominant hummingbird species. The frequency of aggressive behaviors was correlated with the time the hummingbird spent feeding, and bird size. Our results showed that hummingbirds can adopt different strategies to enhance food acquisition; that more aggressive species feeding more than less aggressive species; and that the birds, especially if they were dominant species, visited high quality food resources more often.

KEY WORDS. Behavioral strategies, dominance, food resources, subordination, trapline.

INTRODUCTION

Hummingbirds are specialized birds that consume predominantly nectar, but can also consume small arthropods (Cotton 2007). Individual hummingbirds have been recorded foraging in more than 200 flowers per day in a single plant (Snow and Snow 1986, Sick 1997, Ortiz-Pulido et al. 2012). They reach the highest aerobic metabolic rates among vertebrates during flight, which explains these voracious appetites (Suarez et al. 1990).

The net energy concept postulates that energetic costs during foraging must be lower than energy intake during foraging (Heinrich 1975). In order to obtain the necessary amount of energy, hummingbirds can select and protect the richest food patches available at an area (Loss and Silva 2005). Normally, sugar concentrations in the nectar of the flowers visited hummingbirds vary between 20–25% (Roberts 1996), and experimental manipulations revealed that hummingbirds prefer sugar concentrations higher than 35% (Tamm and Gass 1986, Roberts 1996, López-Calleja et al. 1997). However, this preference was observed only when the nectar was collected during repeated licking (Kingsolver and Daniel 1983). Thus, in experiments with artificial flowers (hummingbird feeders), it is expected that hummingbirds forage more in feeders that had higher sucrose concentration, since energy intake is greater in these feeders and the volume of nectar is large enough to permit repeated licking cycles.

Three behavioral strategies can be adopted by hummingbirds when foraging: (A) dominance/territoriality, when an individual defends a territory containing food resources and excludes competitors from the resource (Feisinger 1976, Feisinger and Colwell 1978, Stiles 1978, Cotton 1998), and (B) intruder/subordinance, when an individual forages in defended patches until it is expelled by the territorial hummingbird (Feisinger and Colwell 1978, Stiles 1978, Barbosa-Filho and Araújo 2013). A third strategy is known as trapline foraging (C), when an individual repeatedly visits a set of plants in routes through different patches, exploiting resources without displaying any territorial behavior (Feisinger and Colwell 1978, Rios et al. 2010, Tello-Ramos et al. 2015). A trapliner hummingbird can either be expelled by territorialists when foraging in one food resource, or it can simply ignore the presence of the territorial individuals by moving across different territories (Feisinger and Colwell 1978, Garrisson and Gass 1999). Thus, a trapliner individual also can eventually act as a subordinate one, performing strategy B, yet an individual that acts according to strategy B may not necessarily adopt a traplining strategy.

Typical behaviors exhibited by territorial hummingbirds are "perching near the food resource" (Loss and Silva 2005, Longo and Fischer 2006), inter- and intraspecific attacks (Loss and Silva 2005), and intense vocalizations and visual displays (Mendonça and dos Anjos 2006). Since the energetic cost of defending a territory can be up to three times higher than the cost of non-aggressively foraging (Gill and Wolf 1975), it is expected that individuals will engage in resource defense only when the fitness benefits of territoriality outweigh its costs (Brown 1964). Territorial individuals commonly have access to more food than subordinate ones (Justino 2009, Rios et al. 2010), and the intensity of the defense should increase with the quality of the defended resource (Justino et al. 2012). Hummingbirds, for example, defend clumped flowers rich in nectar more aggressively than scattered flowers (Temeles et al. 2005). Moreover, body size affects territoriality in hummingbirds, with medium to larger size species exhibiting more territoriality than smaller ones (Feisinger and Colwell 1978, Abrahamczyk and Kessler 2014).

The aim of this study was to evaluate the feeding behavior of hummingbirds in artificial food patches at an Atlantic Forest fragment in Brazil. We described the behavior exhibited by the hummingbirds in the food patches and identified the sugar concentration most visited by the birds. We hypothesized that: 1) most feeding visits would be to the feeders that have 35% sugar-concentration, since this food resource provides more energy to the birds (Tamm and Gass 1986, Roberts 1996); 2) larger and heavier hummingbirds will defend the feeders from smaller and lighter birds, since body size determines dominance in hummingbirds (Antunes 2003, Araújo-Silva and Bessa 2010); 3) smaller hummingbirds will exhibit subordination behaviors

and bigger hummingbirds will exhibit dominance behaviors; and (4) the frequency of aggression behaviors exhibited by the hummingbirds will be correlated with the time spent feeding on the feeders, since dominant birds will have more access to the feeders due to their greater aggressiveness.

MATERIAL AND METHODS

The study was conducted in the Itacolomi State Park, in the city of Ouro Preto, Minas Gerais, southeastern Brazil (20°23′S, 43°30′W), in an Atlantic Forest remnant, from June to December 2012. Four artificial food patches, distant linearly 1.5-2.5 m from each other in a 6 m² area, were constructed in the core of the forest fragment (70m distant from the forest edges), each containing five artificial hummingbird feeders (200 ml plastic feeders Mr Pet®, with three red plastic flowers with short corollas) filled with a sugar-water solution of 5, 15, 25 or 35%. Nectar sugar concentration was computed diluting commercial sucrose in filtered water; e.g., in 35% sugar concentration, 350 g of sucrose was diluted in 750 ml of filtered water. Each food patch contained feeders filled with only one concentration, which remained available all day long. The solution was replaced each morning after the feeders were cleaned.

Behavioral recording sessions occurred continuously from 07:00 to 10:00 a.m. and from 02:00 to 05:00 p.m. each day, totaling 325 hours of observation (54 non-consecutive days in total). The birds were observed at a distance of 10m, using a 10x50 (Nikon TX Extreme) binocular. Hummingbirds were observed ad libitum with continuous recording of behaviors (Altmann 1974), and the birds were identified according to Sigrist (2009). We focused on both territorial and subordinate hummingbirds, recording all behaviors during the entire observation period. Since no hummingbirds were captured and marked, to avoid pseudoreplication, we only collected data on the same hummingbird species after intervals of 30 minutes, counting from the time the individual of that species left, or when two or more individuals of the same species were feeding at the same time on the food patch (we recorded behaviors of more than 10 ± 6 individuals feeding at the same time on the food patch). We evaluated the time spent in each food patch and the behavior exhibited by the hummingbirds in each food patch. An ethogram was built based on 100h of pilot observations and information from the scientific literature (Barçante and Mahecha 2004, Loss and Silva 2005, Toledo and Moreira 2008, Araújo-Silva and Bessa 2010) (Table 1).

Friedman's non-parametric ANOVA with Dunn's post-hoc tests were used to evaluate if the hummingbird species differed in the frequency of the behaviors they exhibited. Both frequency of behaviors and the time spent in the behavior were used in the analyses of the "feeding", "alert" and "vocalizing" behaviors; only the number of observations was used in the analysis of the other behaviors evaluated. Time spent feeding was used to evaluate sugar concentrations most visited by each

Table 1. Ethogram for the hummingbirds recorded at the Itacolomi State Park, Ouro Preto, Minas Gerais, Brazil.

Abbreviation	Behavior	Description
FEE	Feeding	Hummingbird feeds in the artificial feeder, hovering or perched.
EXP	Expelling*	Hummingbird 1 expels hummingbird 2, pursuing it for long or small distances.
FLE	Fleeing**	Hummingbird 1 flees from hummingbird 2, who expelled it.
FIG	Fighting*	Hummingbirds fight using their beak.
FRI	Frightening*	Hummingbird 1 frightens hummingbird 2 simply due to its appearance in the area.
FRIED	Frightened**	Hummingbird 2 was frightened by hummingbird 1 simply due to its appearance in the area.
EXA	Expel attempt*	Hummingbird 1 tries to expel hummingbird 2, but hummingbird 2 continues to feed without caring about the presence of hummingbird 1.
IMP	Impassive	Hummingbird 2 behaves normally when hummingbird 1 tries to expel it from the feeders.
PRS	Persecution with only one individual identified*	Hummingbird 1 pursue hummingbird 2, but only one individual is identified.
AL	Alert*	Hummingbird perched, observing the food patches.
VOC	Vocalizing*	Hummingbird vocalizes in or near the food patches.

*Aggressive behaviors, **Subordinate behaviors.

Table 2. Average lengths (with tails and bills included) and weights of the hummingbird species recorded at Itacolomi State Park, Ouro Preto, according to Sick (1997) and Dunning-Junior (2008).

Species	Mean length (cm)	Mean weight (g)	Social status in this study
Amazilia versicolor	8.5	4.1	Subordinate
Chlorostilbon lucidus	8.5	2.5	Subordinate
Clytolaema rubricauda	12.0	7.9	Dominant
Leucochloris albicollis	10.5	6.3	Subordinate
Phaethornis eurynome	15.5	5.3	Subordinate
Phaethornis spp.	15.5	5.3	Subordinate
Thalurania glaucopis	11.1	4.8	Dominant

hummingbird species. The frequency of aggression behaviors recorded were summed ("expelling", "fighting", "frightening", "expel attempt", "alert" and "vocalizing") and correlated with the time spent feeding on all sugar concentrations (5-35%) and with hummingbird sizes and weights – according to Sick 1997 (Table 2) – using a Spearman's correlation test (Zar 1999). When an individual expelled or frightened another, he was considered the winner of the agonistic encounter. The individual that was expelled or frightened was considered the loser of the agonistic encounter. The sizes and weights of hummingbird species as defined by Sick (1997) were used in this study. All tests were conducted using the software Minitab v.16, at a confidence level of 95%.

RESULTS

Six species of hummingbirds of were recorded visiting the feeders during the study: Brazilian ruby *Clytolaema rubricauda* (Boddaert, 1783), Violet-capped woodnymph *Thalurania glaucopis* (Gmelin, 1788), Scale-throated hermit *Phaethornis eurynome* (Lesson, 1832), White-throated hummingbird *Leucochloris albicollis* (Vieillot, 1818), Versicoloured emerald *Amazilia versicolor* (Vieillot, 1818), Glittering-bellied emerald *Chlorostilbon lucidus* (Shaw, 1812) and an unidentified species of *Phaethornis* spp. [four species of *Phaethornis* Swainson, 1827 occur in the Itacolomi State Park: *P. eurynome*, *P. pretrei* (Lesson & Delattre, 1839), *P. squalidus* (Temminck, 1822) and *P. ruber* (Linnaeus, 1758)

(Ribon 2006); data of more than one *Phaethornis* species could be computed in the results of *Phaethornis* spp., excluding *P. eurynome*, which results were analyzed separately].

The most-visited feeders were those containing a solution of 35% sugar, and the least-visited feeders were those containing a solution of 5% sugar; the frequency of visitations differed between the feeders (F = 177.380, d.f. = 3, p < 0.001, n = 94).

Clytolaema rubricauda, *P. eurynome* and *Phaethornis* spp. visited the feeders with sugar solution of 35% more often (*C. rubricauda*: F = 164.5, d.f. = 3, p < 0.001, n = 4568; *P. eurynome*: F = 41.7, d.f. = 3, P < 0.001, n = 409; *Phaethornis* sp.: F = 9.6, d.f. = 3, p = 0.023, n = 71). *Thalurania glaucopis* visited more often the feeders containing a sugar solution of 25% (F = 154.6, d.f. = 3, p < 0.001, n = 5992). *L. albicollis* and *A. versicolor* visited the feeders containing a sugar solution of 15% more often (*L. albicollis*: F = 28.81, d.f. = 3, p < 0.001, n = 97; *A. versicolor*: F = 19.93, d.f. = 3, p < 0.001, n = 15). *Chlorostilbon lucidus* was the species that less frequently visited the food patches, and no differences were found between the number of visits in each food patch (F = 3.67, d.f. = 3, p = 0.34, n = 3, Fig. 1).

Clytolaema rubricauda won most of the aggressive encounters with other hummingbird species, both in total and in each different sugar solution concentrations, followed by *T. glaucopis* (Table 3). No other species won aggressive encounters (Table 3).

Aggressive behaviors ("expelling", "fighting", and "expel attempt") were exhibited by *C. rubricauda*, *T. glaucopis*, *P. eurynome* and *L. albicollis*. Among the aggressive behaviors, *P. eurynome* and *L. albicollis* exhibited only "expel attempts" against other hummingbirds (Table 4). The other species did not exhibit aggressive behaviors, but displayed subordinate behaviors (Table 4).

Clytolaema rubricauda and *T. glaucopis* behaved similarly, being the most aggressive species observed (Table 4). The behavior "frightened" differed between these species, with *T. glaucopis* being frightened more often (Table 4). The behaviors "expelling" and "fighting" were only exhibited by these species, and *C. rubricauda* expelled more and fought less than *T. glaucopis* (Table 4). *Amazilia versicolor*, *C. lucidus*, *L. albicollis*, *P. eurynome*

Table 3. The outcomes of aggressive encounters between different hummingbird species in relation to different sugar solution.

Winners	Losers						
	Aggressive winning percentages between species						
	A. versicolor	C. lucidus	C. rubricauda	L. albicollis	P. eurynome	Phaethornis sp.	T. glaucopis
C. rubricauda	–	–	–	100.00% (12)	100.00% (44)	100.00% (9)	87.66% (334)
T. glaucopis	100.00% (1)	100.00% (1)	12.34% (47)	100.00% (20)	100.00% (43)	100.00% (10)	–
	Aggressive winning percentages between species in 5% patch						
	A. versicolor	C. lucidus	C. rubricauda	L. albicollis	P. eurynome	Phaethornis sp.	T. glaucopis
C. rubricauda	–	–	–	–	–	–	100.00% (7)
T. glaucopis	100.00% (1)	–	–	–	–	–	–
	Aggressive winning percentages between species in 15% patch						
	A. versicolor	C. lucidus	C. rubricauda	L. albicollis	P. eurynome	Phaethornis sp.	T. glaucopis
C. rubricauda	–	–	–	100.00% (2)	100.00% (3)	–	89.39% (59)
T. glaucopis	–	–	10.61% (7)	100.00% (5)	100.00% (4)	100.00% (1)	–
	Aggressive winning percentages between species in 25% patch						
	A. versicolor	C. lucidus	C. rubricauda	L. albicollis	P. eurynome	Phaethornis sp.	T. glaucopis
C. rubricauda	–	–	–	100.00% (2)	100.00% (9)	100.00% (3)	87.59% (120)
T. glaucopis	–	100.00% (1)	12.41% (17)	100.00% (5)	100.00% (3)	100.00% (3)	–
	Aggressive winning percentages between species in 35% patch						
	A. versicolor	C. lucidus	C. rubricauda	L. albicollis	P. eurynome	Phaethornis sp.	T. glaucopis
C. rubricauda	–	–	–	100.00% (3)	100.00% (27)	100.00% (6)	88.55% (116)
T. glaucopis	–	–	11.45% (15)	100.00% (7)	100.00% (35)	100.00% (4)	–

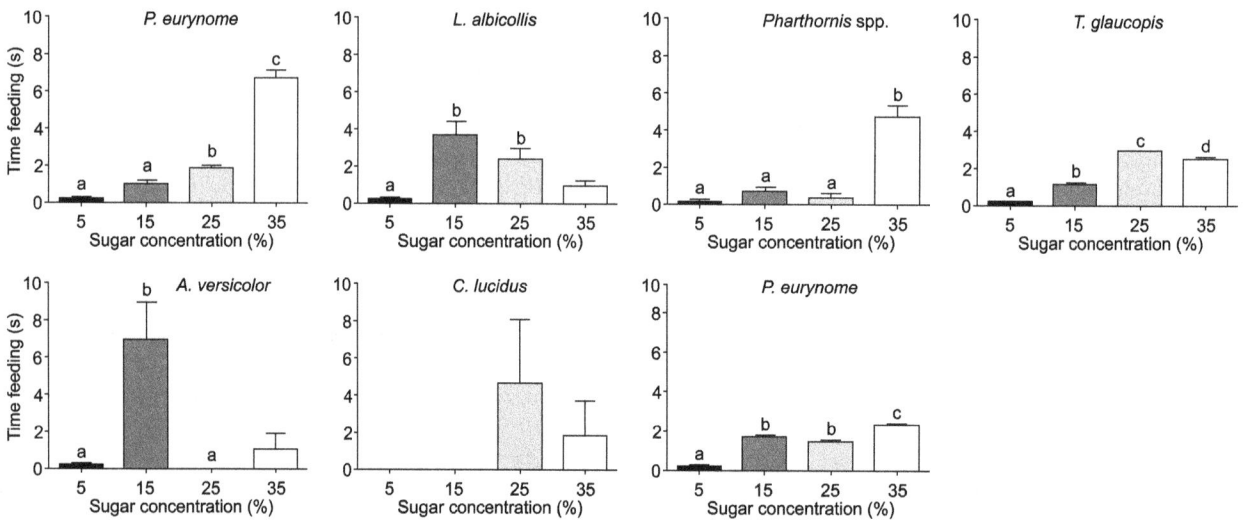

Figure 1. Sugar concentrations most visited by the recorded hummingbird species based on time spent foraging (mean duration of time spent foraging ± SD). Different letters represents statistical differences between sugar solutions as per the results of Dunn's post hoc tests (P-value < 0.05).

and *Phaethornis* spp. also behaved similarly, but these species expressed more subordinate than aggressive behaviors (Table 4).

Thalurania glaucopis got involved in the greatest number of pursuits in which only one bird was identified, and they also expressed more "expel attempts" (Table 4). *Clytolaema rubricauda* stood alert, vocalized, fought and expelled more times than any other species. Furthermore, it stood impassive when faced with

the expel attempts of *T. glaucopis* (Table 4) more often than the other species. *Thalurania glaucopis* fed the most, both in terms of time spent feeding and frequency of feeding, followed by *C. rubricauda* and *P. eurynome*, (Tables 3, 4).

Time spent feeding was positively correlated with the expression of aggressive behaviors (r = 0.86, p < 0.0001) (Fig. 2). Two of the biggest hummingbirds, *C. rubricauda* and *T. glaucopis*,

Table 4. Behaviors (mean number of total recordings ± standard error) exhibited by the six hummingbird species visiting artificial flowers in an Atlantic Forest remnant of Brazil, from June to December 2012.

Behaviors	Species							F	P-value
	A. versicolor	C. lucidus	C. rubricauda	L. albicollis	P. eurynome	Phaethornis spp.	T. glaucopis		
FEE	0.16 ± 0.06 [a]	0.03 ± 0.02 [a]	48.45 ± 2.02 [b]	1.05 ± 0.27 [a]	4.44 ± 0.56 [c]	0.76 ± 0.16 [ac]	63.67 ± 3.71 [b]	383.09	< 0.001
EXP	–	–	16.49 ± 1.65 [a]	–	–	–	7.62 ± 0.70 [a]	325.66	< 0.001
FLE	0.02 ± 0.02 [a]	0.03 ± 0.02 [a]	9.60 ± 0.02 [b]	0.67 ± 0.19 [a]	1.26 ± 0.19 [a]	0.46 ± 0.10 [a]	12.13 ± 0.84 [b]	324.56	< 0.001
FIG	–	–	0.58 ± 0.10 [a]	–	–	–	2.04 ± 0.27 [a]	72.66	< 0.001
FRI	–	–	0.96 ± 0.13 [a]	0.01 ± 0.01 [b]	0.03 ± 0.09 [bc]	–	0.51 ± 0.10 [ac]	59.06	< 0.001
FRIED	0.02 ± 0.02 [a]	–	0.32 ± 0.08 [a]	0.32 ± 0.02 [a]	0.53 ± 0.17 [a]	0.32 ± 0.02 [a]	0.87 ± 0.12 [b]	47.68	< 0.001
EXA	–	–	0.03 ± 0.02 [ab]	0.01 ± 0.01 [a]	0.02 ± 0.02 [ab]	–	0.39 ± 0.07 [b]	17.00	0.01
IMP	–	–	0.36 ± 0.07 [a]	–	0.01 ± 0.01 [b]	–	0.11 ± 0.04 [ab]	17.19	0.01
PRS	–	0.01 ± 0.01 [a]	1.56 ± 0.33 [b]	–	0.01 ± 0.01 [a]	–	1.88 ± 0.30 [b]	77.11	< 0.001
AL	0.02 ± 0.02 [a]	0.03 ± 0.02 [a]	24.87 ± 1.84 [b]	0.34 ± 0.11 [a]	0.10 ± 0.05 [a]	0.02 ± 0.02 [a]	18.40 ± 1.31 [b]	341.38	< 0.001
VOC	0.10 ± 0.06 [a]	0.11 ± 0.04 [a]	2.44 ± 0.29 [b]	0.22 ± 0.06 [a]	0.14 ± 0.04 [a]	–	1.67 ± 0.28 [b]	138.76	< 0.001

F = Friedman's test; N = 326; df = 6. Superscript letters: Different letters mean statistical differences according to the Tukey's post hoc test. Behaviors: FEE = feeding; EXP = expelling; FLE = fleeing; FIG = fighting; FRI = frightening; FRIED = frightened; EXA = expel attempt; IMP = impassive; PRS = persecution with only one individual identified; AL = alert; VOC = vocalizing.

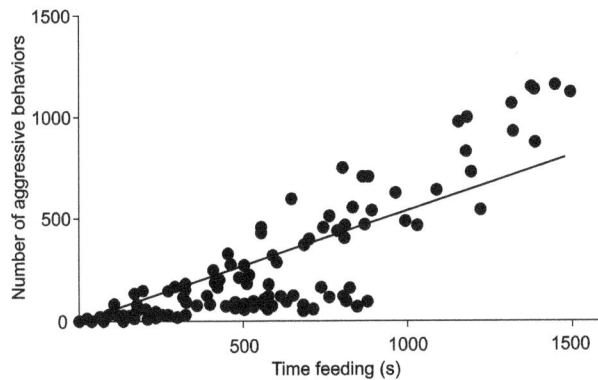

Figure 2. Positive correlation between the time spent feeding and the exhibition of aggressive behaviors by the hummingbirds. For this analysis, we included the time spent feeding by all hummingbird species in all sugar concentrations, and we summed the aggressive behaviors "expelling", "fighting", "frightening", "expel attempt", "alert" and "vocalizing".

expressed more aggressive behaviors than smaller species (r = 0.24, p < 0.05), but they also expressed more submission behaviors than smaller ones (r = 0.17, p < 0.0001) (Figs 3–4). The same results were found for hummingbird weight: two of the heaviest hummingbirds, C. rubricauda and T. glaucopis, expressed more aggressive (r = 0.39, p < 0.001), as well as submission (r = 0.17, p < 0.001) behaviors than lighter species (Figs 5–6).

DISCUSSION

Clytolaema rubricauda and T. glaucopis were considered dominant hummingbirds in this study since they presented the greatest feeding frequencies and were the most aggressive.

Besides, since they expressed the behaviors "alert", "expelling" and "vocalizing" more often than other species, they were also considered territorial. All other hummingbird species in our data fed less and expressed fewer aggressive behaviors, and were therefore considered subordinates.

Thalurania glaucopis and C. rubricauda also exhibited the "fleeing" behavior more often, due to the great number of pursuits within and between individuals of these two species. The frequency of aggressive behaviors exhibited by A. versicolor, L. albicollis, P. eurynome and Phaethornis spp. was statistically lower than by T. glaucopis and C. rubricauda; the former species were expelled from the feeders by the latter dominant species, therefore feeding less in the artificial patches. Dominant species limit the access of subordinate species to food sources (Stiles 1978, Roussau et al. 2014) when defending a territory, but only if the energy gain is higher than the energy loss (Heinrich 1975). The artificial food patches created in this study, especially those with 35% sugar concentration, had enough energy to warrant the expression of territorial behaviors. These behaviors allowed the dominant species to have more access to food resources than the subordinate species. More aggressive and territorial hummingbirds spent more time feeding in the artificial flowers in this study than the less aggressive and subordinate species, confirming our hypothesis.

One of the factors determining aggressive behavior in hummingbirds is body size; the bigger and heavier the hummingbird is, the more dominant it is (Antunes 2003, Loss and Silva 2005, Mendonça and dos Anjos 2006, Rodrigues et al. 2009). In this study, small and light hummingbird species were expelled from the food patches by the bigger and heavier species, except for Phaethornis spp. and P. eurynome (Sick 1997), which were expelled the most from the food patches, even though they are the biggest recorded in the study area.

Justino (2009) and Rios et al. (2010) found that individuals of Phaethornis explored less the food resources they were studying

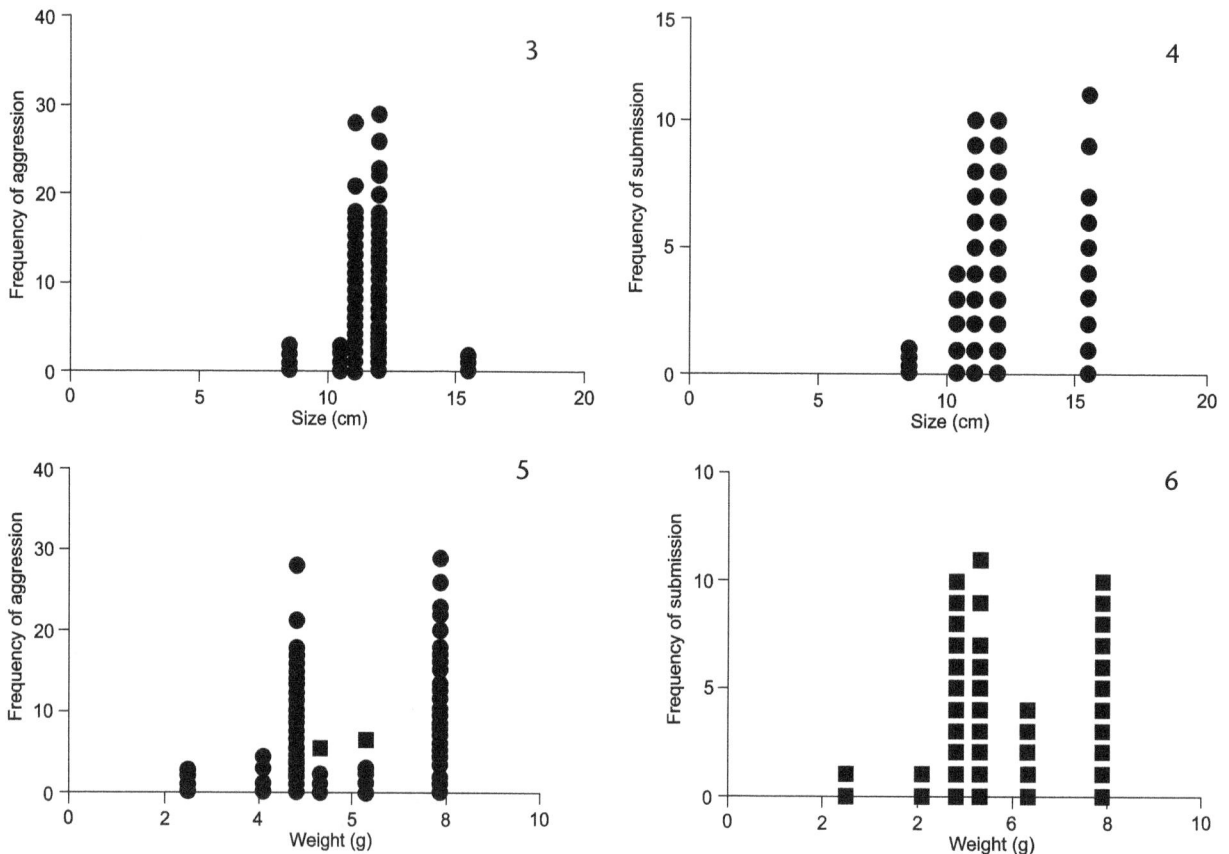

Figures 3–6. Positive correlation between size (cm) and the exhibition of aggressive (3) and submissive (4) behaviors by the hummingbirds. Positive correlation between weight (g) and the exhibition of aggressive (5) and submissive (6) behaviors by the hummingbirds. For both analysis, we included aggressive and submissive behaviors exhibited by all hummingbird species in all sugar concentrations, and we summed the aggressive behaviors "expelling", "fighting", "frightening", "expel attempt", "alert" and "vocalizing", and the submissive behaviors "fleeing" and "frightened".

than other hummingbird genera. This genus is known to use the trapline strategy for food acquisition, where routes are followed with no defined territories (Feisinger and Colwell 1978, Gill 1988, Garrisson and Gass 1999 Temeles et al. 2006, Rios et al. 2010). In this study, *Phaethornis* hummingbirds avoided confrontations with territorial species, hiding whenever dominant individuals arrived in the area. This could explain the low number of species of this genus recorded feeding in this and other studies, even though they are bigger than the dominant species recorded here. It is interesting to observe that even in our limited sample size, the Phaetornithinae in this study showed a preference for the richest sugar solution (35%); their large size probably provided some protection against the dominant species *C. rubricauda* and *T. glaucopis*. They were recorded being frightened many times by the dominant species, but instead of flying away, they hid in the shrubs, remained quiet, and returned to the feeder soon after the dominant species left the area. To the best of our knowledge, this

hide-and-wait strategy had never being recorded for Phaethornithinae before. Thus, our hypothesis of bigger and heavier hummingbirds being the most aggressive were partially corroborated, since not only size seemed to be related to aggressive behavior and time spent feeding, but also the behavioral strategy adopted by the hummingbird, with bigger territorial hummingbirds and bigger trapliners feeding more than smaller submissive ones.

Hummingbirds often show food preferences (Heinrich 1975, Loss and Silva 2005), but not all species do. In this study, five out of seven species fed more on the 25-35% sucrose feeders, showing that hummingbirds prefer the most energy-dense solutions. Barçante and Mahecha (2004) analyzed the interactions between two species in areas with food resources available, and found that the dominant species selected the resources according to their quantity and quality, while the subordinate species chose resources according to the presence or absence of the dominant species, preferring less rich resources that were not guarded. Stiles

(1978) described the same behavior, i.e., subordinate species wait for the dominant to leave the area before feeding, or they sneak in to feed until they are expelled by the dominants; such strategies were also observed in the present study. The fact that *C. lucidus* did not have a preference for a certain sugar concentration, and the great amount of time *A. versicolor* and *L. albicollis* spent feeding at the 15% sugar concentration feeders were probably due to the influence of the dominants, which forced the subordinates to feed on less concentrated sugar solutions or to wait until the dominants were absent before feeding on the richest sucrose feeders. The subordinate species would probably show a preference for the more profitable resource (35% sucrose feeders) if the dominant species were absent, but this hypothesis needs to be tested. The influence of the dominant hummingbirds on the feeding behavior of the submissive hummingbirds was demonstrated by Pimm et al. (1985). Thus, our hypothesis of preference for the richest sugar solution feeders was corroborated, but dominance and subordination of each species were important in the sugar solution choices of hummingbirds.

Van-Sluys and Stotz (1995) proposed that dominant hummingbirds have difficulties defending all resources within large territories, which gives subordinate species a chance to feed there occasionally. In this study, all feeders were located in a 6 m² area. Clearly, 6 m² is not a big area, but facing a virtually infinite food resource, many individuals visited the area at the same time, which made it difficult for the dominant hummingbirds to expel all subordinates. While the dominant was chasing a subordinate, other subordinates would take advantage and feed on the momentarily available food resource, which can explain the visits of subordinate individuals. Even with so many intruders, dominant hummingbirds did not abandon their aggressive behaviors, showing that the artificial food patches were an important energy resource at that time. It is important to state that the presence of the feeders, an unlimited resource of carbohydrates, may have increased the abundance of the hummingbirds in the area, as observed by Sonne et al. (2016) in their study, and that this increase may have influenced the expression of dominant and/or subordinate behaviors by the hummingbirds; larger numbers of hummingbirds around the richest sugar concentration feeders may have led to difficulties in defending the food resource by the dominant individuals.

In conclusion, all three behavioral strategies related to food resources were recorded for this area of Atlantic Forest. *Clytolaema rubricauda* and *T. glaucopis* were the dominant species; *A. versicolor*, *C. lucidus*, and *L. albicollis* are the subordinate species, and *P. eurynome* and *Phaethornis* spp. are the species that used the trapline strategy or acted as subordinate species with an evasive strategy to avoid confrontations with the dominants. The richest sugar solutions, with 25% and 35% sugar concentration, were most visited by the dominant species and by Phaetornithinae species; subordinate species visited less rich food patches. Finally, aggression was directly linked to the time that a hummingbird spent feeding; the more aggressive it was, the more it fed.

ACKNOWLEDGEMENTS

We thank the staff of the Itacolomi State Park for the permission to use their facilities during the study. We would also like to thank UFOP for providing scholarship and logistic support to L. Lanna. CNPq and CAPES provided scholarship support to Y. Antonini (CNPq 306840/2015-4) and R. Oliveira (CAPES/PNPD-1432299), respectively. Finally, we thank R.J. Young for invaluable suggestions on this paper. We appreciate the improvements in English usage made by R. Cramer through the Association of Field Ornithologists' program of editorial assistance.

REFERENCES

Abrahamczyk S, Kessler M (2014) Morphological and behavioural adaptations to feed on nectar: how feeding ecology determines the diversity and composition of hummingbird assemblages. Journal of Ornithology 156: 333–347. https://doi.org/10.1007/s10336-014-1146-5

Altmann J (1974) Observational study of behavior: sampling methods. Behavior 49: 227–267. https://doi.org/10.1163/1568539-74X00534

Antunes AZ (2003) Partilha de néctar de *Eucalyptus* spp., territorialidade e hierarquia de dominância em beija-flores (Aves: Trochilidae) no sudeste do Brasil. Ararajuba 11: 39–44.

Araújo-Silva LE, Bessa E (2010) Territorial behavior and dominance hierarchy of *Anthracothorax nigricolis* Vieillot, 1817 (Aves: Trochilidae) on food resources. Revista Brasileira de Ornitologia 18: 89–96.

Barbosa-Filho WG, Araújo AC (2013) Flowers visited by hummingbirds in an urban Cerrado fragment, Mato Grosso do Sul, Brazil. Biota Neotropica 13: 21–27. https://doi.org/10.1590/S1676-06032013000400001

Barçante L, Mahecha GAB (2004) Efeitos da disponibilidade de recursos alimentares sobre as respostas comportamentais de *Amazilia lactea* e de *Eupetomena macroura* (Apodiformes: Trochilidae). Bios 12: 69–70.

Brown JL (1964) The evolution of diversity in avian territorial systems. Wilson Bulletin 76: 160–169.

Cotton PA (1998) Temporal portioning of a floral resource by territorial hummingbirds. Ibis 140: 647–653. https://doi.org/10.1111/j.1474-919X.1998.tb04710.x

Cotton PA (2007) Seasonal resource tracking by amazonian hummingbirds. Ibis 149: 135–142. https://doi.org/10.1111/j.1474-919X.2006.00619.x

Dunning-Junior JB (2008) CRC Handbook of Avian Body Masses. Boca Raton, CRC Press, 2nd ed., 672p.

Feisinger P (1976) Organization of a tropical guild of nectarivorous birds. EcologicalMonographs 46: 257–291. https://doi.org/10.2307/1942255

Feisinger P, Colwell RK (1978) Community organization among Neotropical nectar-feeding birds. American Zoologist 18: 779–795. https://doi.org/10.1093/icb/18.4.779

Garrisson JSE, Gass CL (1999) Response of a traplining hummingbird to changes in nectar availability. Behavioral Ecology 10: 714-725. https://doi.org/10.1093/beheco/10.6.714

Gill FB (1988) Trapline foraging by Hermit hummingbirds: competition for and undefended, renewable resource. Ecology 69: 1933–1942. https://doi.org/10.2307/1941170

Gill FB, Wolf LL (1975) Economics of feeding territoriality in the golden-winged sunbird. Ecology 56: 333-345. https://doi.org/10.2307/1934964

Heinrich B (1975) Energetics of pollination. Annual Review of Ecology and Systematics 6: 139–170. https://doi.org/10.1146/annurev.es.06.110175.001035

Justino DA (2009) Distribuição e disponibilidade de recursos florais e estratégias de forrageamento na interação entre beija-flores e Palicourea rígida (Rubiaceae). Uberlândia, Master's Dissertation, Universidade Federal de Uberlândia. https://repositorio.ufu.br/handle/123456789/13322 [Accessed: 12/08/2016]

Justino DA, Maruyama PK, Oliveira PE (2012) Floral resource availability and hummingbird territorial behavior on a Neotropical savanna shrub. Journal of Ornithology 153: 189–197. https://doi.org/10.1007/s10336-011-0726-x

Kingsolver JG, Daniel TL (1983) Mechanical determinants of nectar feeding strategy in hummingbirds: energetic, tongue morphology, and licking behavior. Oecologia 60: 214–226. https://doi.org/10.1007/BF00379523

Longo JM, Fischer E (2006) Efeito da taxa de secreção de néctar sobre a polinização e a reprodução de sementes em flores de Passiflora speciosa Gardn. (Passifloriaceae) no Pantanal. Revista Brasileira de Botânica 29: 481–488. https://doi.org/10.1590/S0100-84042006000300015

López-Calleja MV, Bozinovic F, del Río CM (1997) Effects of sugar concentration on hummingbird feeding and energy use. Comparative Biochemistry and Physiology 118: 1291–1299. https://doi.org/10.1016/S0300-9629(97)00243-0

Loss ACC, Silva AG (2005) Comportamento de forrageio de aves nectarívoras em Santa Tereza – ES. Natureza On-line 3: 48–52.

Mendonça LB, dos Anjos L (2006) Flower morphology, nectar features, and hummingbird visitation to Palicourea crocea (Rubiaceae) in the Upper Paraná River foodplain, Brazil. Anais da Academia Brasileira de Ciência 78: 45–57. https://doi.org/10.1590/S0001-37652006000100006

Ortiz-Pulido R, Diaz SA, Valle-Diaz OI, Fisher AD (2012) Hummingbirds and the plants they visit in the Tehuacán-Cuicatlán Biospher Reserve, Mexico. Revista Mexicana de Biodiversidad 83: 152–163.

Pimm SL, Rosenzweig ML, Mitchell W (1985) Competition and food selection: field tests of a theory. Ecology 66: 798–807. https://doi.org/10.2307/1940541

Ribon R (2006) Plano de manejo do Parque Estadual do Itacolomi: avifauna. Ouro Preto, Universidade Federal de Ouro Preto, Relatório Técnico.

Rios PAF, da Silva JB, Moura FBP (2010) Visitantes florais da Aechmea constantinii (Mez) L.B. Sm. (Bromeliaceae) em um rema-nescente da Mata Atlântica do Nordeste Oriental. Biomas 23: 29–36.

Roberts WM (1996) Hummingbirds' nectar concentration preferences at low volume: the importance of time scale. Animal Behaviour 52: 361–370. https://doi.org/10.1006/anbe.1996.0180

Rodrigues MS, Antonini RD, Piratelli A (2009) Ecologia comportamental de beija-flores em Malvaviscus arboreus, em uma área de Mata Atlântica na ilha da Marambaia, Rio de Janeiro, Brasil. In: Anais III Congresso Latino Americano de Ecologia, São Lourenço, MG, p. 1–3.

Rosseau F, Charette Y, Bélisle M (2014) Resource defense and monopolization in a marked population of ruby-throated hummingbirds (Archilochus colubris). Ecology and Evolution 4: 776–793. https://doi.org/10.1002/ece3.972

Sick H (1997) Ornitologia Brasileira. Rio de Janeiro, Nova Fronteira, 862p.

Sigrist T (2009) Avifauna Brasileira. Vinhedo, Editora Avis Brasilis, 608p.

Snow DW, Snow BK (1986) Feeding ecology of hummingbirds in the Serra do Mar, southeastern Brazil. Hornero 12: 286–296.

Sonne J, Kyvsgaard P, Maruyama PK, Vizentin-Bugoni J, Ollerton J, Sazima M, Rahbek C, Dalsgaard B (2016) Spatial effects of artificial feeders on hummingbird abundance, floral visitation and pollen deposition. Journal of Ornithology 157: 573–581. https://doi.org/10.1007/s10336-015-1287-1

Stiles FG (1978) Ecological and evolutionary implications of bird pollination. American Zoology 18: 715–727. https://doi.org/10.1093/icb/18.4.715

Suarez RK, Lighton JRB, Moyes CD, Brown GS, Gass CL, Hochachka PW (1990) Fuel selection in rufous hummingbirds: Ecological implications of metabolic biochemistry. Ecology 87: 9207–9210. https://doi.org/10.1073/pnas.87.23.9207

Tamm S, Gass CL (1986) Energy intake rates and nectar concentration preferences by hummingbirds. Oecologia 70: 20–23. https://doi.org/10.1007/BF00377107

Tello-Ramos MC, Hurly TA, Healy SD (2015) Traplining in hummingbirds: flying short-distance sequences among several locations. Behavioral Ecology 26: 812–819. https://doi.org/10.1093/beheco/arv014

Temeles EJ, Goldman RS, Kudla AU (2005) Foraging and territory economics of sexual dimorphic Purple-throated Caribs (Eulampis jugularis) on three Heliconia morphs. The Auk 122: 187–204. https://doi.org/10.1642/0004-8038(2005)122[0187:FATEOS]2.0.CO;2

Temeles EJ, Shaw KC, Kudla AU, Sander SE (2006) Traplining by purple-throated carib hummingbirds: behavioral responses to competition and nectar availability. Behavioral Ecology and Sociobiology 61: 163–172. https://doi.org/10.1007/s00265-006-0247-4

Toledo MCB, Moreira DM (2008) Analysis of the feeding habits of the swallow- tailed hummingbird, Eupetomena macroura (Gmelin, 1788), in a urban park in southeastern Brazil. Brazilian

Journal of Biology 68: 419–426. https://doi.org/10.1590/S1519-69842008000200027

Van-Sluys M, Stotz DF (1995) Padrões de visitação a Vriesea neoglutinosa por beija- flores no Espírito Santo, Sudeste de Brasil. Bromélia 2: 27–35.

Zar JH (1999) Biostatistical Analysis. Upper saddle River, Prentice Hall, 4th ed., 929p.

Author Contributions: LLL and RMC conducted the experiments and analyzed the data; CSA, RO and YA designed the experiments, analyzed the data and wrote the paper.

Competing Interests: The authors have declared that no competing interests exist.

15

Taxonomic identification using geometric morphometric approach and limited data: an example using the upper molars of two sympatric species of *Calomys* (Cricetidae: Rodentia)

Natália Lima Boroni[1,3], Leonardo Souza Lobo[2,3], Pedro Seyferth R. Romano[3], Gisele Lessa[3]

[1]*Departamento de Zoologia, Universidade Federal de Minas Gerais. 31270-901 Belo Horizonte, MG, Brazil.*
[2]*Laboratório de Processamento de Imagem Digital, Museu Nacional, Universidade Federal do Rio de Janeiro. 20940-040 Rio de Janeiro, RJ, Brazil.*
[3]*Departamento de Biologia Animal, Universidade Federal de Viçosa. 36570-900 Viçosa, MG, Brazil.*
Corresponding author: Natália Lima Boroni (natalia_boroni@hotmail.com)

http://zoobank.org/A73FD3AB-66CC-47D2-BFCA-B4FF8569D0BC

ABSTRACT. The taxonomic identification of micromammals might be complicated when the study material is fragmented, as it is the case with pellets and fossil material. On the other hand, tooth morphology generally provides accurate information for species identification. Teeth preserve notably well, retaining their original morphology, unlike skulls and mandibles, which can get crushed or have missing parts. Here, we explored a geometric morphometrics approach (GM) to identify fragmented specimens of two sympatric *Calomys* Waterhouse, 1837 species – *Calomys tener* (Winge, 1888) and *Calomys expulsus* (Lund, 1841) – using the morphology of intact molars as the basis for identification. Furthermore, we included some specimens of uncertain taxonomic identification to test their affinities and the utility of the shape of the molar to identify incomplete specimens. We evaluated the variations in the shape of the first upper molar (M1) among 46 owl pellets specimens of *Calomys*, including *C. expulsus* (n = 15), *C. tener* (n = 15), and unidentified specimens treated as *Calomys* sp. (n = 16) through GM analysis using 17 landmarks. The data was explored using PCA, PERMANOVA, and Discriminant analyses over the Procrustes residuals matrix were applied to evaluate inter- and intraspecific shape differences. Also, we evaluated whether allometric shape differences could impact the data, but found no evidence of a correlation between size and shape. Our results support that shape differences in the M1 are effective for discriminating between *C. tener* and *C. expulsus*. Moreover, the unidentified specimens do not represent a third shape but could be identified with confidence either as *C. tener* or *C. expulsus*. Our results show that even with fragmentary materials, GM is a feasible and useful tool for exploring inter-specific shape differences and assisting in taxonomic identification as a complement to traditional qualitative description of diagnostic features in poorly preserved specimens.

KEY WORDS. Landmarks, morphology, owl pellets, Sigmodontinae, taxonomy.

INTRODUCTION

The complex morphology of the molars (with the cones, flexes, and lophs) is a source of information for the study of cricetid rodent taxonomy, as these structures provide diagnostic characteristics for subfamilies (Reig 1977). Over the years, it has been demonstrated that geometric morphometrics is a useful tool for systematics, for example taxonomic identification by analysis of molar shape differences among rodents, especially of fossil material (Polly and Head 2004, Kryštufek and Janžekoviè

2005, Macholán 2006, Marcolini et al. 2009). Enamel hardness protects the molars so they often are the only intact structure of a fragmented skull and form the only known elements of several extinct taxa in the fossil record (Reig 1977).

Among the genera with complicated taxonomy and great morphological similarity between some species is *Calomys* Waterhouse, 1837 (Almeida et al. 2007). They are small cricetid rodents distributed mainly in areas of dry vegetation, with wide distribution in South America (Bonvicino et al. 2010). The Brazilian species can be separated into two major groups, based on skull and body size measurements: a group of larger-bodied *Calomys* that includes *Calomys callosus* (Rengger, 1830), *Calomys expulsus* (Lund, 1840), *Calomys tocantinsi* Bonvicino, Lima & Almeida, 2003, *Calomys callidus* (Thomas, 1916), and *Calomys cerqueirai* Bonvicino, Oliveira & Gentile, 2010; and a group of smaller-bodied individuals with two species, *Calomys tener* (Winge, 1888) and *Calomys laucha* (G. Fisher, 1814). Despite this morphometric clustering, the smaller body size group is not monophyletic and *Calomys laucha* shares a more recent common ancestor with *C. expulsus* than *C. tener* does (Almeida et al. 2007). In contrast to their general morphological and morphometric similarity, the karyotype differs greatly between species of *Calomys*, making cytogenetic studies jointly with molecular data useful for species discrimination within this genus (Bonvicino and Almeida 2000, Salazar-Bravo et al. 2013, Almeida et al. 2007, Bonvicino et al. 2010).

In Brazil, *C. tener* and *C. expulsus* are both widely distributed (Salazar-Bravo 2015). The first occurs mainly in the Cerrado and Atlantic Forest borders; and the second occurs in the Caatinga and Cerrado (Bonvicino et al. 2008). Sympatry between these species is common, especially in the central region of Brazil (Bonvicino et al. 2010, Salazar-Bravo 2015). The morphometry, karyotype, distribution, and ecological differences between these two species had been described by Bonvicino and Almeida (2000), but there is little data on the dental morphology of either species. These species have the same diploid number, but differ in the fundamental number (Bonvicino and Almeida 2000). Morphologically, these two species can be distinguished by their size, and some cranial measurements such as the lengths of the skull and molar series (Bonvicino et al. 2010), but owing to the size variation and ontogenetic development, species identification might be inaccurate (Hingst-Zaher et al. 2000).

Although karyotypes and gene sequences may be useful for discriminating among *Calomys* species (Bonvicino and Almeida 2000, Almeida et al. 2007, Bonvicino et al. 2010), these data are rarely available from fossil and subfossil material. Even considering that *C. tener* is a little smaller than *C. expulsus* in some cranial characters, is difficult to separate both species when the material available for study is incomplete. Fragmented material is often found in owl pellets, fossils and sub fossils, and *Calomys* remains are very common in these samples throughout South America (e.g., Pardiñas et al. 2000, 2002, Salles et al. 2006, Scheibler and Christoff 2007).

Calomys and other members of the tribe Phyllotini generally share simplified molars and complete loss of the mesoloph and mesolophid (Hershkovitz 1962), but discrete dental characters useful for identification of *C. tener* and *C. expulsus* are lacking (see Hershkovitz 1962). In the absence of such features and in face of the difficulties to identify either species with fragmentary remains, the variation between them need to be studied and verified using alternative quantitative tools. Many studies, not only involving rodents, have explored the use of morphometric methods with superposition of forms for iden-

tifying different taxa, including molar analysis (e.g., Rohlf and Slice 1990, Bookstein 1991, Rohlf 1999, Becerra and Valdecasas 2004, Macholán 2006, Marcolini et al. 2009, Matthews and Stynder 2011). With this methodology, a complete identification is not always possible, but at least one can reduce the number of steps and time necessary for a correct identification (Becerra and Valdecasas 2004). In the Neotropics, however, studies investigating the usefulness of geometric morphometrics of the molar for taxonomic identification are unusual, and only skull and post-cranial elements are generally employed (e.g., Corti et al. 2001, Cordeiro-Estrela et al. 2006, 2008, Morgan 2009, Astúa et al. 2015).

The aim of present study is to explore geometric morphometric analysis to identify fragmented materials (modern and fossil) of small vertebrates based on molars; and to assess whether this method allows for accurate identifications. For this we applied this technique in one area: species-level identification of *Calomys* (*C. expulsus* and *C. tener*) based on their upper molars.

MATERIAL AND METHODS

The *Calomys* specimens analyzed were from owl pellets collected in the Natural Monument Peter Lund, Cordisburgo, in the central karst region of Minas Gerais, the Bambuí group, Brazil (Fig. 1). At the Natural Monument Peter Lund the pellets were collected inside the Salitre cave (19°07′17″S, 44°28′24″W) during the Park Management Plan.

In the Köppen climate classification system, the regional climate is Aw tropical humid, characterized by hot, rainy summers and dry winters (Travassos 2010). The average annual temperature in Cordisburgo is 22°C and the average annual precipitation ranges from 1250 to 1500 mm (Travassos 2010). In the karst area of Cordisburgo, much of the Cerrado has been replaced by agriculture and silviculture of *Eucalyptus* spp. (Travassos 2010). In the surrounding area and limestone outcrops, there is a semi-deciduous forest conditioned by the type of rock and climate, and it is possible to identify riparian and gallery forests along major drainages (Travassos 2010).

For species identification, we analyzed complete and fragmented skulls, maxillae and mandibles, examining the cranial sutures and morphology of the teeth. However, some fragments did not allow a consistent identification to the species level, thus restricting classification to the level of genus. The nomenclature of the species was based on that described by Patton et al. (2015). The molar teeth nomenclature was based on Reig (1977). All collected materials were deposited in the Mastozoology collection of the Museu de Zoologia João Moojen, Universidade Federal de Viçosa, Brazil (MZUFV 3861, MZUFV 3862, MZUFV 3863).

With the fragmented skull, the best characteristics used to differentiate *C. tener* from *C. expulsus* was the presence of alisphenoid strut, which is observed only in *C. tener* (Salazar-Bravo 2015) (Suppl. material 1), and the length of the molar series, which is on average 3.4 mm (range 3.1–3.9 mm) in *C.*

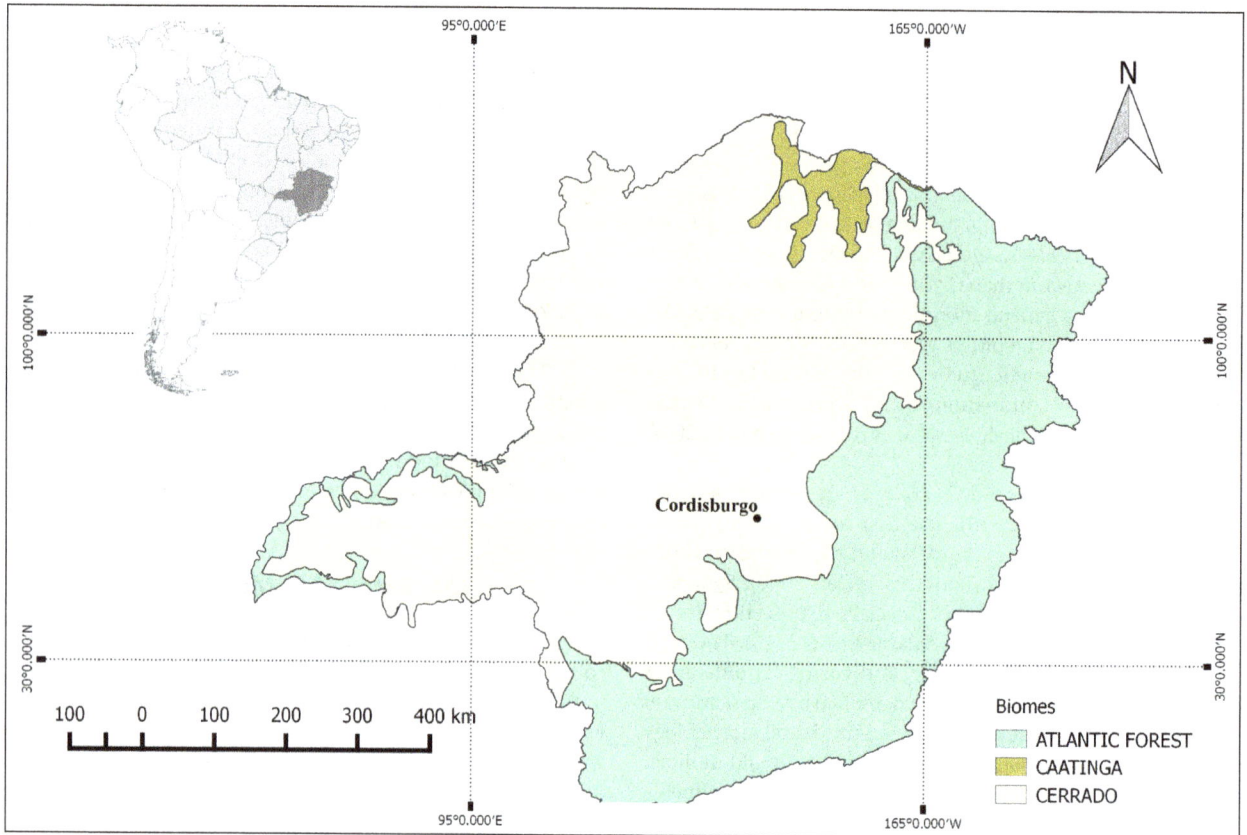

Figure 1. Map of Minas Gerais (Brazil), with the study location, municipality of Cordisburgo.

tener and on average 4.0 mm (range 3.8–4.2 mm) in *C. expulsus* (Bonvicino and Almeida 2000, Bonvicino et al. 2010). Due to an overlap in the measurements of the length of the molar series and difficulties to observe the presence of alisphenoid strut in all skulls due to fragmentation of the posterior part of the skull, some specimens were only identified as *Calomys* sp.

A total of 124 skulls were analyzed and separated into classes of dental wear, including 32 of *C. tener*, 57 of *C. expulsus*, and 35 of *Calomys* sp. (Suppl. material 2). For all analyses only the first right molar of *Calomys* skulls was analyzed, all of which had been previously identified as being of the same age based on wear class (see Suppl. materials 2).

The wear category class 2 showed the most numerically balanced sample between *C. tener*, *C. expulsus*, and *Calomys* sp. (Table 1) and, for that reason, was chosen for the morphometric analysis (see Suppl. material 3 for further information about the specimens analyzed). This subset included 46 specimens (15 of *C. tener*, 15 of *C. expulsus* and 16 of. *Calomys* sp.) Comparison of specimens from the same age category eliminated this potential source of age-related variation in enamel morphology due to tooth wear caused by chewing. Moreover, it minimized

Table 1. Specimens examined and the respective sample sizes by dental wear category.

	Calomys tener	*Calomys expulsus*	*Calomys* sp.
Class 1	7	10	8
Class 2	15	15	16
Class 3	9	20	8
Class 4	–	11	2
Total	32	57	35

the allometric effect in the data, since the size of the specimens showed little variation.

The landmarks were digitalized using TpsDig v.2.17 (Rohlf 2015). We selected 17 landmarks corresponding to type II (points of maximum curvature, *sensu* Bookstein 1991), spanning the enamel folds connecting the main cusps of the first molar, which are described as follows (Fig. 2). Landmark 1, anterior extremity of anterolabial conule; Landmark 2, posterior extremity of anteroflexus; Landmark 3, anterior extremity of anterolabial conule; Landmark 4, lateral extremity of anterolabial conule;

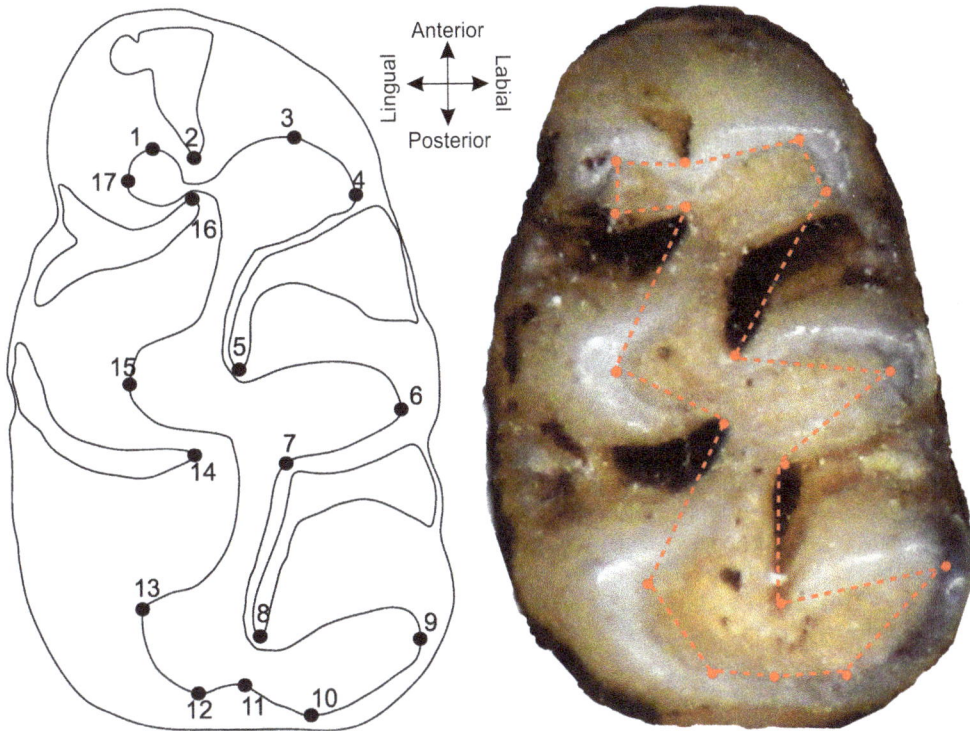

Figure 2. Landmarks of the molar used in this study. For landmarks description, see text.

Landmark 5, posterior extremity of protoflexus; Landmark 6, lateral extremity of protocone; Landmark 7, anterior extremity of hypoflexus; Landmark 8, posterior extremity of hypoflexus; Landmark 9, lateral extremity of hypocone; Landmark 10, posterior extremity of hypocone; Landmark 11, hypocone in contact with the metacone; Landmark 12, posterior extremity of metacone; Landmark 13, lateral extremity of metacone; Landmark 14, medial extremity of metaflexus; Landmark 15, lateral extremity of paracone; Landmark 16, medial extremity of paraflexus; and Landmark 17, lateral extremity of anterolingual conule.

The multivariate analyses were performed using the MorphoJ v. 1.05f (Klingenberg 2011). After digitizing all landmarks using TpsDig, the centroid sizes of all specimens were calculated from the original coordinate matrix. The centroid size is the square root of the summed squared distances of each landmark to the centroid and can be used as a measure of general size of specimens (Zelditch et al. 2004). The matrices of coordinates were then superimposed using Procrustes standardization (the generalized least square method, GLS) to remove differences in size, orientation, and position between specimens (Zelditch et al. 2004). All further statistics were performed using the Procrustes residuals to analyze differences in shape and the centroid size values to evaluate differences in size.

For increased reliability of data, an analysis of error in digitization of anatomical landmarks was performed following the protocol of Adriaens (2007) using the TpsSmall v. 1.26 (Rohlf 2015). For this analysis, 12 specimens that were duplicated to evaluate the accuracy in digitization of anatomical landmarks were used. The results of the analysis of error are presented in Suppl. material 4.

From this experimental design, a preliminary principal component analysis (PCA) was performed using only species anatomically distinguishable, namely, 15 specimens of *C. tener* and 15 specimens of *C. expulsus*. This PCA was conducted to visualize the projection of individuals from the two axes of greatest variation, thereby detecting the distribution pattern in the graph. Then, we included the 16 specimens that could not be definitively identified as *C. tener* or may be *C. expulsus*. This addition aimed to increase the sample, totaling to 46 specimens, and explore the projection of *Calomys* sp. in the two axes of the PCA in the presence of distinguishable specimens. The second goal of adding specimens of *Calomys* sp. was to classify the unidentified specimens to one or another expected species (*C. tener* or *C. expulsus*). The wireframe, which helps in comparing the shapes of the specimens plotted in the PCA, was used whenever necessary to describe and discuss the shape of the specimens.

After each *Calomys* spp. was assigned to *C. tener* or *C. expulsus* through the pattern observed in the PCA, a Discriminant analysis (DA) was performed using MorphoJ v. 1.05f (Klingenberg 2011) and with 10.000 replications in the permutation test. As the Procrustes distances did not meet assumptions of normality, homoscedasticity and homogeneity of covariance (Suppl. material 5), the differences between the molar morphology of *C. tener* and *C. expulsus* were tested by PERMANOVA with adjustment of Euclidean distance (Anderson 2005). This analysis is analogous to ANOVA, however is a non-parametric test used to perform a comparison between two or more groups. In this test the significance is computed by permutation of group membership and $p < 0.05$ as the criterion of significance. "*Calomys* sp." potentially identified as *C. tener* or *C. expulsus* into the PC1 vs. PC2 individual projections were also tested by PERMANOVA. Both analyzes were performed using 10.000 replications in permutation test using PAST v.3.11 (Hammer et al. 2001).

Lastly, a pooled regression within each species (multivariate regression) was performed to evaluate whether the differences in the size of the tooth were correlated to the pattern of differentiation of the PCA projection. For this analysis, we used the centroid size of specimens as the independent variable and Procrustes coordinates as the dependent variable. Also, we performed a permutation test with 10,000 rounds to evaluate whether the dependence of shape (Procrustes distances) on the size (centroid size) of the tooth is significant.

RESULTS

The first PCA was performed only with *C. expulsus* and *C. tener* together. The PC1 and PC2 totaled 29.5% and 13.8% of the total variance, respectively, explaining 43.3% of the shape variation within the sample. It was possible to visualize either species separated in PC1 vs PC2 individual projections (Suppl. material 6). PERMANOVA test showed significant difference between *C. expulsus* and *C. tener* (p-value = 0.001 F = 6.079). The second PCA, which includes also the *Calomys* sp. specimens, showed a similar pattern as that of the previous PCA, with PC1 and PC2 holding respectively 28.7% and 12.3%, amounting 41% of the variation (Fig. 3). Discriminant analysis of first upper molar Procrustes coordinates classified 100% of the specimens into the correct groups; when the cross-validation technique was applied, 82.6% of specimens were recovered into the correct groups (Table 2). The permutation test using the Procrustes distance values (0.0484, p-value = < 0.0001) and Mahalanobis distances (6.4366, T-square = 0.0008, p-value = 0.0006) corroborated a distinction in the shape of the molar between species. The result of the PERMANOVA test for *Calomys* sp. (identified as *C. tener* or *C. expulsus* by PCA) together with the other specimens of *C. expulsus* and *C. tener* was significant (p-value = 0.0001 F = 8.283).

The variation expressed by PC1 showed differences in shape of the occlusal between both species. These differences are illustrated based on the wireframe view of mean shape (Fig. 4) and disparate shapes (Fig. 5) of *C. tener* and *C. expulsus*. In general, *C. tener* showed a more retracted shape in the lingual-labial axis, due to an expansion of the cusps in *C. expulsus* compared to *C. tener*; flexus retraction and smaller

aperture angles of the lingual flexus in *C. expulsus* compared to *C. tener*. The anterolingual conule of *C. tener* was retracted in the anterior-posterior axis compared to *C. expulsus*, was located more anteriorly, and was more oblique in relation to the anteromedian flexus. When compared with *C. expulsus*, *C. tener* had a retraction of the anterolabial conule in the labial-lingual axis and suffered another retraction in relation to the anterior and labial extremities (between the anatomical landmarks 3 and 4). The paraflexus of *C. tener* compared to *C. expulsus* was expanded in the anterior-posterior axis, and the paracone was retracted in the labial-lingual axis. The protoflexus of *C. tener*, when compared with that of *C. expulsus*, was expanded and the protocone was retracted on both the orientation axes. The metacone of *C. tener*, when compared with that of *C. expulsus*, was retracted in both the axes. The hypoflexus of *C. tener*, in relation to *C. expulsus*, was expanded and the hypocone was retracted in its anterior part. The posterior region of the molar of *C. tener*, in relation to *C. expulsus*, was retracted in the lingual region (landmarks 10, 11, and 12) and anterior-posterior axis.

CT 02 (*C. tener*) can be distinguished from the other individuals by observing the shape differences explained by PC2 and illustrated in Fig. 6. CT 02 differed from the mean value for PC2 scores owing to the following morphological disparities: expansion of the anterior-posterior axis and slight retraction of the labial-medial axis of the anterolabial conule (landmarks 3 and 4); expansion of anterior-posterior axis in the protoflexus (landmark 5) (the anterior extremity of hypoflexus (landmark 7) did not expand proportionally), which caused a narrowed protocone in CT 02; slight retraction of posterior extremity of hypocone; and lingual expansion of metacone (landmark 13) and posterior displacement of anterior extremity of metaflexus (landmark 14).

To evaluate whether tooth size influences the patterns of variation in oclusal surface, we performed a multivariate regression with the Procrustes coordinates residual (shape) on the centroid size (size measurement). This regression showed a weak relationship that accounted for just 5.3% of the shape variation predicted by the size (Fig. 7), pursuant the permutation test accepting the null hypothesis of independence between the variables (p-value = 0.2118).

Table 2. Discriminant analysis results (classification/misclassification table) for *Calomys expulsus* and *Calomys tener*.

	Discriminant function – Allocated to		Total	Percentage
	Calomys expulsus	*Calomys tener*		
Calomys expulsus	22	0	22	100
Calomys tener	0	24	24	100
	Cross-validation – Allocated to			
	Calomys expulsus	*Calomys tener*		
Calomys expulsus	19	3	22	86.4
Calomys tener	5	19	24	79.2

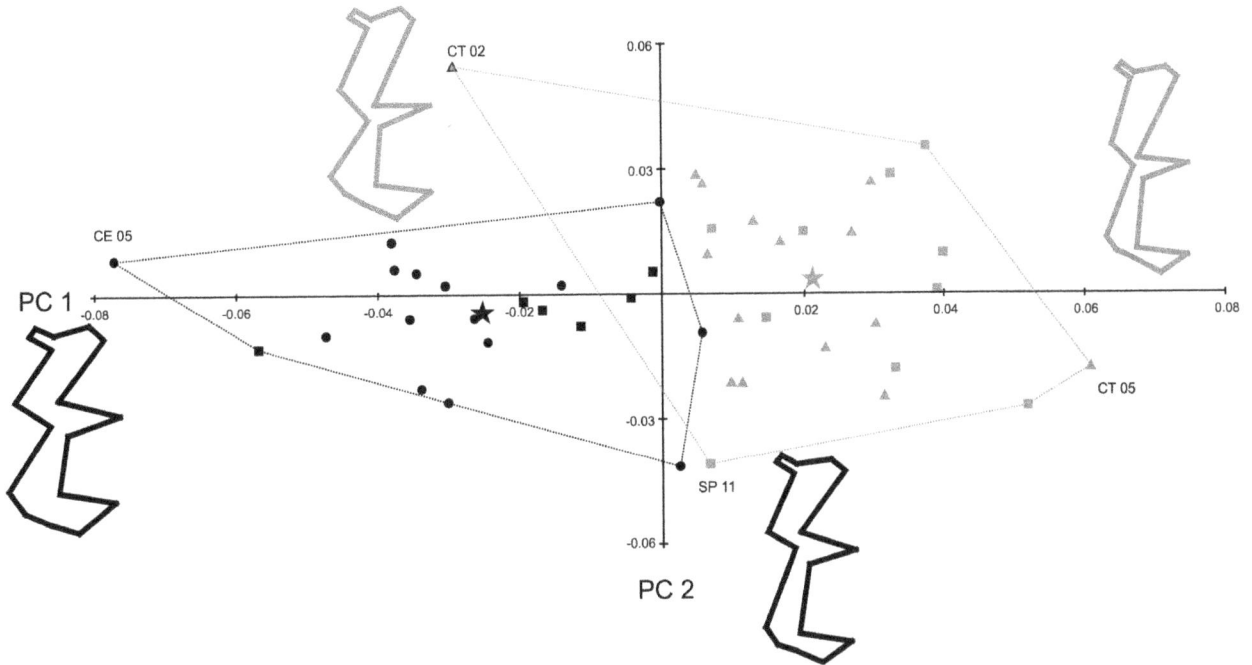

Figure 3. PCA showing the individual projections of *Calomys* sp., *C. expulsus* and *C. tener* in the two major axis (PC1 28.7%, PC2 12.3% of the variance). The wireframes illustrate shape differences between most different specimens: CE 09, CT 05, CT 02, and SP 11. Black circles: *C. expulsus*; Gray triangle: *C. tener*; Black square: *Calomys* sp. identified as *C. expulsus*; Gray square: *Calomys* sp. identified as *C. tener*; Gray triangle with black edge: specimen CT 02; Stars: mean shape for each species.

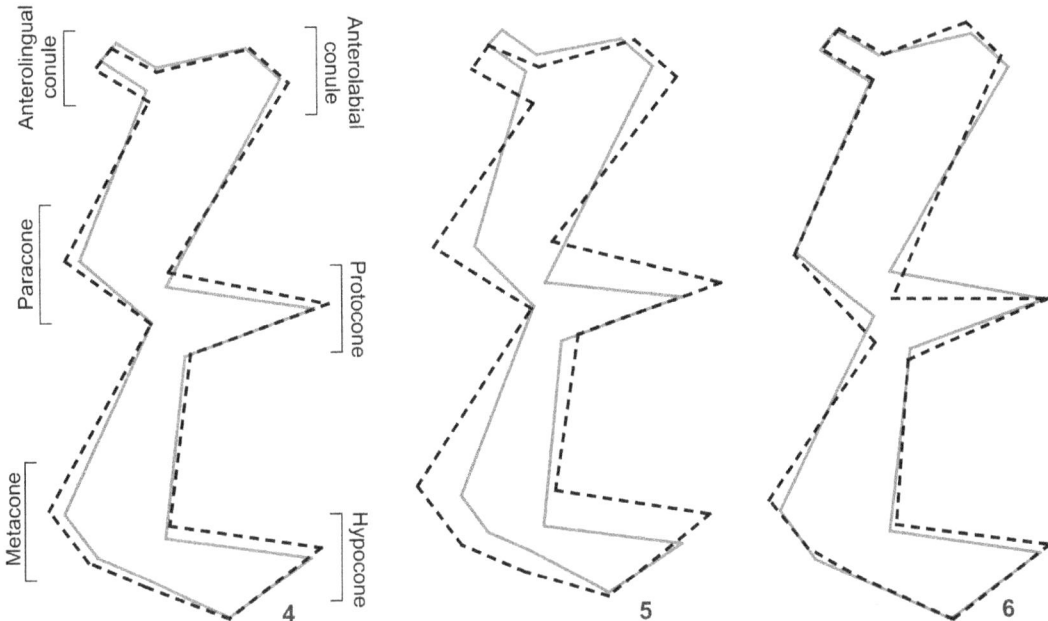

Figures 4–6. Shape differences wireframes in PC1. Gray solid lines indicate *C. tener* (positive PC1 values) and black dotted lines indicate *C. expulsus* (negative PC1 values), see Fig. 3 for individual projections. (**4**) Differences between the mean shapes of *C. tener* and *C. expulsus*; (**5**) differences between the disparate shape of *C. tener* represented by CT 05 and *C. expulsus* represented by CE 09; (**6**) shape differences wireframes in PC2. Gray solid lines indicate the mean of shape of PC2 value and black dotted lines indicate specimen CT 02 (*C. tener* outlier).

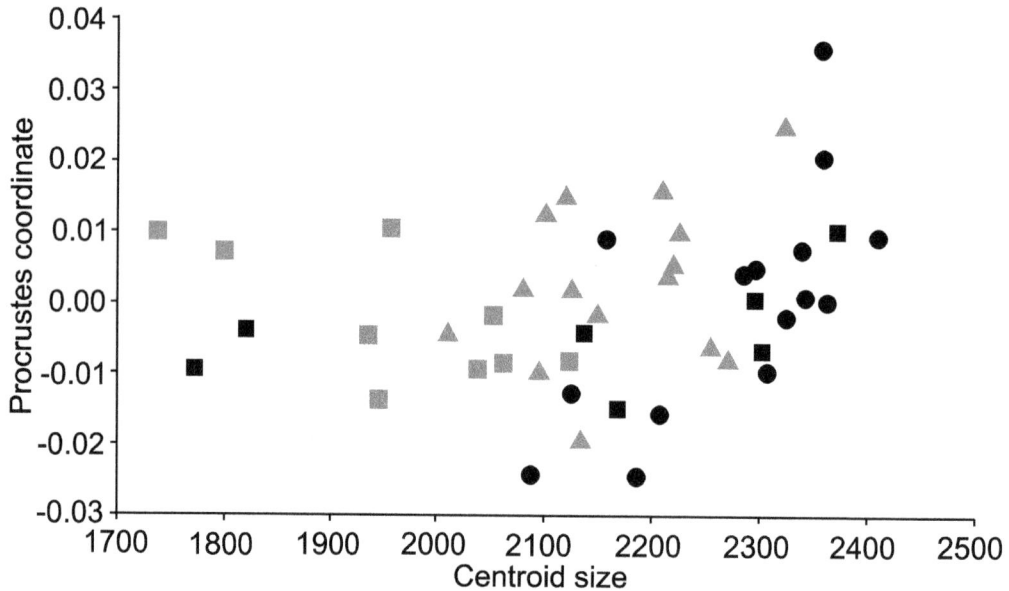

Figure 7. Pooled regression within the two species, between shape (dependent variable) and centroid size (independent variable). Black circles: *C. expulsus*; Gray triangle: *C. tener*; Black square: *Calomys* sp. identified as *C. expulsus*; Gray square: *Calomys* sp. identified as *C. tener*.

DISCUSSION

Calomys tener and *C. expulsus* are grouped into different clades (Almeida et al. 2007), however, neither species has discrete craniodental characters for species diagnosis (Hershkovitz 1962, Cordeiro-Estrela et al. 2006), and commonly used morphometric measures also overlapped (see Bonvicino et al. 2010). Nevertheless, the analyses performed in the present study indicate that the presence of the alisphenoid strut can be used to identify either species.

The overlap areas in morphometric cranial analyses using orthogonal matrix transformation techniques as PCA, irrespective of whether traditional or geometric, have also been identified in other species of other genera and in *Calomys* (Cordeiro-Estrela et al. 2006, 2008, Astúa 2009, Bonvicino et al. 2010, Martínez and Di Cola 2011). Furthermore, the results obtained in the present study for M1 shape are in agreement with those reported in other studies that showed a divergence in the shape of *Calomys* skull, but with small areas of intersection (Cordeiro-Estrela et al. 2006, 2008).

The main criterion for distinguishing *C. tener* from *C. expulsus* by traditional morphometrics is the size of the body (Bonvicino and Almeida 2000). *Calomys expulsus* can be differentiated from *C. tener* by having a larger body, smaller ear, proportionally smaller tail (generally), more robust skull, smaller braincase, length of molar row close to 4.0 mm in *C. expulsus*, 3.0 mm in *C. tener* (Winge 1888, Bonvicino and Almeida 2000). However, Cordeiro-Estrela et al. (2006) analyzed the skull of *C.*

expulsus and *C. tener* using geometric morphometrics and noted that allometry is not the main source of variation between the two species, despite its contribution to their differences. The main morphological differences between these two species are related to the patterns of shape variation (Cordeiro-Estrela et al. 2006), as observed in the present study through molar analysis.

As mentioned earlier, *C. expulsus* and *C. tener* occur in sympatry, and were differentiated in the present study through the presence of alisphenoid strut and length of maxillary molars. We observed a high similarity between our analysis and previous identification based on both characteristics, with more than 80% accuracy in DA. However, our results indicate that tooth size characters are not useful to diagnose either species. Thus, we suggest that the use of size characters (e.g., Bonvicino and Almeida 2000, Bonvicino et al. 2010) to distinguish *C. tener* and *C. expulsus* should be avoided. Cordeiro-Estrela et al. (2008) arrived at a similar conclusion considering the skull centroid size of *C. laucha* and *Calomys musculinus* (Thomas, 1913).

In our results, CT02 deserves mention. This specimen was identified as *C. tener* because it had alisphenoid strut and a 3.7 mm long molar. However, the shape of this specimen, at least observing the differences explained by PC1, reveals more similarity with *C. expulsus*. Since CT 02 is a fragmentary specimen, our conclusion regarding its identification is limited. This specimen could be *C. tener* with larger molars or perhaps *C. expulsus* with alisphenoid strut and small molars. On the other hand, another possibility is that CT02 is a *C. cerqueirai*, a species described from near the investigation area of the present study (see Bonvicino et al. 2010). However, we are not able to confirm this hypothesis, since *C. cerqueirai* shows no discrete cranial or dental characters or difference in size, when compared with *C. expulsus* (Bonvicino et al. 2010). As described by Bonvicino et al. (2010), the main differences between *C. cerqueirai* to *C. expulsus* are a Cinnamon-Brown overall dorsal coloration, head paler than dorsum, small ears with short brownish hairs, and sharply bicolored tail. Also, some descriptive cranial measurements, as palatal bridge, breath of incisive foramen, breadth of first maxillary molar, are overlapped. Unfortunately, we could not add *C. cerqueirai* specimens in our analysis and there are only twelve specimens deposited in other institutions (Bonvicino et al. 2010, Mesquita and Passamani 2012, Colombi and Fagundes 2015.) But since neither species (*C. cerqueirai* and *C. expulsus*) has alisphenoid strut (Salazar-Bravo 2015), the most likely identification of specimen CT02 is outliner *C. tener*.

We observed a pattern of differentiation in the oclusal shape of class 2, among sympatric specimens of *C. expulsus* and *C. tener*. In this class, the cusps of *C. tener* (as protocone and hipocone) were more retracted, whereas the cusps of *C. expulsus* are expanded in the lingual-labial axis (when compared with *C. tener*); the flexus (as paraflexus and protoflexus) are expanded and the aperture angles are larger in *C. tener* than in *C. expulsus*. Different molar morphologies in rodents can represent adaptations to different diets and can be relevant to avoid intraspecific competition (Parra et al. 1999, Renaud et al. 1999). This could mean that the differences in the shape of the molar, observed between *C. tener* and *C. expulsus*, observed in this study, could reflect different diets. This hypothesis needs testing, since the diet of *C. expulsus* and *C. tener* has not been published in the literature. A study on the outline of the first molar of extinct murine rodents demonstrated that different diet groups could be distinguished through geometric morphometric analysis (Cano et al. 2013). However, in marmots and some Didelphidae it has been demonstrated that diet is weakly correlated with molar shape (Caumul and Polly 2005, Chemisquy et al. 2015).

The possibility to perform taxonomic identifications based on molar shape, principally in rodents, is a very useful, especially when working with fossil materials (Macholán 2006, Barrón-Ortiz et al. 2008, Marcolini et al. 2009, Matthews and Stynder 2011). Matthews and Stynder (2011) distinguished *Aethomys* fossil specimens from extant specimens based on molar shape using geometric morphometric analysis; however, they emphasized that although the shape of the molar is important, a small sample size might hamper a complete analysis. Limitations in analyses owing to insufficient sample size are common, particularly in paleontological samples (Matthews and Stynder 2011), as it was in ours. Is important to point out that geometric morphometrics and features from the molars can be used in other types of studies beyond taxonomic identifications, as for example: correlating changes in molar shape with paleoclimate aspects (McGuire 2010), phylogenetic (Caumul and Polly 2005), morphological evolution and environmental variations (Renaud and Van Dam 2002).

Many studies using small mammal fossils or owl pellets for taxonomic identification are restricted to the level of genus (e.g. Salles et al. 1999, 2006, Bonvicino and Bezerra 2003, Scheibler and Christoff 2007, Rocha et al. 2011). With a good taxonomic identification it is possible to go beyond listing the species that occur at an area. The possibility to identify unknown specimens with an accurate morphology-based taxonomic assignments permits us to gain a better understanding of geographic distribution, migration and environmental response (Polly and Head 2004). Approaches using different quantitative, statistics and morphological analyzes can provide more scientifically accurate and reliable results (Polly and Head 2004).

This is the first study using molars found in owl pellets for taxonomic identification through geometric morphometrics in South America. Our results were effective for the identification of dubious *Calomys* sp. specimens. All of them were identified either as *C. tener* or *C. expulsus* based on the individual projections of the PC1. The use of size as a diagnostic character is not effective for molar differentiation of *C. tener* and *C. expulsus*, as the regression test showed no evidence of correlation between size and shape. These results demonstrated the huge potential of geometric morphometric analysis as a tool in the rodent taxonomic identification using molars, especially in works with fragmented material (modern or fossil). This technique can be extensively useful principally by paleontologists, who have to identify isolated morphological elements, as molars of small vertebrates. This method can be also applied to other genera (e.g. *Oligoryzomys* Bangs, 1900 and *Akodon* Meyen, 1833) and to other groups such as bats and marsupials, wherever accurate taxonomic identifications of morphologically similarity and sympatric species are needed.

ACKNOWLEDGMENTS

We are grateful to Ulyses Pardiñas for the initial support in the identification of the owl pellets. Rodolfo Stumpp and the reviewers for all relevant comments. We thank the staff of Peter Lund National Monument for allowing us to use their facilities for the collection of samples. We are also grateful to Prof. Weyder Santana for permitting us to use the stereoscopic microscope and camera. The Federal Forest Institute (IEF) granted permits for the collection of specimens. We also thank Coordenação de Aperfeiçoamento de Pessoal de Nível Superior (CAPES) for the student scholarship. This study was supported by Fundação de Amparo à Pesquisa do Estado de Minas Gerais (FAPEMIG APQ-02262-12), grants to Gisele Lessa and was developed during 2013 at the Graduation Program in Animal Biology of the Federal University of Viçosa.

REFERENCES

Adriaens D (2007) Protocol for error testing in landmark based geometric morphometrics. http://www.fun-morphometrics.

morph.ugent.be/Miscel/Methodology/Morphometrics.pdf [Accessed: 18/03/2016]

Almeida FC, Bonvicino CR, Cordeiro-Estrela P (2007) Phylogeny and temporal diversification of Calomys (Rodentia, Sigmodontinae): Implications for the biogeography of an endemic genus of the open/dry biomes of South America. Molecular Phylogenetics and Evolution 42: 449–466. https://doi.org/10.1016/j.ympev.2006.07.005

Anderson MJ (2005) Permutational multivariate analysis of variance. Auckland, Department of Statistics, University of Auckland.

Astúa D (2009) Evolution of scapula size and shape in didelphid marsupials (Didelphimorphia: Didelphidae). Evolution 63: 2438–2456. https://doi.org/10.1111/j.1558-5646.2009.00720.x

Astúa D, Bandeira I, Geise L (2015) Cranial morphometric analyses of the cryptic rodent species Akodon cursor and Akodon montensis (Rodentia, Sigmodontinae). Oecologia Australis 19: 143–157. https://doi.org/10.4257/oeco.2015.1901.09

Barrón-Ortiz CR, Riva-Hernández G, Barrón-Corvera R (2008) Morphometric analysis of equid cheek teeth using a digital image processor: A case study of the Pleistocene Cedazo local fauna equids, Mexico. Revista Mexicana de Ciencias Ggeológicas 25: 334–345

Becerra JM, Valdecasas AG (2004) Landmark superimposition for taxonomic identification. Biological Journal of the Linnean Society 81: 267–274. https://doi.org/10.1111/j.1095-8312.2003.00286.x

Bonvicino CR, Almeida FC (2000) Karyotype, morphology and taxonomic status of Calomys expulsus (Rodentia: Sigmodontinae). Mammalia 64: 339–351. https://doi.org/10.1515/mamm.2000.64.3.339

Bonvicino CR, Bezerra AMR (2003) Use of regurgitated pellets of Barn Owl (Tyto alba) for inventory small mammals in the Cerrado of central Brazil. Studies on Neotropical Fauna and Environment 38: 1–5. https://doi.org/10.1076/snfe.38.1.1.14030

Bonvicino CR, Oliveira JA, D'andrea OS (2008) Guia dos roedores do Brasil, com chaves para gêneros baseadas em caracteres externos. Rio de Janeiro, Centro Pan-Americano de Febre Aftosa, OPAS/OMS.

Bonvicino CR, Oliveira JA, Gentile R (2010) A new species of Calomys (Rodentia: Sigmodontinae) from Eastern Brazil. Zootaxa 25: 19–25.

Bookstein FL (1991) Morphometric tools for landmark data: geometry and biology. Cambridge University Press, New York.

Caumul R, Polly PD (2005) Phylogenetic and environmental components of morphological variation: Skull, mandible, and molar shape in Marmots (Marmota, Rodentia). Evolution 59: 2460–2472. https://doi.org/10.1111/j.0014-3820.2005.tb00955.x

Cano GA, Hernández FM, Álvarez-Sierra M (2013) Dietary ecology of Murinae (Muridae, Rodentia): A geometric morphometric approach. PLOS ONE 8: 1–7. https://doi.org/10.1371/journal.pone.0079080

Chemisquy MA, Prevosti FJ, Martin G, Flores DA (2015) Evolution of molar shape in didelphid marsupials (Marsupialia: Didelphidae): analysis of the influence of ecological factors and phylo-

genetic legacy. Zoological Journal of the Linnean Society 173: 217–235. https://doi.org/10.1111/zoj.12205

Colombi VH, Fagundes V (2015) First record of Calomys cerqueirai (Rodentia: Phyllotini) in Espirito Santo (Brazil) with description of the 2n = 36, FNA = 66 karyotype. Mammalia 79: 479–486. https://doi.org/10.1515/mammalia-2014-0076

Cordeiro-Estrela P, Baylac M, Denys C, Marinho-Filho J (2006) Interspecific patterns of skull variation between sympatric Brazilian vesper mice: Geometric morphometrics assessment. Journal of Mammalogy 87: 1270–1279. https://doi.org/10.1644/05-MAMM-A-293R3.1

Cordeiro-Estrela P, Baylac M, Denys C, Polop J (2008). Combining geometric morphometrics and pattern recognition to identify interspecific patterns of skull variation: Case study in sympatric Argentinian species of the genus Calomys (Rodentia: Cricetidae: Sigmodontinae). Biological Journal of the Linnean Society 94: 365–378. https://doi.org/10.1111/j.1095-8312.2008.00982.x

Corti M, Aguilera M, Capanna E (2001) Size and shape changes in the skull accompanying speciation of South American spiny rats (Rodentia: Proechimys spp.). Journal of Zoology 253: 537–547. https://doi.org/10.1017/S0952836901000498

Hershkovitz P (1962) Evolution of neotropical Cricetine rodents (Muridae) with special reference to the phyllotine group. Fieldiana Zoology 46: 1–524. https://doi.org/10.5962/bhl.title.2781

Hammer Ø, Harper DAT, Ryan PD (2001) PAST: Paleontological Statistics Software Package for education and data analysis. Palaeontologia Electronica 4: 1–9.

Hingst-Zaher E, Marcus L, Cerqueira R (2000) Application of geometric morphometrics to the study of post-natal size and shape changes in the skull of Calomys expulsus. Hystrix 11: 99–113. https://doi.org/10.4404/hystrix-11.1-4139

Klingenberg CP (2011) MorphoJ: An integrated software package for geometric morphometrics. Molecular Ecology Resources 11: 353-357. https://doi.org/10.1111/j.1755-0998.2010.02924.x

Kryštufek B, Janžekovič F (2005) Relative warp analysis of cranial and upper molar shape in rock mice Apodemus mystacinus sensu lato. Acta theriologica 50: 493–504. https://doi.org/10.1007/BF03192642

Macholán M (2006) A geometric morphometric analysis of the shape of the first upper molar in mice of the genus Mus (Muridae, Rodentia). Journal of Zoology 270: 672–681. https://doi.org/10.1111/j.1469-7998.2006.00156.x

Marcolini F, Piras P, Martin RA (2009) Testing evolutionary dynamics on first lower molars of Pliocene Ogmodontomys (Arvicolidae, Rodentia) from the Meade Basin of Southwestern Kansas (USA): A landmark-based approach. PALAIOS 24: 535–543. https://doi.org/10.2110/palo.2008.p08-114r

Matthews T, Stynder DD (2011) An analysis of the Aethomys (Murinae) community from Langebaanweg (Early Pliocene, South Africa) using geometric morphometrics. Palaeogeography, Palaeoclimatology, Palaeoecology 302: 230–242. https://doi.org/10.1016/j.palaeo.2011.02.003

Martínez JJ, Di Cola V (2011) Geographic distribution and phenetic skull variation in two close species of Graomys (Rodentia,

Cricetidae, Sigmodontinae). Zoologischer Anzeiger 250: 175–194. https://doi.org/10.1016/j.jcz.2011.03.001

McGuire J (2010) Geometric morphometrics of vole (*Microtus californicus*) dentition as a new paleoclimate proxy: Shape change along geographic and climatic clines. Quaternary International 212: 1–8. https://doi.org/10.1016/j.quaint.2009.09.004

Mesquita AO, Passamani M (2012) Composition and abundance of small mammal communities in forest fragments and vegetation corridors in Southern Minas Gerais, Brazil. Revista de Biologia Tropical 60: 1335–1343. https://doi.org/10.15517/rbt.v60i3.1811

Morgan CC (2009) Geometric morphometrics of the scapula of South American caviomorph rodents (Rodentia: Hystricognathi): form, function and phylogeny. Mammalian Biology-Zeitschrift für Säugetierkunde 74: 497–506. https://doi.org/10.1016/j.mambio.2008.09.006

Pardiñas UF, D'Elía G, Ortiz PE (2002) Sigmodontinos fósiles (Rodentia, Muroidea, Sigmodontinae) de América del Sur: estado actual de su conocimiento y prospectiva. Mastozoología Neotropical 9: 209–252.

Pardiñas UF, Moreira GJ, Garcia-Esponda CM, Santis LJ (2000) Deterioro ambiental y micromamíferos durante el holoceno en el nordeste de la estepe patagônica. Revista Chilena de Historia Natural I73: 9–21. https://doi.org/10.4067/S0716-078X2000000100002

Parra V, Loreau M, Jaeger JJ (1999) Incisor size and community structure in rodents: two tests of the role of competition. Acta Oecologica 20: 93–101. https://doi.org/10.1016/S1146-609X(99)80021-6

Patton JL, Pardiñas UFG, D'elía G (2015) Mammals of South America, vol. 2. The University of Chicago Press, Chicago. https://doi.org/10.7208/chicago/9780226169606.001.0001

Polly PD, Head JJ (2004) Maximum-likelihood identification of fossils: taxonomic identification of Quaternary marmots (Rodentia, Mammalia) and identification of vertebral position in the pipesnake *Cylindrophis* (Serpentes, Reptilia). In: Elewa AMT (Ed.) Morphometrics applications in biology and paleontology. Springer, New York, 197–221. https://doi.org/10.1007/978-3-662-08865-4_14

Reig OA (1977) A proposed unified nomenclature for the enamelled components of the molar teeth of the Cricetidae (Rodentia). Journal of Zoology 181: 227–241. https://doi.org/10.1111/j.1469-7998.1977.tb03238.x

Renaud S, Michaux J, Mein P, Aguilar JP, Auffray JC (1999) Patterns of size and shape diferentiation during the evolutionary radiation of the European Miocene murine rodents. Lethaia 32: 61–71. https://doi.org/10.1111/j.1502-3931.1999.tb00581.x

Renaud S, Van Dam J (2002) Influence of biotic and abiotic environment on dental size and shape evolution in a Late Miocene lineage of murine rodents (Teruel Basin, Spain). Palaeogeography, Palaeoclimatology, Palaeoecology 184: 163–175. https://doi.org/10.1016/S0031-0182(02)00255-9

Rocha RG, Ferreira E, Leite YLR, Fonseca C, Costa LP (2011) Small mammals in the diet of barn owls, *Tyto alba* (Aves: Strigiformes) along the mid-Araguaia River in central Brazil. Zoologia 28: 709–716. https://doi.org/10.1590/S1984-46702011000600003

Rohlf FJ (1999) Shape statistics: Procrustes superimposition and tangent spaces. Journal of Classification 16: 197–223. https://doi.org/10.1007/s003579900054

Rohlf FJ (2015) The tps series of software. Hystrix, the Italian Journal of Mammology 26: 9–12. https://doi.org/10.4404/hystrix-26.1-11264.

Rohlf FJ, Slice D (1990) Extensions of the Procrustes method for the optimal superimposition of landmarks. Systematic Zoology 39: 40–59. https://doi.org/10.2307/2992207

Salazar-Bravo J, Pardiñas UF, D'Elía G (2013) A phylogenetic appraisal of Sigmodontinae (Rodentia, Cricetidae) with emphasis on phyllotine genera: systematics and biogeography. Zoologica Scripta 42: 250–261. https://doi.org/10.1111/zsc.12008

Salazar-Bravo J (2015) Genus *Calomys*. In: Patton J, Pardiñas UFJ, D'elía G (Eds) Mammals of South America, vol. 2. The University of Chicago Press, Chicago, 481.

Salles LO, Cartelle C, Guedes PG, Boggiani P, Janoo A, Russo CAM (2006) First report on Quaternary mammals from the Serra da Bodoquena, Mato Grosso do Sul, Brazil. Boletim do Museu Nacional 521: 1–12

Salles LO, Carvalho GS, Weksler M, Sicuro FL, Abreu F, Camardella AR, Guedes PG, Avilla LS, Abrantes EAP, Sahate V, Costa ISA (1999) Fauna de mamíferos do Quaternário de Serra da Mesa (Goiás, Brasil). Publicações Avulsas do Museu Nacional 78: 1–15

Scheibler DR, Christoff AU (2007) Habitat associations of small mammals in southern Brazil and use of regurgitated pellets of birds of prey for inventorying a local fauna. Revista Brasileira de Biologia 67: 619–625. https://doi.org/10.1590/S1519-69842007000400005

Travassos LE (2010) Considerações sobre o carste da região de Cordisburgo, Minas Gerais, Brasil. Belo Horizonte, Tradição Plana.

Winge H (1888) Jordfundne og nulevende Gnavere (Rodentia) fra Lagoa Santa, Minas Geraes, Brasilien. E Museo Lundii 1: 1–200.

Zelditch LM, Swiderski DL, Sheets HD, Fink WL (2004) Geometric Morphometrics for Biologists: A Primer. Elsevier Academic Press, San Diego, 436 pp.

Author Contributions: NLB and LL conducted the experiments, analyzed the data and wrote the paper. PSRR analyzed the data and wrote the paper. GL reviewed the paper.

Competing Interests: The authors have declared that no competing interests exist.

16

Review of *Coeliaria* (Coleoptera: Coccinellidae: Chnoodini)

Julissa M. Churata-Salcedo[1], Lúcia M. Almeida[1]

[1]*Laboratório de Sistemática e Bioecologia de Coleoptera, Department of Zoology, Universidade Federal do Paraná. Caixa Postal 19030, 81581-980 Curitiba, PR, Brazil.*
Corresponding author: Lúcia M. Almeida (lalmeida@ufpr.br)

http://zoobank.org/500319E5-EFA3-4B84-8188-B8BE475B7D4A

ABSTRACT. *Coeliaria* Mulsant, 1850 is revised based on the external morphology and genitalia of the adults and is distinguished from the other Chnoodini by the following characters: dorsal surface pubescent; antenna 11-segmented; hypomera with rounded fovea; tibia flat and angulated; abdominal postcoxal line incomplete, recurved and with oblique line. A new species, *Coeliaria castanea* **sp. nov.**, from Brazil, and two new combinations, are proposed: *Coeliaria bernardinensis* **comb. nov.** and *C. luteicornis* **comb. nov.**, expanding the distribution of the genus to Bolivia and Paraguay.

KEY WORDS. Coccinellinae, Neotropical Region, taxonomy.

INTRODUCTION

Coccinellidae Latreille, 1807 is an ecologically and morphologically diverse group of predators that are often used in biological control programs of insect pests (Hodek and Honěk 1996).

Since the establishment of Coccinellidae as a family, several authors have proposed classification systems for it. Beginning in the second half of the nineteen-century, Mulsant (1846, 1850) proposed a generic classification based on the pubescence patterns. He was followed Crotch (1874) and Chapuis (1876), who divided the family into "Aphidiphages" and "Phytophages". Casey (1899) recognized 16 tribes, and other authors worked on an improved classification system for the family (Weise 1895, Sicard 1907, 1909). Korschefsky (1931, 1932), in his catalog, recognized three subfamilies and twenty tribes in Coccinellidae, whereas Bouchard et al. (2011) considered only two: Microweiseinae and Coccinellinae. In addition Bouchard et al. (2011) revalidated the name Chnoodini, which was accepted and followed by Seago et al. (2011), González (2013) and Krüger et al. (2016). Nedved and Kovář (2012), however, continued using the name Exoplectrini in their chapter on the phylogeny and classification of Coccinellidae, where they listed 20 genera for the tribe. Although the name Exoplectrini was used by Nedved and Kovář (2012) after the work of Bouchard et al. (2011) to name the tribe, Chnoodini is the correct name for the group, since it was first used by Mulsant (1850) as "Chnoodiens" (Principle of priority)

and later Latinized and used by Sicard (1909). Currently, in the Neotropical Region, Chnoodini includes *Chnoodes* Chevrolat, 1849, *Coeliaria* Mulsant, 1850, *Dapolia* Mulsant, 1850, *Dioria* Mulsant, 1850, *Exoplectra* Chevrolat, 1844, *Gordonita* González, 2013, *Incurvus* González, 2013, *Neorhizobius* Crotch, 1874, *Sidonis* Mulsant, 1850 and *Siola* Mulsant, 1850.

Mulsant (1850) described *Coeliaria* for *Exoplectra erythrogaster* Mulsant, 1850 based on the presence of "a deep fovea in the hypomera".

Crotch (1874) briefly redescribed *Coeliaria* and indicated the presence of thoracic foveae and epipleura subfoveolate in *C. erythrogaster*. Chapuis (1876) treated *Coeliaria* as a subgenus of *Exoplectra*, indicating that the thoracic fovea of *C. erythrogaster* differs from the condition found in the other species of *Exoplectra*, which have a flat epipleura. Gemminger and Harold (1876), Korschefsky (1932) and Blackwelder (1945) listed *C. erythrogaster* in the monotypic genus, in their catalogs/checklist. Gordon (1994) and Fürsch (1990a,b) placed *Coeliaria* in Exoplectrini, and González (2013) recorded *C. erythrogaster* from Paraguay.

The Brazilian species of *Exoplectra* were reviewed by Costa et al. (2008), who provisionally removed *E. bernardinensis* and *E. luteicornis* from the genus.

In this paper, *Coeliaria* is revised based on its external morphology and genitalia. One new species from Brazil and two new combinations are proposed, expanding the size and distribution records of the genus.

MATERIAL AND METHODS

The specimens examined were provided by the California Academy of Sciences, California, USA (CAS); Coleção Entomológica Pe. J.S. Moure, Universidade Federal do Paraná, Curitiba, Paraná, Brazil (DZUP); Museu de Ciências Naturais, Fundação Zoobotânica, Rio Grande do Sul, Porto Alegre, Brazil (MCN); Museé d'Histoire Naturelle de Lyon, France (MHNL); Museu Nacional do Rio de Janeiro, Rio de Janeiro, Brazil (MNRJ); Museu de Zoologia da Universidade de São Paulo, São Paulo, Brazil (MZSP); United States National Collection, Smithsonian Institution, Washington, DC, USA (USNM); and the Zoological Museum, University of Copenhagen, Denmark (ZMUC).

Parts that were dissected from specimens (mouthparts, antennae, legs, abdomen and genitalia) were stored in microvials with glycerin. The microvials were pinned together with the respective specimen. Photographs were taken using a Sony Cyber-shot (DSC-W300) digital camera coupled to a Zeiss Stemi SV6 compound stereomicroscope and a Zeiss Stereo Discovery Standard 20 microscope.

The terminology used in the descriptions follows Costa et al. (2008) and Krüger et al. (2016).

The labels of the type material are arranged in sequence from top to bottom, with the data for each label within double quotes (" "); slashes (/) separate the rows, and the information between brackets ([]) provides additional details recorded on the labels.

TAXONOMY

Coeliaria Mulsant, 1850

Figs 1–12

Coeliaria Mulsant, 1850: 1042 (description); Crotch 1874: 283 (description); Chapuis 1876: 242 (description); Gemminger and Harold 1876: 3801 (catalog); Korschefsky 1932: 229 (catalog); Blackwelder 1945: 451 (checklist); Gordon 1987: 34 (catalog); Fürsch 1990a: 4 (catalog); Fürsch 1990b: 9 (catalog); Gordon 1994: 683 (taxonomy); Fürsch 2007: 1 (catalog); Costa et al. 2008: 366 (citation); González 2013: 64 (distribution).

Type species. *Exoplectra erythrogaster* Mulsant, 1850 (original designation).

Diagnosis. *Coeliaria* is a Neotropical genus that resembles *Exoplectra* Mulsant, 1850, and *Gordonita* González, 2013, by the angulation of the tibia. *Coeliaria* is distinguished from *Exoplectra* by the presence of a fovea in the hypomera; pronotum with the inner angles more prominent and emarginated. In *Gordonita*, the fovea of the hypomera is small and deep; the body is elongated and depressed. *Coeliaria* is distinguished from the other genera of Chnoodini by the following combination of characters: Body black or dark, without spots; dorsal surface pubescent; antenna 11-segmented; labrum truncated; hypomera with rounded fovea; tibia flat and angulated; abdomen with five visible sternites (females) or six sternites (males); abdominal postcoxal line incomplete, recurved, oblique line present.

Redescription. Body rounded or oval, convex, with yellowish or whitish pubescence, fine and dense, with punctuation fine and sparse. Integument brownish or black, with green, bluish, or bronze metallic reflections, without spots. Ventral surface reddish, brown, or black. Head black or brownish; clypeus merged with forehead, without fronto-clypeal suture, expanded laterally and with rounded front edge, distinctly emarginated; eyes divided by the gena (Figs 1, 2) partially covered by the pronotum. Antennae 11-segmented with conspicuous club (Fig. 3). Labrum transverse, truncate anteriorly (Fig. 4); mandibles asymmetric, robust with apex bifid (Figs 5, 6); maxillae with last segment of palpus distinctly securiform; labium with short bristles on ligule (Fig. 7). Pronotum transverse, narrower than elytra, anterior border emarginated with lateral margin straight, rounded above and subsinuous posteriorly. Hypomera with large, deep and rounded fovea (Fig. 9). Prosternal process with rounded apex, without carina (Fig. 8). Elytra expanded with projected humeral callus, with anterior margin truncate; elytral epipleuron wide, deeply excavated for reception of femoral apex, with carina parallel in the inner margin, curved at base (Fig. 10). Legs with relatively wide femora, excavated for reception of tibiae; tibiae flattened, with acute angulation on outer margin (Figs 11, 12); claws bifid. Abdomen with five (females) or six (males) visible ventrites, with descending post-coxal line, attached to the posterior edge of the first ventrite, with oblique line.

Male genitalia. Tegmen with penis guide and symmetrical parameres; penis slender, with developed penis capsule.

Female genitalia. Coxites elongated and sub-triangular; spermatheca C-shaped, simple.

Key to the species of *Coeliaria* Mulsant, 1850

1. Body black or dark brown, with reflections of different colors 2
1'. Body black, without reflections, elytra very expanded, 5.67 to 10.0 mm (Figs 13–16) ....*C. erythrogaster* Mulsant, 1850
2. Body longer than wide, black, with bronze or blue reflections, 3.08 to 4.0 mm (Figs 38–41) ....*C. luteicornis* (Mulsant, 1850), comb. nov.
2'. Body as long as wide, black or brown, with reflections of other colors 3
3. Bluish reflections, body 3.0 to 4.0 mm .... *C. bernardinensis* (Brèthes, 1925), comb. nov.
3'. Brown reflections, body 5.0 mm (Figs 47–50) ....*C. castanea* sp. nov.

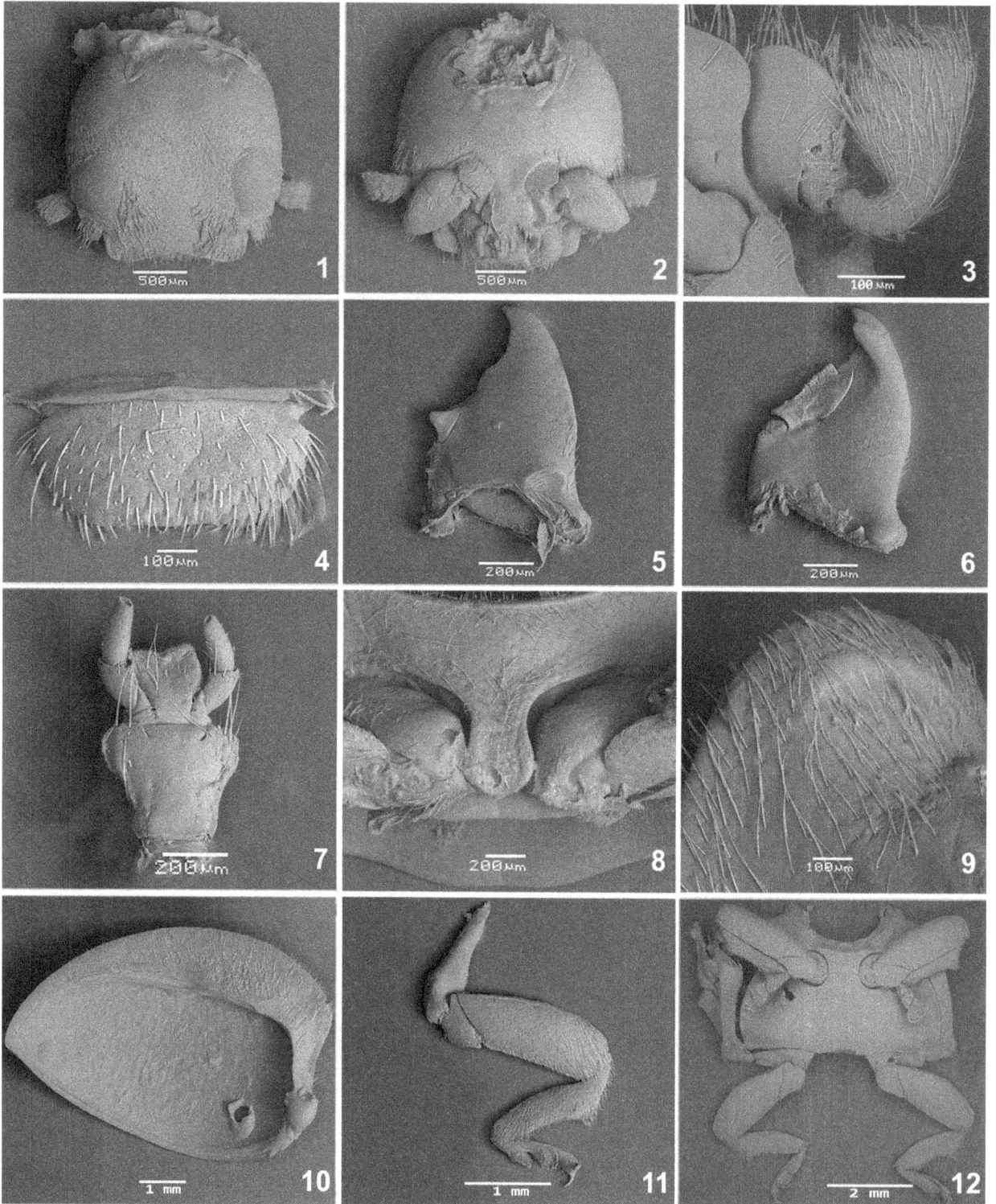

Figures 1–12. Scanning electron microscopy of *Coeliaria erythrogaster*: (1) head, dorsal view, (2) head, ventral view; (3) antenna, lateral view; (4) labrum; (5) mandibles, dorsal view, (6) mandibles, ventral view; (7) labium; (8) prosternal process; (9) hypomera; (10) elytra, ventral view; (11) leg, dorsal view; (12) metathorax, ventral view.

Coeliaria erythrogaster Mulsant, 1850

Figs 13–28, 55

Coeliaria erythrogaster Mulsant, 1850: 1042 (description); Crotch 1874: 283 (description); Gemminger and Harold 1876: 3801 (catalog); Korschefsky 1932: 229 (catalog); Blackwelder 1945: 451 (checklist); Gordon 1987: 34 (catalog); Fürsch 1990b: 9 (systematic); Costa et al. 2008: 367 (revision, Brazilian species).
Exoplectra erythrogaster Mulsant, 1850: 916 (original description); Mariconi and Zamith 1959: 261 (biology); Mariconi and Zamith 1960: 229 (biology).

Redescription. Male. Length 5.67–10.00 mm, width 5.42–9.17 mm. Body hemispherical, rounded and convex, with fine pubescence, short, thick, whitish or yellowish, with thin, sparse punctuation; integument black and dark brown (Figs 13–16, 24–26). Head dark, antennae and mouthparts reddish (Fig. 14). Pronotum transverse, narrower than elytra, with anterior margin emarginate, lateral margin straight, rounded anteriorly, and posterior margin subsinuous (Fig. 15), hypomera with deep, rounded fovea. Prosternal process with rounded apex, without carina, longer than wide, strangulated at base (Fig. 8). Scutellum black. Elytra dark, without spots and strongly expanded (Figs 15, 25); epipleurae wide, narrowing towards apex, fovea shallow for reception of femoral apex (Figs 10, 14). Meso- and metaventrite black. Legs black, with flattened femora and tibiae, with acute angulation on outer margin (Figs 11, 14). Abdomen with incomplete post-coxal line, attached to posterior margin of first ventrite, with oblique line (Figs 17, 27). Genitalia symmetrical; tegmen with penis guide symmetrical, broad at base, narrowing at apex; parameres slightly wide, with short bristles, slightly larger than penis guide (Figs 18, 19). Penis sclerotized, J-shaped, with rounded apex, penis capsule with highly developed inner arm (Fig. 20).

Female. Similar to the male. Coxites longer than wide, sub-triangular, with long bristles; mammiliform style (Fig. 22). Spermatheca C-shaped, with acute apex (Fig. 21).

Type material. It was only possible to examine photographs of the syntype (Figs 24–28) deposited in the Musée des Confluences, Lyon, France (MHNL). In 1970, R.D. Gordon indicated, on a label, that the specimen as the Lectotype, but this designation was not published. Here we designate this specimen as the Lectotype.

Material examined. Bolivia. Santa Cruz: Roboré, 28-II to 1-III-1954, C. Gans-F. Pereira leg., 1 specimen (DZUP 188194). Brazil. Goiás: Faz. Cachoeirinha, Jatai, X. 1962, Exp. Dep. Zool., 1 specimen (MZSP); Minas Gerais: Faz dos Campos, XII-1920, Col. J.F. Zikán, 1 specimen (DZUP 146675); Espírito Santo: Sta. Teresa, 12-X-64, C. Elias leg., 1 specimen (DZUP 185643); 19-X-64, C. Elias leg., 2 specimens (DZUP 188181, 185641); 26-X-64, C. Elias leg., 1 specimen (DZUP 185642); São Paulo: Marília, 1.XI.945, Coll. H. Zellibor, 2 specimens (MNRJ); São Paulo, Mus. Pragense, Korschefsky Collection 1952, 1 specimen (USNM); (Jabaquara), 10.XII. 45, Coll. H. Zellibor, 1 specimen (MNRJ); Rio de Janeiro: Itatiaia, I-1929, Coll. J.F. Zikán, 1 specimen (DZUP 185612); (Corcovado), VII. 1958, Alvarenga and Seabra, Coll. M. Alvarenga, 1 male (DZUP 188193);

18-IX-61, J.S. Moure, Alvarenga and Seabra, 1 male (DZUP 288378); X-1961, Seabra and Alvarenga leg., 2 specimens (DZUP 185644, 185645); 8.X.1962, Alvarenga and Seabra, Coll. M. Alvarenga, 1 female (DZUP 288374); XI.1962, Alvarenga and Seabra, Coll. M. Alvarenga, 1 specimen (DZUP 188192); X.1966, Alvarenga and Seabra, Coll. M. Alvarenga, 1 specimen (DZUP 288373); XI.1967, Alvarenga and Seabra, Coll. M. Alvarenga, 1 specimen (DZUP 288375); 30.X.1975, M.A. Monné, 1 specimen (MNRJ); 1.X.1976 M.A. Monné, 1 specimen (MNRJ); 7.X.1976 M.A. Monné, 1 specimen (MNRJ); XI.1955, Alvarenga and Seabra, 1 female (MNRJ); X.1958, Alvarenga and Seabra, 1 specimen (MNRJ); XI.1958/ Alvarenga and Seabra", 3 specimens (MNRJ); XII.1958, Alvarenga and Seabra, 1 female (DZUP 288376); Nova Friburgo, IV.2005, E.J. Grossi col., 1 specimen (DZUP 132000); Paraná: Arapongas, XII.1951, A. Maller, 1 female (DZUP 185640); Santa Catarina: Seara (Nova Teutônia), 27°11'8", 52°23'1", Fritz Plaumann, 1.X.1949, 1 specimen (FPNT); XI.1953, 1 specimen (FPNT); Corupá, Hansa Humbolt, Oct. 1944, 1 specimen (CAS); 2-I-38, 1 specimen (DZUP 188185); Mus. Westerm., 1 specimen (ZMUC).

Geographical distribution. Bolivia, Brazil and Paraguay (Fig. 55).

Remarks. Coeliaria erythrogaster, first described in Exoplectra, was characterized by presenting a fovea in the hypomera, differentiating it from all other species of Exoplectra. Since then it was considered the only species of Coeliaria. In addition to this character, C. erythrogaster has strongly expanded elytra, pubescence very dense and uniform, and is larger than the other species of the genus.

Biological data. Costa Lima (1950) published a report of the metamorphosis of C. erythrogaster larvae, and mentioned that it was covered by abundant waxy secretions. Mariconi and Zamith (1959, 1960) described the larvae and adult, as well as biological aspects of E. erythrogaster preying on Mimosicerya hempeli (Cockerell, 1899) (Hemiptera: Margarodidae) on the plant Cassia fistula Linnaeus (Fabaceae) in Piracicaba, São Paulo, Brazil. In their description of the adult they mentioned various bare areas, apparently with shorter and thinner pubescence. This description is consistent with Crotch (1874), who mentioned that recently collected specimens have gray pubescence, giving the impression of bare spots, which is very peculiar. According to those authors, the larvae are completely covered by white secretions that form a mass of conspicuous flaky wax, which extends 30 to 35 mm in length.

Coeliaria bernardinensis (Brèthes, 1925), comb. nov.

Figs 29–37, 55

Exoplectra bernardinensis Brèthes, 1925: 8 (original description); Korschefsky 1932: 227 (catalog); Denier 1939: 581 (list); Blackwelder 1945: 450 (checklist); Costa et al. 2008: 365 (revision, Brazilian species).

Redescription. Male. Length 3.08–4.00 mm, width 2.58–3.83 mm. Body rounded, with fine pubescence, thick,

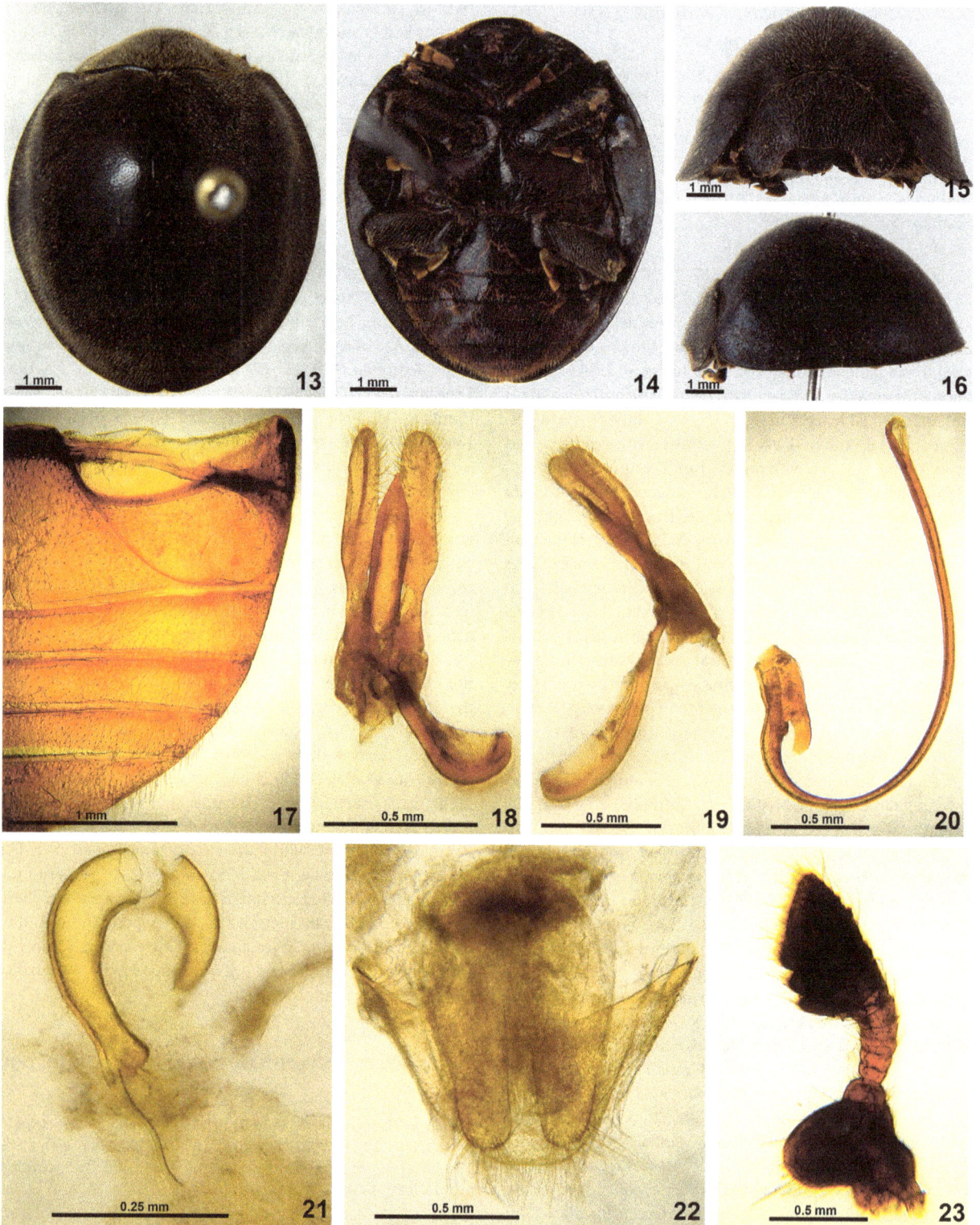

Figures 13–23. *Coeliaria erythrogaster*: (13) dorsal view, (14) ventral view, (15) frontal view, (16) lateral view, (17) abdomen. Male genitalia: tegmen (18) dorsal view, (19) lateral view; (20) penis; female genitalia (21) spermatheca, (22) coxites. (23) Antenna.

Figures 24–28. *Coeliaria erythrogaster*, type material from Musée des Confluences, Lyon, France (MNHL): (24) dorsal view, (25) frontal view, (26) lateral view, (27) abdomen, (28) labels.

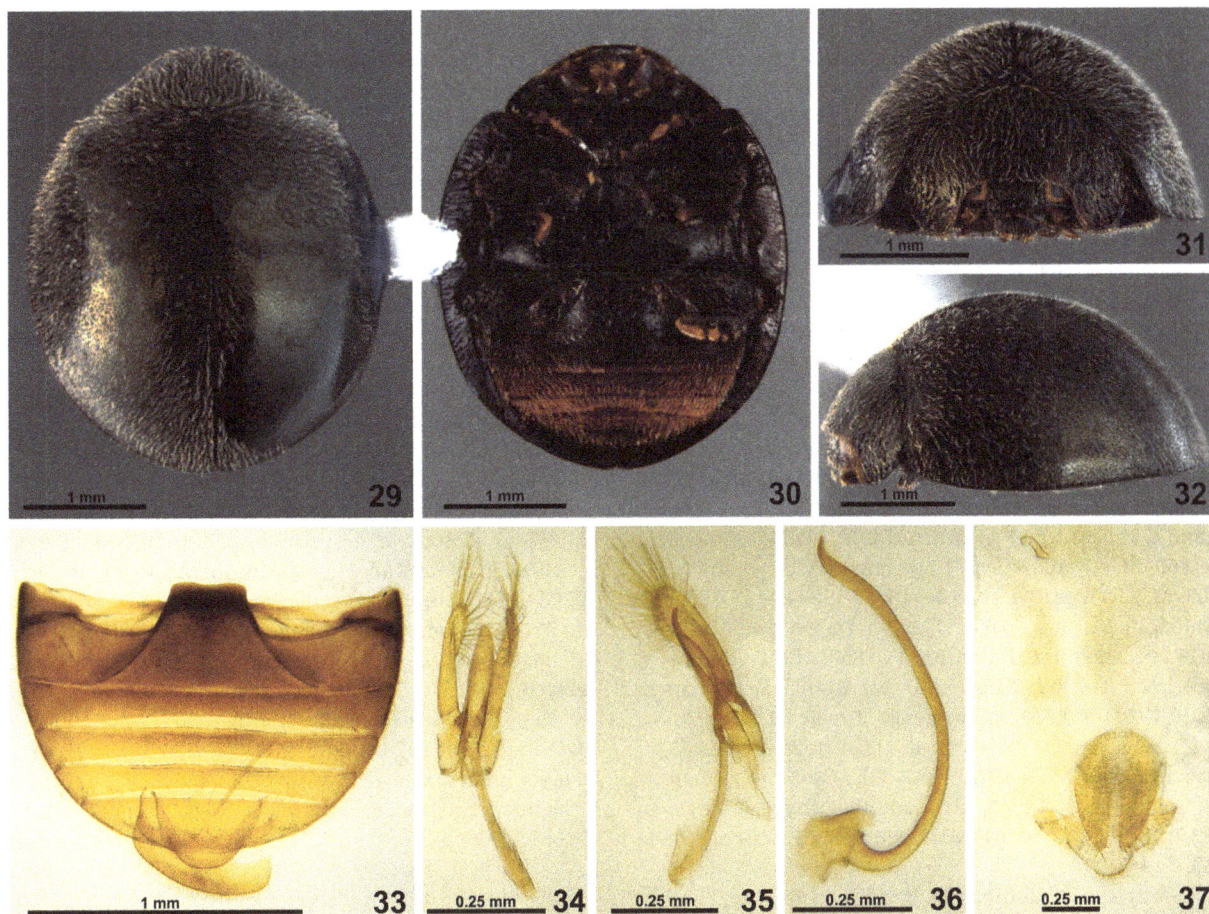

Figures 29–37. *Coeliaria bernardinensis* comb. nov.: (29) dorsal view, (30) ventral view, (31) frontal view, (32) lateral view, (33) abdomen. Male genitalia: tegmen (34) dorsal view, (35) lateral view; (36) penis; (37) female genitalia (coxites and spermatheca).

whitish, with thin, sparse punctuation; integument black with blue or green metallic reflections (Figs 29–32). Head, antennae and mouthparts dark or reddish (Fig. 30). Pronotum transverse, narrower than elytra, with emarginated anterior margin, straight lateral margin, rounded anteriorly, and subsinuous posterior margin (Fig. 31), hypomera with deep fovea. Prosternal process with sub-quadrangular apex, without carina, as wide as long. Scutellum black. Elytra dark without spots (Figs 29, 32); epipleura narrowing towards apex, with fovea for reception of femoral apex (Fig. 30). Meso- and metaventrite black. Legs black, with flattened femora and tibiae with acute angulation at the outer margin. Abdomen with central region of first ventrite blackish, others lighter, with oblique line (Fig. 33). Genitalia symmetrical: tegmen with symmetrical penis guide, wide at base and narrow at apex, which is slightly recurved; parameres with long bristles slightly larger than penis guide (Figs 34, 35). Penis sclerotized, with acuminate and recurved apex, developed penis capsule (Fig. 36).

Female. Similar to male. Coxites elongated, longer than wide, sub-triangular with long bristles (Fig. 37). Spermatheca more or less recurved, C-shaped, with undeveloped ramus (Fig. 37).

Type material. According to Gordon (1987) the holotype would be found in British Museum of Natural History, London, England. However, Horn and Kahle (1935–1937) indicated that the type material was deposited in the Museo Argentino de Ciencias Naturales "Bernardino Rivadavia" (MACN), Buenos Aires, Argentina, but it was not possible to study this material, because MACN does not lend material for study.

Material examined. Brazil. *Alagoas*: Maceió, VI. 1993, Lima, I.M.M., 1 female (DZUP 131993); *Goiás*: Dianópolis, 11–14.I.1962, J. Bechyné col., 1 specimen (DZUP 192029); Corumbá de Goiás, 5. II.1962, J. Bechyné col., 1 specimen (DZUP 192042); Jataí, Cerrado, Faz. Nova Orlandia, I. 964, Martins, Morgante and Silva, 1 specimen (DZUP 192033); *Minas Gerais*: Belo Horizonte (Campus-UFMG) 17.VI.81, C13, N. S. Domingos, 1 male (DZUP 192032); 07.V.81, B17, N.S. Domingos, 1 specimen (DZUP 192037); 06.XII.82, B17, N.S. Domingos, 1 male (DZUP 192030); *Espírito Santo*: Conceição da Barra (BR 16), 21/IX/68, C. and C.T. Elias leg., 1 specimen (DZUP 192036); Guarapari, IX-1960, M. Alvarenga leg., 1 specimen (DZUP 192043); IX-1960, M. Alvarenga leg., 1 male (DZUP 192045); IX-1960, M. Alvarenga leg., 1 male (DZUP 192025); IX-1960, M. Alvarenga leg., 1 specimen (DZUP 192040); IX-1960, M. Alvarenga leg., 1 specimen (DZUP 192035); XI-61, M. Alvarenga, 1 specimen (DZUP 192044); XI-61, M. Alvarenga, 1 male (DZUP 192034); XI-61, M. Alvarenga, 1 specimen (DZUP 192038); XI-61, M. Alvarenga, 1 specimen (DZUP 192039); *Mato Grosso*: Chapada dos Guimarães, 20-I-1961, MT/J. and B. Bechyné, 1 specimen (DZUP 187185); 23-I-1961, MT/J. and B. Bechyné, 1 specimen (DZUP 146701).

Geographical distribution. Brazil and Paraguay (Fig. 55).

Remarks. *Coeliaria bernardinensis* was first described as *Exoplectra bernardinensis*, but is herein transferred to *Coeliaria*

based on the presence of a deep fovea in the hypomera, which characterizes *Coeliaria*. It differs from the other species of *Coeliaria* by its small size and pattern of genitalia.

Coeliaria luteicornis (Mulsant, 1850), comb. nov.

Figs 38–46, 55

Exoplectra luteicornis Mulsant, 1850: 919 (original description); Crotch 1874: 284 (synonymy); Korschefsky 1932: 227 (catalog); Blackwelder 1945: 450 (checklist); Costa et al. 2008: 365, 373, 374 (revision, Brazilian species).

Redescription. Male. Length 5.50–6.25 mm, width 4.42–5.25 mm. Body ovate, with fine, short, thick, whitish or yellowish pubescence; with thin, sparse punctuation; integument black with blue metallic reflections (Figs 38–41). Head, antennae and mouthparts dark, reddish (Fig. 39). Pronotum transverse, narrower than elytra, with emarginated anterior margin, straight lateral margin, rounded anteriorly, and with posterior margin subsinuous (Fig. 40), hypomera with rounded fovea (Fig. 39). Prosternal process sub-quadrangular, apex without carina (Fig. 39). Scutellum black. Elytra dark, without spots (Figs 38–41); epipleura wide, narrowing towards apex, with slight fovea for reception of femoral apex (Fig. 39). Meso- and metaventrite black. Legs black with flattened femora, and tibiae with acute angulation at the outer margin (Fig. 39). Abdomen reddish, with oblique line (Figs 39, 41). Genitalia symmetrical; tegmen with penis guide wide at base and narrowing at apex; parameres wide, with short bristles, slightly larger than penis guide (Figs 43, 44). Penis sclerotized, J-shaped, with rounded apex, developed penis capsule (Fig. 45).

Female. Similar to male. Coxites sub-triangular, with long bristles; spermatheca C-shaped with highly developed ramus (Fig. 46).

Type material. It was not possible to study the type material, which, according to Gordon (1987) is deposited in the Muséum National d'Histoire Naturelle, Paris, France. We were not able to locate this specimen there.

Material examined. Brazil. *São Paulo*: Mairiporã, 4–13. I. 1967, C. Costa col., 1 specimen (DZUP 188384); *Paraná*: P. Grossa, Pedreira, Coleção F. Justus Jor, 2 specimens (DZUP 192084, 288380); *Santa Catarina*: Seara (Nova Teutônia), XI.1951, F. Plaumann col., 1 female (DZUP 192101); XI.1951, F. Plaumann col., 4 specimens (DZUP 192055, 192066, 192082, 192083); X.1965, F. Plaumann col., 1 specimen (DZUP 192100); XI.1965, F. Plaumann col., 8 specimens (DZUP 192077–192078, 192080–192081, 192085–192086–192087, 192099); I.1966, F. Plaumann col., 2 specimens (DZUP 192053, 192065); XI.1966, F. Plaumann col., 6 specimens (DZUP 192052, 192059–192060–192061, 192064, 192079); 27°11'B. 52°23'L, Fritz Plaumann, I.1974, 6 specimens (DZUP 192057, 288382–288383, 192058, 288385–288386); V.1974, 2 specimens (DZUP 192046, 192026); X.1974, 1 female (DZUP 192056); X.1974, 2 specimens (DZUP 192047, 288387); X.1974, 4 specimens (DZUP 192048, 192051,

Figures 38–46. *Coeliaria luteicornis* comb. nov.: (38) dorsal view, (39) ventral view, (40) frontal view, (41) lateral view, (42) abdomen. Male genitalia: tegmen (43) dorsal view, (44) lateral view; (45) penis; (46) female genitalia (coxites and spermatheca).

288384,192049); XI.1974, 5 specimens (DZUP 192050,192062, 288381, 192063, 288388); Nov 74, 1 specimen (DZUP 192028); XII.1974, 1 specimen (DZUP 192054); Fritz Plaumann, 1 female (DZUP 192027); *Rio Grande do Sul*: Derrubadas, 27.X. 2003, I. Heydrich col., 1 specimen (MCN 228613); 28.X. 2003, I. Heydrich col., 1 specimen (MCN 227521); 30.X.2003, A. Barcellos col., 27°14′14.7″S, 53°58′46.0″W, 1 specimen (MCN 227530); 31.X.2003, L. Moura col., 1 specimen (MCN 227520); 20.X. 2004, A. Barcellos col., 1 specimen (MCN 231547); 20.X.2004, L. Moura col., 1 specimen (MCN 231575); 21.X. 2004, L. Moura col., 5 specimens (MCN 231544, 231593, 231581, 231592, 231589); 21.X. 2004, R. Ott col., 2 specimens (MCN 231566, 231596); 21.X. 2004, L. Podgaiski col., 1 specimen (MCN 231556); 22.X. 2004, I. Heydrich col., 2 specimens (MCN 231582, 230461); 22. X. 2004, R. Ott col., 1 specimen (MCN 231579); 22.X. 2004, A. Barcellos col., 2 specimens (MCN 230462, 230463); 22.X. 2004, L. Moura col., 1 specimen (MCN 231610); 22.X. 2004, L. Podgaiski col., 1 specimen (MCN 230479).

Geographical distribution. Brazil (Fig. 55).

Remarks. *Coeliaria luteicornis* shows a deep fovea in the hypomera, characteristic of *Coeliaria*; additionally, it has an oval body with whitish or yellowish pubescence. The male genitalia is similar to that of *C. erythrogaster*, but the female differs in the shape of the spermatheca. The most distinctive characteristic of this species is its oval body.

Figures 47–54. *Coeliaria castanea* sp. nov.: (47) dorsal view, (48) ventral view, (49) frontal view, (50) lateral view, (51) abdomen. Female genitalia (52) coxites, (53) spermatheca; (54) labels.

Coeliaria castanea sp. nov.

http://zoobank.org/BC6551EB-703D-4B3E-82AA-B8A2177A6729

Figs 47–54, 55

Description. Holotype female. Length 5.58 mm, width 5.25 mm. Body rounded, integument light brown, white pubescence (Fig. 47–50). Pronotum transverse, narrower than elytra, anterior margin emarginated yellowish, lateral margin rounded, posterior margin subsinuous (Fig. 49), hypomera with rounded fovea (Fig. 48). Prosternal process sub-quadrangular, apex without carinae, as long as wide (Fig. 48). Ventral color lighter than dorsal (Fig. 48). Scutellum brown. Elytra brownish, without spots (Figs 47–50); epipleura wide, narrowing to apex, with fovea for reception of femoral apex. Meso- and metaventrite brownish (Fig. 48). Legs brownish, flattened femora and tibiae, with acute angulation at the outer margin. Abdomen yellowish, with oblique line (Fig. 51).

Genitalia symmetrical; coxites very elongated, three times longer than wide, styles with long bristles (Fig. 52). Spermatheca C-shaped, with sclerotized infundibulum (Fig. 53).

Male. Unknown.

Material examined. The holotype female is labeled as follows: Brazil, *Santa Catarina* "Nova Teutônia/SC, Brasil/XI. 1966/F. Plaumann col.", "♀", "HOLOTIPO/*Coeliaria castanea*" Churata-Salcedo & Almeida [red label], 1 specimen "DZUP/186709" [DZUP] (Fig. 54). The holotype is double mounted, and is in good condition (genitalia on microvial with glycerin).

Etymology. The species epithet, *castanea*, is a reference to the color pattern of this species.

Geographical distribution. Brazil (Santa Catarina) (Fig. 55).

Remarks. *Coeliaria castanea* sp. nov. resembles *C. erythrogaster* by having very conspicuous, deep and rounded fovea, but differs in the brownish color and by the shape of the female genitalia.

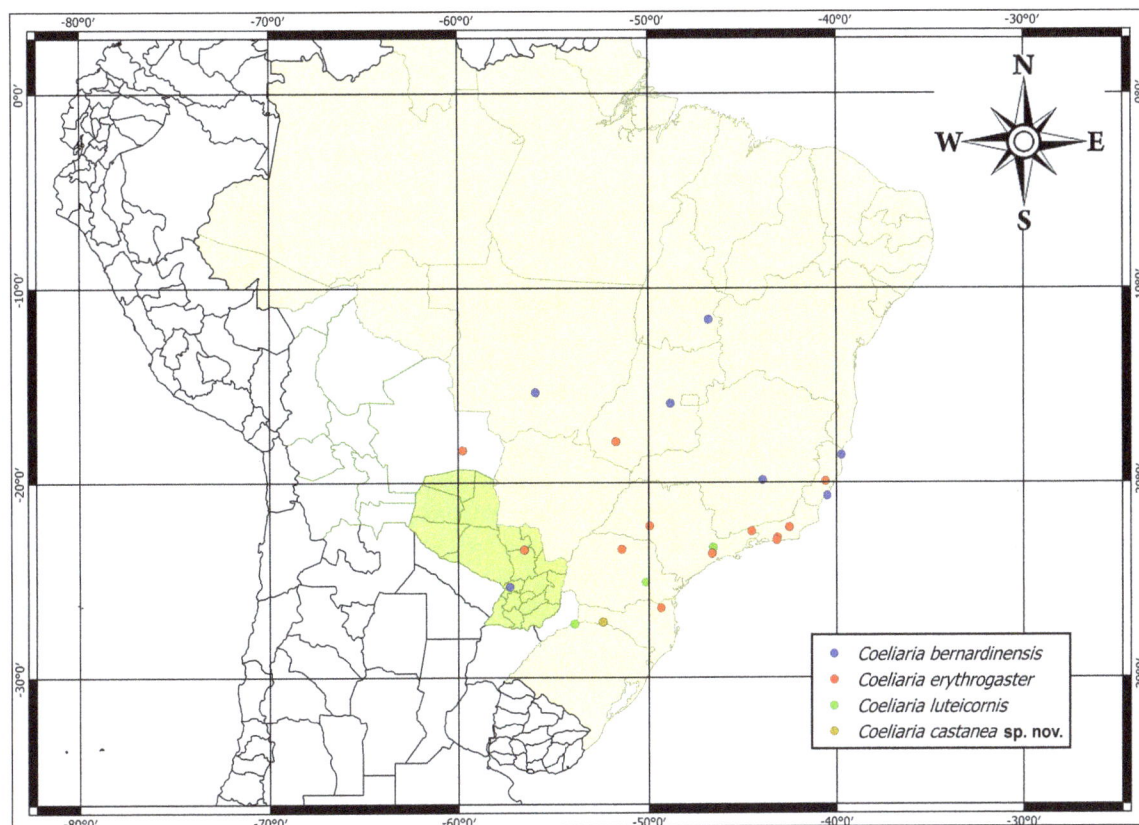

Figure 55. Map showing the known geographical distribution of the species of *Coeliaria*.

ACKNOWLEDGMENTS

We thank Coordenação de Aperfeiçoamento de Pessoal de Nível Superior for the Master's fellowship to JMCS (CAPES 1347909/2014), Conselho Nacional de Desenvolvimento Científico e Tecnológico for the research fellowship to LMA (CNPq 309764/2013-0), the Centro de Microscopia Eletrônica (UFPR) and Roy Funch who helped with the English of the previous version of this manuscript.

REFERENCES

Blackwelder RE (1945) Checklist of the Coleopterous insects of Mexico, Central America, the West Indies, and South America. Bulletin of the United States National Museum 185: 1–188. https://doi.org/10.5479/si.03629236.185.3

Bouchard P, Bousquet Y, Davies AE, Alonso-Zarazaga MA, Lawrence JF, Lyal CHC, Newton AF, Reid CAM, Schmitt M, Ślipiński A, Smith ABT (2011) Family-group names in Coleoptera (Insecta). ZooKeys 88: 1–972. https://doi.org/10.3897/zookeys.88.807

Brèthes J (1925) Coccinellides du British Museum. Nunquam Otiosus IV, 1–10.

Casey TD (1899) A revision of the American Coccinellidae. Appendix II: on South American Coccinellidae. Journal of the New York Entomological Society 7: 168–169.

Chapuis F (1876) Famille des phytophages des érotyliens des endomychides et des coccinellides. Tomo 12. In: Lacordaire Jt, Chapuis F (Eds) Histoire naturelle des insectes. Genera des Coléoptères. Roret, Paris, 424 pp.

Chevrolat LA (1844) *Exoplectra*. In: d'Orbigny CD (Ed.) Dictionaire Universel d'Histoire Naturelle. L. Houssiaux, Paris [1861], vol. 5, 545 pp.

Chevrolat LA (1849) *Chnoodes*. In: d'Orbigny CD (Ed.) Dictionnaire Universel d'Histoire Naturelle. L. Houssiaux, Paris [1861], vol. 3, 612 pp.

Costa Lima A (1950) Nota sobre a larva de uma joaninha (Coleoptera, Coccinellidae). Revista de Entomologia 21: 592–593.

Costa AV, Almeida LM, Corrêa GH (2008) Revisão das espécies brasileiras do gênero *Exoplectra* Chevrolat (Coleoptera, Coccinellidae, Exoplectrinae, Exoplectrini). Revista Brasileira de Entomologia 52: 365–383. https://doi.org/10.1590/S0085-56262008000300010

Crotch GR (1874) A revision of the Coleopterous Family Coccinellidae. E.W. Janson, London, 311 pp. https://doi.org/10.5962/bhl.title.8975

Denier CL (1939) De Coccinellidis Brethesianis. Typorum Speciarum Recensio. Physis 17: 569–587.

Fürsch H (1990a) Taxonomy of Coccinellids. Coccinella 2: 4–6.

Fürsch H (1990b) Valid genera and subgenera of Coccinellidae. Coccinella 2: 7–18.

Fürsch H (2007) Taxonomy of Coccinellids. Coccinella 6: 1–3.

Gemminger M, Harold B (1876) Chrysomelidae (Par II.), Languridae, Erotylidae, Endomychidae, Coccinellidae, Corylophidae, Platypsyllidae. Família LXXII: Coccinellidae, Tom. 12. In: Gemminger M, Harold B (Eds) Catalogus Coleopterorum hucusque descriptorum synonymicus et systematicus, 3740–3818.

González G (2013) Gordonita n. gen. y otros aportes al conocimiento de los Chnoodini de América del Sur (Coleoptera: Coccinellidae). Boletín de la Sociedad Entomológica Aragonesa 53: 63–79.

Gordon RD (1987) A catalogue of the Crotch collection of Coccinellidae (Coleoptera). Occasional Papers on Systematic Entomology 3: 1–46.

Gordon RD (1994) South American Coccinellidae (Coleoptera). Part III: Definition of Exoplectrinae Crotch, Azyinae Musant, and Coccidulinae Crotch; a taxonomic revision of Coccidulini. Revista Brasileira de Entomologia 38: 681–775.

Hodek I, Honěk A (1996) Ecology of Coccinellidae. Kluwer Academic Publishers, Dordrecht, 464 pp. https://doi.org/10.1007/978-94-017-1349-8

Horn W, Kahle I (1935–1937) Über entomologische Sammlungen, Entomologen & Entomo-Museologie. Berlin, Dahlem, 533 pp.

Korschefsky R (1931) Coccinellidae II. In: Junk W, Schenkling S (Eds) Coleopterorum Catalogus. W. Junk, Berlin, Pars 118, 1–224.

Korschefsky R (1932) Coccinellidae II. In: Junk W, Schenkling S (Eds) Coleopterorum Catalogus. W. Junk, Berlin, Pars 120: 225–659.

Krüger TC, Castro-Guedes CF, Almeida LM (2016) Two new species of Chnoodes Chevrolat (Coleoptera: Coccinellidae) from Brazil. Zootaxa 4078: 269–283. https://doi.org/10.11646/zootaxa.4078.1.24

Latreille PA (1807) Genera Crustaceorum et Insectorum Secundum Ordinem Naturalem in Familias Disposita, Iconibus Exemplis

que Plurimus Explicata. Paris, vol. 3, 258 pp.

Mariconi FAM, Zamith APL (1959) Notas sobre uma cochonilha e seu predador. O Biológico 25: 258–265.

Mariconi FAM, Zamith APL (1960) Contribuição para o conhecimento da Mimosicerya hempeli (Cockerell, 1899) (Homoptera, Margarodidae) e de seu predador Exoplectra erythrogaster Mulsant, 1851 (Coleoptera, Coccinellidae). Anais da Escola Superior de Agricultura "Luiz de Queiroz", 223–238.

Mulsant E (1846) Histoire Naturelle des Coléoptères de France. Sulcicolles, sécuripalpes. Paris, [From Horn.], 280p.

Mulsant E (1850) Species des Coléoptères trimères sécuripalpes. Annales des Sciences Physique et Naturelles d'Agriculture et d'Industrie, Lyon, 1104 pp. https://doi.org/10.5962/bhl.title.8953

Nedvěd O, Kovář I (2012) Phylogeny and classification. In: Hodek I, Emdem HF, Honěk A (Eds) Ecology and Behaviour of the Ladybird Beetles. Wiley, Blackwell, Oxford, 1–12. https://doi.org/10.1002/9781118223208.ch1

Seago A, Giorgi J, Li J, Ślipiński A (2011) Phylogeny, classification and evolution of ladybird beetles (Coleoptera: Coccinellidae) based on simultaneous analysis of molecular and morphological data. Molecular Phylogenetics and Evolution 60: 137–151. https://doi.org/10.1016/j.ympev.2011.03.015

Sicard A (1907) Revision des Coccinellides de la Faune Malgache (I). Annales de la Société Entomologique de France 76: 425–482.

Sicard A (1909) Revision des Coccinellides de la Faune Malgache. Annales de la Société Entomologique de France 78: 63–165.

Weise J (1895) Neue Coccinelliden, sowie Bemerkungen zu bekannten Arten. Annales de la Société Entomologique de Belgique 39: 120–146.

Author Contributions: JMCS and LMA participated equally in the preparation of this article.

Competing Interests: The authors have declared that no competing interests exist.

17

Two new Brazilian species of Chelodesmidae of the genera *Iguazus* and *Tessarithys* (Diplopoda: Polydesmida)

Rodrigo S. Bouzan[1], João Paulo P. Pena-Barbosa[1], Antonio D. Brescovit[1]

[1]*Laboratório Especial de Coleções Zoológicas, Instituto Butantan. Avenida Vital Brasil 1500, 05503-090 São Paulo, SP, Brazil.*
Corresponding author: Rodrigo S. Bouzan (rodrigobouzan@outlook.com)

http://zoobank.org/739E57D7-535D-4EF3-A6C9-49623563B31E

ABSTRACT. Two new species of Chelodesmidae from the Brazilian northeast are described, *Iguazus robustus* **sp. nov.**, from the state of Paraíba, and *Tessarithys exacuminatus* **sp. nov.**, from the states of Pernambuco and Sergipe. *Iguazus robustus* **sp. nov.** differs from other species of the genus by having a constriction in the zone of the gonopodal acropodite tip and an extra branch at the tip of the acropodite. *Tessarithys exacuminatus* **sp. nov.** differs from the other species of the genus by the large and ascending subterminal dorsal branch of the prefemoral process of the gonopod. Brief reviews of the taxonomy, geographic distribution and a key for males of the respective genera are provided.

KEY WORDS. Brazil, Chelodesminae, millipedes, Neotropical, taxonomy.

INTRODUCTION

Among the Diplopoda, with almost 5,000 described species in 31 families, the Polydesmida Leach, 1814 is the most speciose order (Brewer et al. 2012). Within the order, Chelodesmidae constitutes the second largest family, with about 800 described species. Members of this family occur in western Africa (Prepodesminae Cook, 1896) and South America (Chelodesminae Cook, 1895), according to Hoffman (1980). The systematics of Chelodesminae was reviewed by Hoffman (1980), who placed some genera and species in tribes. The remaining species were not reviewed by Hoffman (1980) or described after his classification. As a result, of the 171 genera and ca 800 species in Chelodesmidae, 89 genera and 455 species are currently not assigned to a tribe.

During the examination of the Diplopoda collection of the Instituto Butantan, Brazil, two new species of Chelodesminae from the Brazilian Northeast were found. The first belongs to *Iguazus* Chamberlin, 1952 and the other to *Tessarithys* Hoffman, 1990. Neither genera has been assigned to a tribe within the Chelodesmidae.

Iguazus was proposed by Chamberlin (1952) for a single species, *I. ornithopus* (Brölemann, 1902), described from Cerqueira César, state of São Paulo, Brazil. Hoffman (1965)

noticed that the genus *Hoffmanodesmus* Schubart, 1962 is a junior synonym of *Iguazus*. He then transferred *H. roseofasciatus* Schubart, 1962, from Porto Real do Colégio, state of Alagoas, Brazil, to *Iguazus*.

Hoffman (1990) proposed the genus *Tessarithys* for three species, *T. neoecobius* Hoffman, 1990, the type species, from Senhor do Bonfim, and *T. machaerophorus* (Schubart, 1956), from Juazeiro (both localities are in the state of Bahia, Brazil), and *T. soledadinus* (Attems, 1931) labeled only as "Soledad, Brasilien", (probably Soledade, in the state of Paraíba, Brazil). The last two species were removed from *Leptodesmus* when Hoffman (1971) reorganized the systematic structure of the genus.

Many members of various groups within the Chelodesmidae share character states in their external morphology, and are mostly differentiated by variations in the structures of the male gonopod. These differences in the gonopod indicate that it has undergone rapid divergent evolution in the genitalia, as evidenced in numerous other arthropod groups (Eberhard 2010). Thus, discrimination among species in many chelodesmid groups depends on the analysis of the male gonopod.

In this work we describe the new species *Iguazus robustus* sp. nov. from Araruna, state of Paraíba and *Tessarithys exacuminatus* sp. nov. from São Caetano, state of Pernambuco and provide a map of their records.

MATERIAL AND METHODS

Morphological observations and illustrations were made using a Leica MZ12 stereomicroscope with a camera lucida. Photographs were taken with a Leica DFC 500 digital camera mounted on a Leica MZ16A stereomicroscope. Extended focal range images were composed with Leica Application Suite version 2.5.0. All measurements are in millimeters. The terminology of the gonopodal structures follows Pena-Barbosa et al. (2013) and Hoffman (1990), while the terminology used to describe somatic traits follows Attems (1898), Brölemann (1900) and Pena-Barbosa et al. (2013). The type material was deposited in the collection of the Instituto Butantan, São Paulo (IBSP, curator: A. D. Brescovit).

Museum acronyms: FMNH, Field Museum of Natural History, Chicago, USA; IBSP, Instituto Butantan, São Paulo, Brazil; MZSP, Museu de Zoologia, Universidade de São Paulo, São Paulo, Brazil; NHMW, Naturhistorisches Museum, Wien, Austria.

TAXONOMY

Chelodesmidae Cook, 1895
Chelodesminae Cook, 1895

Iguazus Chamberlin, 1952

Camptomorpha (*non* Silvestri, 1897), Attems 1938: 73 (in part, *ornithopus* Brölemann); Schubart 1943: 147 (in part, *ornithopus* and *phoenicopterus*).
Iguazus Chamberlin, 1952: 568. Type species: *I. leius* Chamberlin 1952, by original designation.
Hoffmanodesmus Schubart, 1962: 255. Type species: *H. ornithopus* (Brölemann, 1902), by original designation; synonymized by Hoffman 1965: 219.
Iguazus: Hoffman 1965: 219.

Diagnosis. Males of *Iguazus* differ from other chelodesmid genera by the following combination of characters: gonopodal acropodite slender, unbranched and sinuously curved, prefemoral process massive in form of a narrow blade or branch, with one or two secondary processes arising proximally from the midlength region (Figs 1-3).

Distribution. Known from the states of Paraíba, Alagoas and São Paulo, Brazil and in Misiones, Argentina.

Composition. Three species, *Iguazus ornithopus* (Brölemann, 1902), *I. roseofasciatus* (Schubart, 1962), *I. robustus* sp. nov.

Iguazus ornithopus (Brölemann, 1902)

Leptodesmus ornithopus Brölemann, 1902: 87, figs 90-92 (One male and two females syntypes from Cerqueira César, 49°16'59"W, 23°03'85"S, São Paulo, deposited in MZSP, not examined).
Camptomorpha ornithopus: Attems 1938: 73.
Camptomorpha phoenicopterus Schubart, 1943: 147, figs 46-47 (Male holotype from Itapura, São Paulo, deposited in MZSP,

not examined); synonymized by Schubart 1962: 255 under *ornithopus*.
Iguazus leius Chamberlin, 1952: 568 fig. 17 (Male holotype, one male and six female paratypes from Iguazu Falls, Misiones, deposited in FMNH #274, not examined). The type lot contains fragments of a single specimen, gonopods missing (Sierwald et al. 2005). Synonymized by Hoffman 1965: 221.
Hoffmanodesmus ornithopus: Schubart 1962: 254.
Iguazus ornithopus: Hoffman 1965: 221, fig. 1.

Distribution. State of São Paulo, Brazil and Misiones, Argentina.

Iguazus roseofasciatus (Schubart, 1962)

Hoffmanodesmus roseofasciatus Schubart, 1962: 255, fig. 2 (Male holotype, one male and two female paratypes from Porto Real do Colégio, 36°83'78"W, 10°18'53"S, Alagoas, deposited in MZSP, not examined).
Iguazus roseofasciatus: Hoffman 1965: 221.

Distribution. State of Alagoas, Brazil.

Iguazus robustus sp. nov.

http://zoobank.org/CE310199-8C38-444C-9A0D-644298CDBED5

Figs 1-3, 7-12, 19-21, 28

Diagnosis. Males of this species differ from those of other species of the genus by having a constriction in the zone of the gonopodal acropodite tip (Fig. 1, arrow) and an extra branch at the tip of the acropodite.

Description. Female: Unknown. Male (Holotype, IBSP 4397): Head reddish with a yellow labrum, Tömösvary organ suboval in shape. Antennae reddish brown, terminal antennomere with invaginations between the four apical sense cones. Body reddish brown and paranota tip yellow, gradually losing the brown color towards the posterior body rings, reddish brown color in the mid-body ring restricted to the anterior and posterior edge and yellow filling the remaining portion (Figs 19-21). Body rings: tegument smooth; alignment of paranota ventrally curved; paranota with posterior edges acutely produced from body ring 5; ozopore centrally situated on body ring 5; and posteriorly situated on the other body rings; ozopore arrangement at the edge of paranota: 5, 7, 9, 10, 12, 13, 15-19 (following the standard polydesmid pore formula). Penultimate body ring with reduced paranota (Fig. 12, arrows). Stigma oval and elongated. Coxae of leg pair 2nd with rectangular shaped genital papilla (Fig. 7, arrows). Sternite of body ring 5 with a pair of elongated projections (Fig. 8). Sternite of body ring 8 presenting two pairs of pointed projections in the zone anterior to the coxae (Fig. 9, arrows). Gonopod aperture on seventh body ring: elliptical, posterior edge without folds. Legs whitish yellow, with rounded ventro-apical process on the prefemur and with an apical-ventral

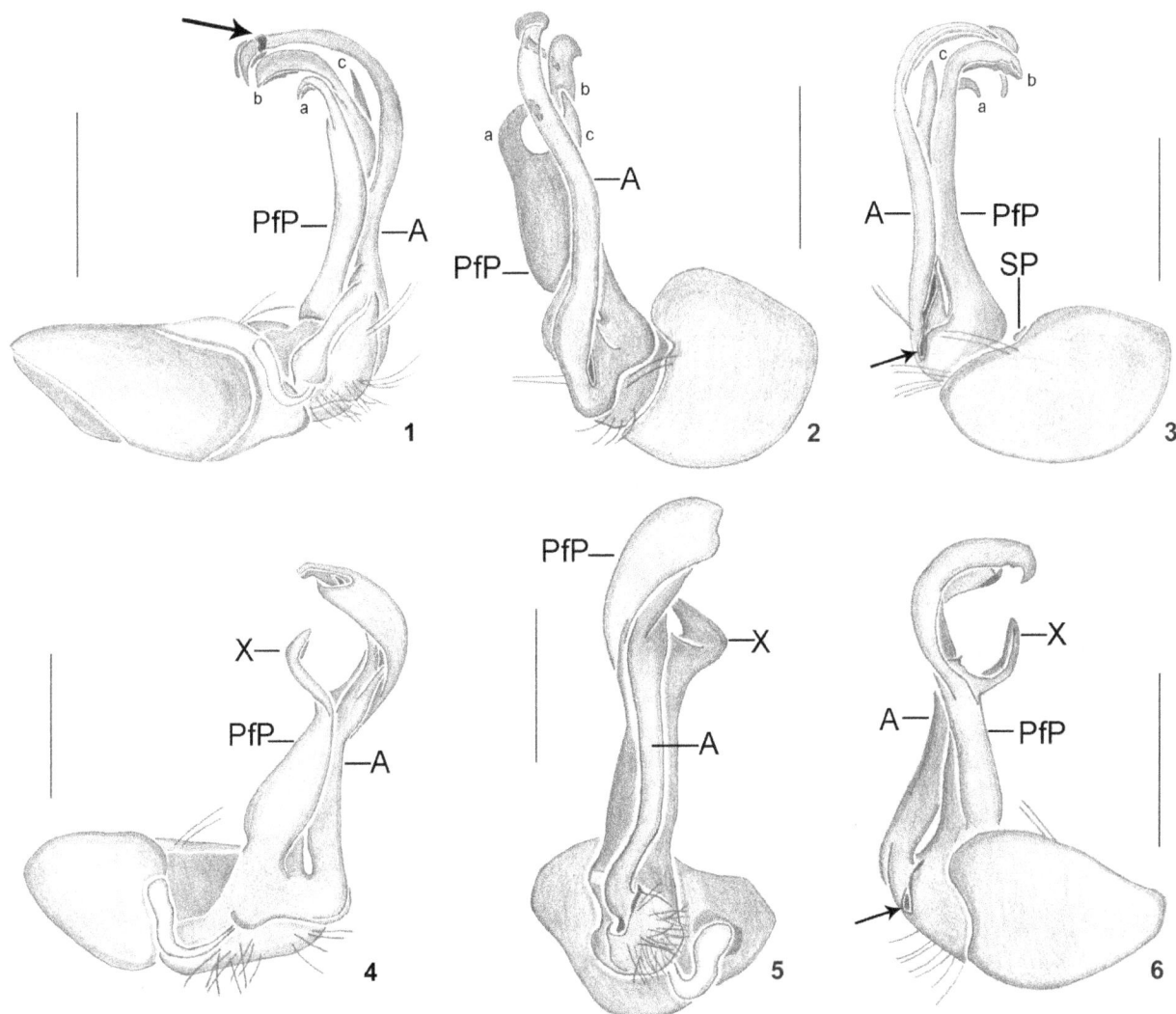

Figures 1–6. (1–3) *Iguazus robustus* sp. nov., left gonopod: (1) mesal view (arrow = insertion point, zone of acropodite); (2) ventral view; (3) ectal view (arrow = groove near the first portion of the acropodite). (4–6) *Tessarithys exacuminatus* sp. nov., left gonopod: (4) mesal view; (5) ventral view; (6) ectal view (arrow = groove near the first portion of the acropodite). Scale bars = 0.5 mm. (a) Process A, (A) acropodite, (b) process B, (c) process C, (PfP) prefemoral process, (SP) spiniform projection on gonopod coxae, (X) subterminal dorsal branch.

membranous projection on the tibia (Figs 10–11), leg modifications are present in all pairs except the last. Telson yellowish with dark brown edges. Total length: 39. Collum 2.08 long, 5.46 wide. Antennomere length: 0.26; 1.14; 1.23; 1.15; 1.27; 1.00; 0.21. Gonopod aperture 1.66 long, 2.56 wide. Telson 0.85 long.

Gonopods (Figs 1–3): gonopod coxae equivalent to about half the length of the telopodite and prominent in ectal view (Fig. 3). Coxae with two bristles in the distal dorsal side. Spiniform projection present. Cannula (Fig. 1): hook-shaped. Prefemoral region ventrally developed and short, 1/3 the size of telopodite. In ectal view, presence of a conspicuous groove near the beginning of the acropodite (Fig. 3, arrow). Prefemoral process (PfP) long, massive,

blade-like. In the middle portion, the prefemoral process divides into three different projections: the first projection, mesal view, sickle-shape (Figs 1–3, a); the mid-projection, the largest among them, boat-shaped (Figs 1–3, b); the last is the spine-shaped lower projection (Figs 1–3, c). Acropodite (A) elongated and slender, carrying the seminal groove; acropodite is unbranched and sinuously-curved (Figs 1–3); its distal portion, mesal view, displays a constriction, that resembles a cingulum where a moveable branch is attached, that results in a pointed tip and in a blade which seems to have the function of protecting this acute blade on the lateral side, thus the tip of acropodite is constituted for two branches (Figs 1, 3).

Type material. Male holotype from Parque Estadual Pedra da Boca, Araruna (6°45'95"S, 35°67'78"W, 228 m), Paraíba, Brazil, 01-02.VI. 2012, I.L.F. Magalhães & J.L. Chavari col., deposited in IBSP 4397.

Distribution. Known only from the type locality (Fig. 28).

Etymology. The species epithet, *robustus*, is a reference to the massive prefemoral process and derives from the Latin word "robustus", "robusta".

Key to males of *Iguazus*

1	Prefemoral process trifurcated..2	
1'	Prefemoral process bifurcated (Schubart 1962: fig. 2) ....*I. roseofasciatus*	
2	Apical portion of acropodite with two branches (Fig. 1).... ....*I. robustus*	
2'	Apical portion of acropodite single branch (Hoffman, 1965: fig. 1)...*I. ornithopus*	

Tessarithys Hoffman, 1990

Tessarithys Hoffman, 1990: 159–166. Type species: *T. neoecobius* Hoffman, 1990, by original designation.

Diagnosis. Males of *Tessarithys* differ from other chelodesmid genera by the combination of the following characters: sternite of body ring 5 with four projections (Fig. 14). Legs with apical-ventral projection on the tibia (Fig. 17). Gonopodal prefemoral process exceeding the acropodite apex and forming a distinct sheath on the ectal side, also displaying a subterminal dorsal branch on the middle portion. Acropodite are divided into two slender and acuminate branches (Figs 4–6).

Distribution. Known from the states of Pernambuco to Bahia.

Composition. Four species, *Tessarithys neoecobius* Hoffman, 1990, *T. machaerophorus* (Schubart, 1956), *T. soledadinus* (Attems, 1931), *T. exacuminatus* sp. nov.

Tessarithys neoecobius Hoffman, 1990

Tessarithys neoecobius Hoffman, 1990: 161, figs 7 (Male holotype and three female paratypes from Senhor do Bonfim, 40°18'68"W, 10°45'97"S, Bahia, deposited in MZSP, not examined).

Distribution. State of Bahia, Brazil.

Tessarithys machaerophorus (Schubart, 1956)

Leptodesmus machaerophorus Schubart, 1956: 424, figs 5-6 (Male holotype, two males and six female paratypes from Joazeiro, 40°50'58"W, 09°42'78"S, Bahia, deposited in MZSP, not examined).
Tessarithys machaerophorus: Hoffman 1990: 1965.

Distribution. State of Bahia, Brazil.

Tessarithys soledadinus (Attems, 1931)

Pseudoleptodesmus soledadinus Attems, 1931: 30, fig. 43-45 (Male holotype, labeled only "Soledad", Brazil, deposited in NHMW, not examined).
Leptodesmus (Pseudoleptodesmus) soledadinus: Attems 1938: 43.
Leptodesmus soledadinus: Schubart 1946: 196.
Tessarithys soledadinus: Hoffman 1990: 163.

Distribution. Labeled only as from "Soledad", Brazil.

Tessarithys exacuminatus sp. nov.

http://zoobank.org/B511E597-DFD1-45C0-AAC9-F95D21001B5A

Figs 4–6, 13–18, 22–27, 29

Diagnosis. Males of this species differ from those of other species of the genus by the large and ascending subterminal dorsal branch of the prefemoral process of the gonopod (Fig. 4).

Description. Female: Unknown. Male (Holotype, IBSP 4431): Head dark reddish with a yellow labrum, Tömösváry organ suboval in shape. Antennae reddish brown, terminal antennomere with invaginations between the four apical sense cones. Body purple, coloration of paranota tip differing from body ring 5, reddish yellow (Figs 22–24). Body rings: tegument slightly rough. Alignment of paranota: ventrally curved (Fig. 18); paranota with posterior edges acutely produced from body ring 5; ozopore centrally situated on body ring 5, and posteriorly situated on the others; ozopore arrangement at the edge of paranota: 5, 7, 9, 10, 12, 13, 15–19 (following the standard polydesmid pore formula). Penultimate body ring with reduced paranota. Stigma oval and elongated. Coxae of 2nd leg pair with rectangular shaped genital papilla (Fig. 13, arrows). Sternite of body ring 4 with a pair of small rounded projections (Fig. 13). Sternite of body ring 5 with two pairs of elongated projections (Fig. 14). Sternite from body ring 8 presenting two pairs of pointed projections in the zone anterior to the coxae (Fig. 15, arrows). Gonopod aperture on seventh body ring: transversal oval, posterior edge without folds. Legs reddish, with a membranous apical-ventral projection on the tibia (Fig. 16–17); leg modifications are present in all pairs except on the last. Telson purple. Total length: 42.5 (Fragmented). Collum 2.03 long, 5.50 wide. Antennomere length: 0.41; 1.17; 1.28; 1.04; 1.23; 1.01; 0.20. Gonopod aperture 1.40 long, 2.77 wide. Telson 1.58 long.

Gonopods (Figs 4–6): gonopod coxae equivalent to about half the length of the telopodite and prominent in ectal view (Fig. 6). Coxae with two bristles in the distal dorsal side. Spiniform projection absent. Cannula (Fig. 4): hook-shaped. Prefemoral region short, ventrally developed, 1/3 the size of telopodite. Presence of a conspicuous groove near the first portion of the acropodite, in the ectal view (Fig. 6, arrow). Prefemoral process (PfP) long, parallel to the acropodite (A); displays a long and acuminated subterminal dorsal branch on the middle portion (Figs 4–6, X).

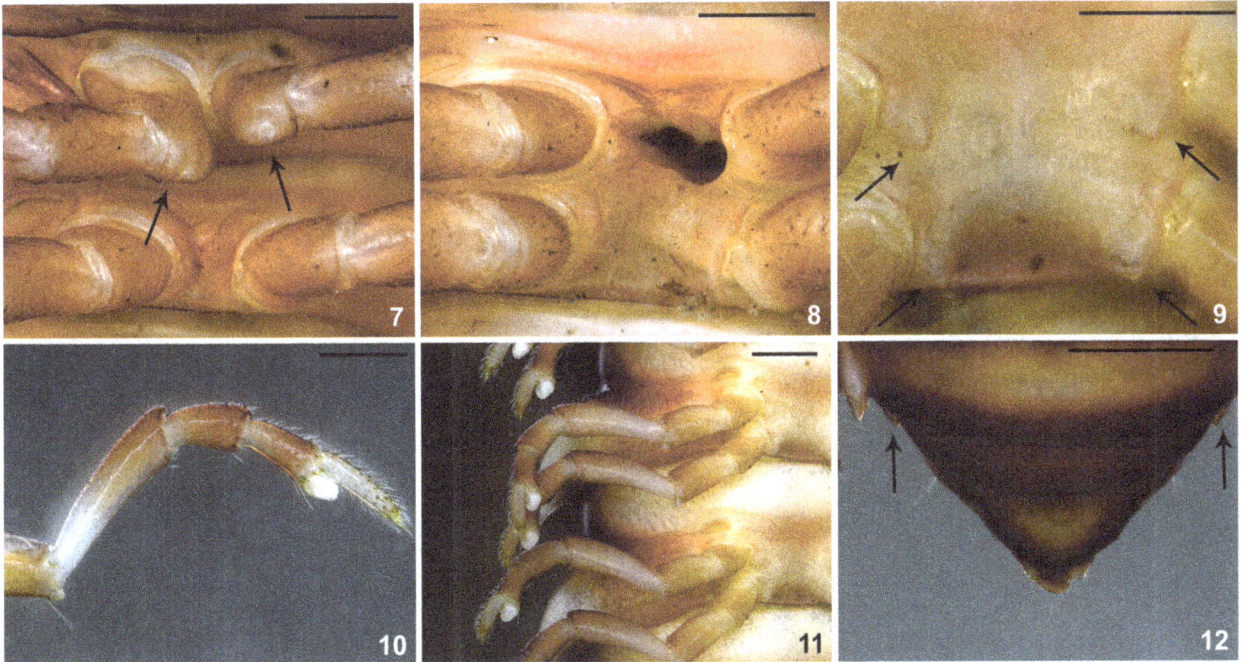

Figures 7–12. *Iguazus robustus* sp. nov.: (7) Sternite, body ring 3 (arrow = genital papilla), body ring 4; (8) Sternite of body ring 5; (9) Sternite on body ring 8 (arrow = pairs of projections); (10) Leg, lateral view; (11) Leg, ventral view; (12) Penultimate body ring (arrow = reduced paranota).

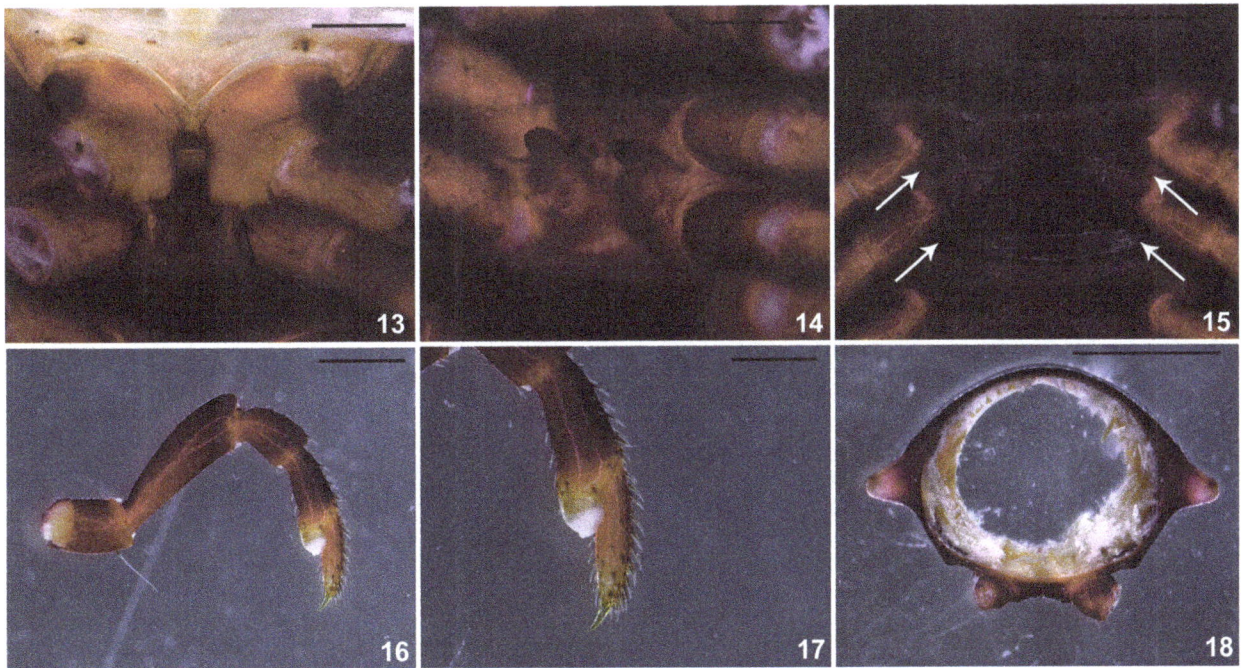

Figures 13–18. *Tessarithys exacuminatus* sp. nov.: (13) Sternite, body ring 3 (arrow = genital papilla), body ring 4; (14) Sternite of body ring 5; (15) Sternite on body ring 8 (arrow = pairs of projections); (16) Leg, lateral view; (17) Tibia, membranous projection apico-ventral, detail; (18) Paranota, ventral.

Figures 19–27. (19–21) *Iguazus robustus* sp. nov., body: (19) first body rings; (20) midbody body rings; (21) last body rings. (22-27) *Tessarithys exacuminatus* sp. nov., body (holotype): (22) first body rings; (23) midbody body rings; (24) last body rings; body (paratype); (25) first body rings; (26) midbody body rings; (27) last body rings.

The final portion of the prefemoral process passes on the back of the acropodite, forming a distinct protection on the lateral side, with two blades forking in the terminal portion (Fig. 4). Acropodite (A) long and slender, with solenomere and one additional branch (Fig. 4). Spermatic groove mostly visible in mesal view except at the base of the acropodite where it diverts to the ectal side.

Variation. The body of the paratype (IBSP 4434) shows reddish brown coloration, with the edges of the paranota whitish yellow, tegument smooth and with a median band present, weakly developed (Figs 25–27).

Type material. Holotype: one male (IBSP 4431) from Reserva Particular do Patrimônio Natural Pedra do Cachorro (8°14'22.9"S, 36°11'13.7"W), São Caetano, Pernambuco, Brazil, 26.V.2012, I.L.F. Magalhães & J.L. Chavari coll. Paratypes: one male (IBSP 4434) same data as holotype and one male (IBSP 4634) from Unidade de Conservação Monumento Natural Grota do Angico, Poço Redondo (9°80'65"S, 37°68'36"W), Sergipe, Brazil. I. 2013, R. G. Faria coll.

Distribution. Known only from the type locality (Fig. 29).

Etymology. In reference to the acuminated subterminal dorsal branch, labelled "X" in figs 7-9 in Hoffman 1990.

Key to *Tessarithys* males

1   Prefemoral process: terminal section consisting of a single branch 2
1' Prefemoral process: terminal section consisting of two branches 3
2   Presence of small denticles beyond the base of the subterminal dorsal branch of prefemoral process (Hoffman 1990: fig. 9)....................................*T. soledadinus*
2' Base of subterminal dorsal process of prefemoral process smooth (Hoffman 1990: fig. 7) .... *T. neoecobius*
3   Subterminal dorsal branch pointing upward (Fig. 4) .... *T. exacuminatus*
3' Subterminal dorsal branch pointing downward (Hoffman 1990: fig. 8).. *T. machaerophorus*

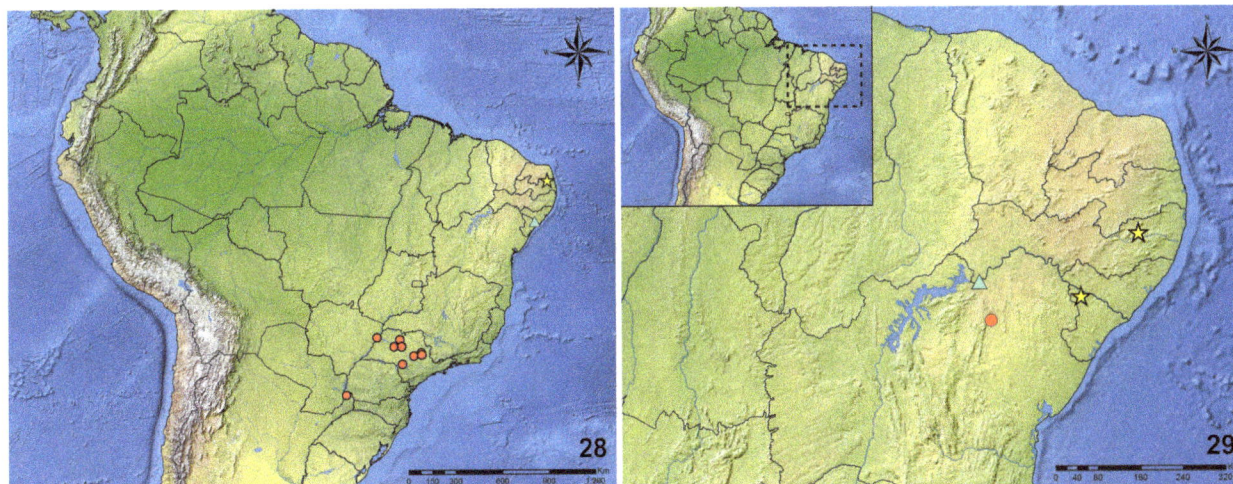

Figures 28–29. (28) Distribution map. *Iguazus ornithopus* = red circles; *I. roseofasciatus* = blue triangle; *I. robustus* sp. nov. = yellow star. Locality data for *I. ornithopus* and *I. roseofasciatus* taken from the literature. (29) Distribution map. *Tessarithys neoecobius* = red circle; *T. machaerophorus* = blue triangle; *T. soledadinus* = ?; *T. exacuminatus* sp. nov. = yellow stars. Locality data for *T. neoecobius* and *T. machaerophorus* taken from the literature.

ACKNOWLEDGMENTS

This work was supported by CNPq/PIBIC-IC (105077/2015-1) grant to RSB; CNPq (301776/2004-0) grant to ADB and CNPq (143049/2011-9) grant to JPB. We are grateful to the reviewers and the editor for their valuable comments. We are also grateful to Luiz F.M. Iniesta (IBSP) for his helping during this work. The version of the manuscript was improved by critical readings from Ross Martin Thomas.

REFERENCES

Attems CG (1898) System der Polydesmiden I. Theil. Denkschriften der Kaiserlichen Akademie der Wissenschaften zu Wien, Mathematisch-Naturwissenschaftliche Klassen 67: 221–482.

Attems CG (1931) Die Familie Leptodesmidae und andere Polydesmiden. Zoologica 30: 1–150.

Attems CG (1938) Polydesmoidea II. Families Leptodesmidae, Platyrhacidae, Oxydesmidae, Gomphodesmidae. Das Tierreich 69: 1–487.

Brewer MS, Sierwald P, Bond JE (2012) Millipede taxonomy after 250 years: classification and taxonomic practices in a mega-diverse yet understudied arthropod group. Plos One 7: 1–12. https://doi.org/10.1371/journal.003724

Brölemann HW (1900) Dous myriapodos notáveis do Brazil, Notas Myriapodologicas. Boletim do Museu Paraense 3: 65–71.

Brölemann HW (1902) Myriapodes du Musée de São Paulo. Revista do Museu Paulista 5: 35–237. https://doi.org/10.5962/bhl.part.9824

Chamberlin RV (1952) Some American polydesmid millipeds in the collection of Chicago Museum of Natural History. Annals of the Entomological Society of America 45: 553–584. https://doi.org/10.1093/aesa/45.4.553

Eberhard WG (2010) Rapid divergent evolution of genitalia. In: Leonard J, Cordoba-Aguilar A (Eds) The evolution of primary sexual characters in animals. Oxford University Press, New York, 40–78.

Hoffman RL (1965) Chelodesmid studies I. The status of the generic name *Hoffmanodesmus* Schubart (Diplopoda: Polydesmida). Papéis Avulsos de Zoologia 17: 219–223.

Hoffman RL (1971) Chelodesmid studies V. Some new, redefined, and resurrected Brasilian genera. Arquivos de Zoolologia 20: 225–277. https://doi.org/10.11606/issn.2176-7793.v20i4p225-277

Hoffman RL (1980) Classification of the Diplopoda. Muséum d'histoire naturelle, Geneva, 237 pp.

Hoffman RL (1990) Chelodesmid studies XXII. Synopsis of *Tessarithys*, a new genus of Brazilian millipeds (Diplopoda: Chelodesmidae). Papéis Avulsos de Zoologia 37: 159–166.

Pena-Barbosa JPP, Sierwald P, Brescovit AD (2013) On the largest chelodesmid millipedes: taxonomic review and cladistic analysis of the genus *Odontopeltis* Pocock, 1894 (Diplopoda; Polydesmida; Chelodesmidae). Zoological Journal of the Linnean Society 169: 737–764. https://doi.org/10.1111/zoj.12086

Schubart O (1943) Espécies novas das famílias Strongylosomidae e Leptodesmidae da ordem Proterospermophora do interior dos estados de São Paulo e de Mato-Grosso. Papéis avulsos do Departamento de Zoologia 3: 127–164.

Schubart O (1946) Contribuição ao conhecimento do gênero *Leptodesmus* (Leptodesmidae, Diplopoda). Anais da Academia Brasileira de Ciências 18: 165–202.

Schubart O (1956) Leptodesmidae Brasileiras. IV: Espécies novas da Bahia (Diplopoda, Proterospermophora). Revista Brasileira de Biologia 16: 421–428.

Schubart O (1962) Leptodesmidae Brasileiras. IX: Sobre algumas espécies do gênero *Camptomorpha* (Proterospermophora, Diplopoda). Revista Brasileira de Biologia 22: 251–261.

Sierwald P, Bond JE, Gurda GT (2005) The millipede type specimens in the collections of the Field Museum of Natural History (Arthropoda: Diplopoda). Zootaxa 1005: 1–64. https://doi.org/10.11646/zootaxa.1005.1.1

Author Contributions: RSB produced all images (drawings, photos and maps) and description; RSB, JPPPB and ADB examined material and wrote the text.

Competing Interests: The authors have declared that no competing interests exist.

18

Morphological and genetic diversity in *Callithrix* hybrids in an anthropogenic area in Southeastern Brazil (Primates: Cebidae: Callitrichinae)

Adrielle M. Cezar[1], Leila M. Pessôa[1], Cibele R. Bonvicino[2,3]

[1]*Laboratório de Mastozoologia, Departamento de Zoologia, Instituto de Biologia, Universidade Federal do Rio de Janeiro. Avenida Brigadeiro Trompowski, Ilha do Fundão, 21941-590 Rio de Janeiro, RJ, Brazil.*
[2]*Laboratório de Biologia e Parasitologia de Mamíferos Silvestres Reservatórios, Instituto Oswaldo Cruz. Avenida Brasil 4365, Manguinhos, 21040-360 Rio de Janeiro, Brazil.*
[3]*Divisão de Genética, Instituto Nacional do Câncer. Rua André Cavalcanti 37, Centro, 20231-050 Rio de Janeiro, RJ, Brazil.*
Corresponding author: Leila M. Pessôa (pessoa@acd.ufrj.brr)

http://zoobank.org/D0A6F9E3-E613-40CC-B180-170074DFCF39

ABSTRACT. Two species of *Callithrix*, *C. jacchus* (Linnaeus, 1758) and *C. penicillata* (É. Geoffroy, 1812), are considered invasive in Rio de Janeiro. This study determined the genetic and morphological diversity and verified the species involved in the hybridization of 10 individuals from the municipalities of Silva Jardim (N = 9) and Rio das Ostras (N = 1). We compared the external morphology and skull of *C. jacchus* (N = 15) and *C. penicillata* (N = 14) specimens deposited in the collection of the National Museum of Rio de Janeiro (MN- UFRJ). Phylogenetic (maximum likelihood and Bayesian inference) and phylogeographical analyses (network analysis) were performed based on cytochrome b sequences. These analyses included hybrids from the metropolitan region of Rio de Janeiro (N = 3), *C. penicillata* (N = 2), *C. jacchus* (N = 2), *C. geoffroyi* (N = 2), *C. kuhlii* (N = 2), *C. aurita* (N = 1), and as outgroups, *Mico emiliae* (N = 1) and *Saguinus mystax* (N = 1). The pelage and skull characters of most hybrids were more closely related to *C. jacchus*. Skull morphometric analysis revealed an intermediate state for the hybrids. Phylogenetic analyses revealed a high similarity between the hybrids and *C. penicillata*. Six haplotypes of hybrids were identified. Network analysis including them and *C. penicillata* recovered the topology generated by phylogenetic analysis. The results corroborate that *C. jacchus* and *C. penicillata* participate in the hybridization process. There was no geographic structure between hybrids from the coastal lowlands and from the metropolitan region of Rio de Janeiro.

KEY WORDS. Atlantic forest, introduced species, marmosets, morphometry, phylogeny.

INTRODUCTION

Callithrix Erxleben, 1777 has six species, all endemic to Brazil. The distribution of *Callithrix* species is closely associated with the Atlantic Forest. *Callithrix jacchus* (Linnaeus, 1758) and *Callithrix penicillata* (É. Geoffroy, 1812) have the largest natural geographical distribution within the genus. They are found in the Atlantic Forest and Caatinga of northeastern Brazil and in the Cerrado of central and northeastern Brazil (De Vivo 1991, Rylands et al. 1996). They are phylogenetically very close, and it has been hypothesized that their most recent common ancestor lived about 700 thousand years ago, in the Atlantic Forest, Cerrado, and Caatinga. A subsequent vicariant speciation event isolated the ancestor of *C. penicillata* in the Cerrado or Caatinga (Buckner et al, 2014 Malukiewicz et al. 2014.

Species of *Callithrix* are commonly called marmosets. The range of *Callithrix* species is allopatric, with some species contacting at the limits of their distribution. However, the ranges of natural species are being altered due to habitat destruction and to anthropogenic introduction of marmoset species outside their natural geographical bounds. As a result of such anthropogenic alterations, *C. jacchus* and *C. penicillata* are often found in sympatry with several other *Callithrix* species and in the natural ranges of other primates (Rylands et al. 1993, 2009, Ruiz-Miranda et al. 2000).

Callithrix jacchus and *C. penicillata* are found in the state of Rio de Janeiro, Brazil, both in forested and disturbed areas. Their introduction and settlement in the state are the result of illicit domestic and international trafficking of primates. Although the history of their introduction into the coastal lowlands is uncertain, the distribution of marmosets is increasing towards the north of the state of Rio de Janeiro in the lowlands at an estimated rate of 1.2 km per year (Ruiz-Miranda et al. 2000). There, they are found in forest fragments of costal lowlands where the golden lion tamarin, *Leontopithecus rosalia* (Linnaeus, 1766), is naturally distributed. The interaction between *C. jacchus* and *C. penicillata* and the native populations of golden lion tamarins is problematic because the ecology of these primates is similar, which may lead to competition for food and territory and disease transmission (Ruiz-Miranda et al. 2000.

Callithrix jacchus and *C. penicillata* are differentiated by the colors of the body pelage and of the auricular tufts, and by the insertion of tufts in the ear. *C. penicillata* has black auricular tufts arranged in front of the ear (pre-auricular) whereas *C. jacchus* has white auricular tufts arranged around the ear (circum-auricular) (Hershkovitz 1977). Most studies on the identification of hybrids consider only the pelage and mitochondrial DNA, excluding cranial morphological characters (e.g. Alonso et al. 1987, Fuzessy et al. 2014).

This study compares genetic hybrids from the coastal lowland with five species of the genus *Callithrix* and other hybrids from Rio de Janeiro's metropolitan region, in order to evaluate morphological and genetic diversity, identify the species involved in hybridization, and verify the geographic structure. To this end, we performed cranial and pelage morphological analyses and molecular phylogenetic estimation using the cytochrome b mitochondrial gene (MT-CYB).

MATERIAL AND METHODS

The sample studied herein comprised 39 individuals: ten *Callithrix* hybrids from two municipalities in the coastal lowlands of the state of Rio de Janeiro, Silva Jardim (N = 9) and Rio das Ostras (N = 1); and *C. penicillata* (N = 14) and *C. jacchus* (N = 15) from localities near the type localities. This sampling strategy ensured that we handled samples of each species separately, thereby avoiding the presence of hybrids. This strategy also limited the sample size. The pure specimens analyzed were identified by the pelage description provided by Hershkovitz (1977). Hybrids were identified by the presence of intermediate characters. Voucher numbers for the specimens analyzed are available in the Appendix 1.

Auricular tuft color and disposition, and general pelage color of all individuals were analyzed to estimate the variation in pelage color and specific patterns for each species and for the hybrids.

A stereoscopic microscope was used to analyze the qualitative characters of *C. jacchus* and *C. penicillata* skulls. Differences between species and between each species and the hybrids were identified.

Ten linear cranial measurements 'were taken from the marmosets, with a digital caliper (mm). The first six were defined by Natori (1994) and De Vivo (1991) and the last four in this study: (PL) prosthion to lambda, (EE) euryon to euryon, (iFO) inside

frontomalare orbitale to frontomalare orbitale, (BB) bicondylar breadth, (CM) mesial surface of the left upper canine to distal surface of the left second upper molar, (ZA) zygomatic arch breadth, (oFO) outside frontomalare orbitale to frontomalare orbitale, (MS) mandibular symphysis height, (lFM) foramen magnum length, and (bFM) foramen magnum breadth.

To analyze the morphological differences between the studied *Callithrix* species and the hybrids, we calculated the mean, standard deviation, maximum and minimum values of morphological measurements described above.

Student's t-test and one-way ANOVA were used to identify differences between species and between each species and the hybrids. Principal component analysis (PCA) was performed to reveal patterns of variation between species and hybrids, and to visualize differences among them. Discriminant function analysis (DFA) was used to verify if the a priori classification of each individual as *C. jacchus*, *C. penicillata*, or hybrid using qualitative characters was correct. Analyses were performed in Statistica 8 (Statistica Software Inc.) and R 3.2.4.

DNA samples were obtained from tissue samples extracted from hybrid specimens and from the species *Callithrix penicillata*, *C. aurita*, *C. kuhlii*, *Mico rondoni*, and *Saguinus mystax* (Table 1). The latter two were used as outgroups. The DNA was extracted following a phenol-chloroform protocol (Sambrook and Russell 2001). Primers for L14724 (Irwin et al. 1991) and Cytb rev (Casado et al. 2010) were used to amplify MT-CYB. The PCR product of the MT-CYB gene was purified and sequenced using the same PCR primers and the internal primers Sot in1 and Sot in2 (Cassens et al. 2000); Alo aot F and Alo aot R (Menezes et al. 2010); and Citb alo (Bonvicino et al. 2001). The product was labeled with XL and BigDye Terminator v3.1 Cycle Sequencing Kit (Applied Biosystems). Sequencing was carried out in an ABI 3130 xl platform. To improve our data, sequences of *C. jacchus* (accession numbers: AF295586 and AY434079), *C. penicillata* (accession number: KR817256.1) and *C. geoffroyi* (accession number: HM368005) were obtained from GenBank online database (www.ncbi.nlm.nih.gov/genbank).

The sequences were analyzed and edited in the software ChromasPro (Mccarthy 1998) and manually aligned in MEGA 5.0 (Tamura et al. 2011).

Genetic distances were estimated with complete deletion using the Kimura 2-parameter model. The MEGA 5.0. Model Generator 0.85 (Keane et al. 2006) identified the best-fitting model for nucleotide substitution using second-order Akaike Information Criteria (AIC) (Akaike 1973).

DNAsp 5 (Librado and Rozas 2009) was used to estimate haplotype and nucleotide diversity. NETWORK was used to reconstruct a median-joining (MJ) network (Bandelt et al. 1999).

Maximum likelihood (ML) and Bayesian inference (BI) phylogenetic trees were built. The ML analysis was inferred using a TN93 + I nucleotide substitution model (Tamura and Nei 1993) and the bootstrap analysis was based on 1000 replicates with PhyML (Guindon et al. 2010). MRBAYES 3.2 (Huelsenbeck and Ronquist 2001) was used to build the Bayesian tree using a TN93 + I model.

Table 1. Samples used in the phylogenetic analyses.

ID	Species	Locality
PRG1415	Callithrix hybrid	RJ, Silva Jardim
PRG1416	Callithrix hybrid	RJ, Silva Jardim
PRG1417	Callithrix hybrid	RJ, Silva Jardim
PRG1454	Callithrix hybrid	RJ, Silva Jardim
PRG1456	Callithrix hybrid	RJ, Silva Jardim
PRG1702	Callithrix hybrid	RJ, Silva Jardim
PRG1703	Callithrix hybrid	RJ, Silva Jardim
PRG1706	Callithrix hybrid	RJ, Silva Jardim
PRG1708	Callithrix hybrid	RJ, Silva Jardim
TDX005	Callithrix hybrid	RJ, Rio das Ostras
ZOOSP01031991	C. penicillata	Zoológico de São Paulo
KR817256.1	C. penicillata	Unavaliable
CRB2587	C. aurita	SP, Cunha
CPRJ1016	C. kuhlii	Centro de Primatologia do Rio de Janeiro
CRB561	Mico rondoni	RO: Ariquemes
CPRJ1621	Saguinus mystax	Centro de Primatologia do Rio de Janeiro
CPRJ452	C. kuhlii	Centro de Primatologia do Rio de Janeiro
CRB3094	Callithrix hybrid	RJ, Rio de Janeiro
CRB3095	Callithrix hybrid	RJ, Rio de Janeiro
LBCE18252	Callithrix hybrid	RJ, Rio de Janeiro
AF295586	C. jacchus	Unavaliable
AY434079	C. jacchus	Unavaliable
HM368005	C. geoffroyi	Germany, Dresden Zoo

RESULTS

Morphological analyses

Callithrix penicillata had blackish pre-auricular tufts, a dark brown neck and throat area, and grey striated hairs on the back with an orange medial band and a basal black band. C. jacchus had white circum-auricular tufts and, as C. penicillata, grey striated hairs on the back with an orange medial band and a basal black band. The general color of the body was grayish for C. jacchus, and ranged from shades of gray to brown for C. penicillata. The general color of the pelage varied widely (in light brown tones) for the hybrids. Most hybrids had white auricular tufts, similar in color but not in ear disposition to those of C. jacchus. The tufts were arranged anterior and lateral to the ear and were broken in some parts. One of the individuals had pre-auricular tufts like those of C. penicillata, but with a grayish color. Two individuals lacked tufts because they were young.

Only one cranial qualitative character could be identified as showing distinct patterns between C. penicillata and C. jacchus. Namely, the presence/absence of a space in the upper jaw after the second molar (Figs 1–4). The space was absent in C. penicillata (i.e. the maxilla ends abruptly after the last molar) and present in C. jacchus. Hybrids exhibited a pattern equivalent to that seen in C. jacchus. Table 2 summarizes the mean and standard deviation of each linear cranial measurement for all Callithrix species and hybrids. Three characters were significantly different between

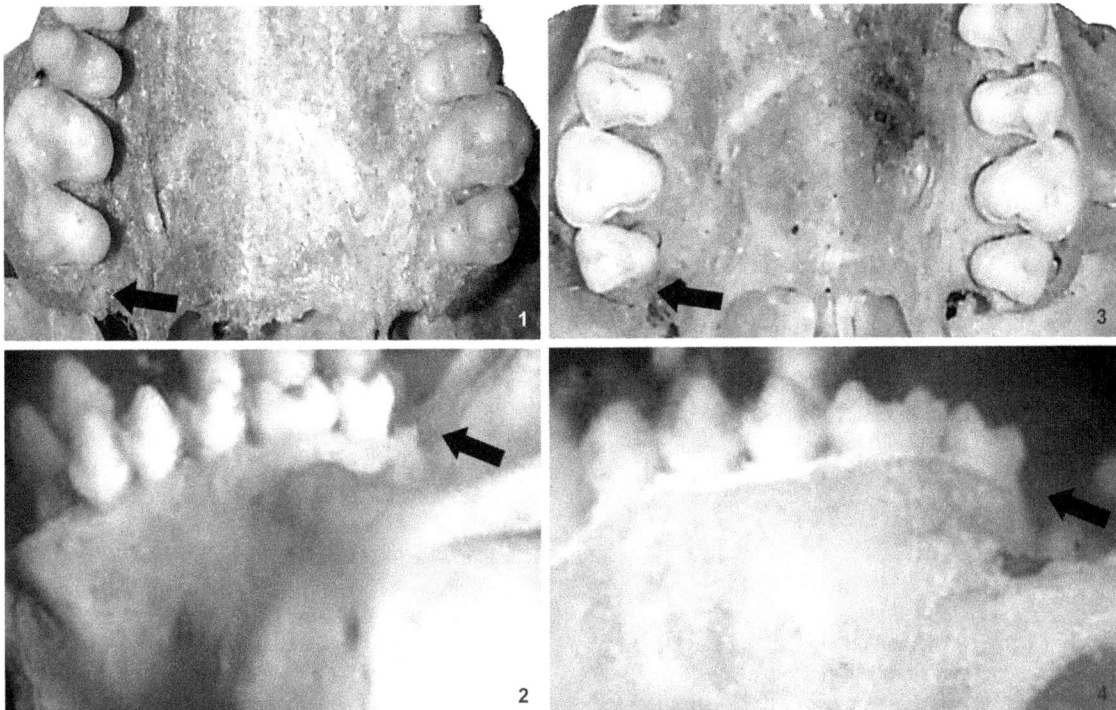

Figures 1–4. Qualitative character differentiating Callithrix jacchus (1–2) from C. penicillata (3–4): presence/absence of a space in the upper jaw after the second molar, indicated by arrows.

Table 2. Approximate mean and standard deviation (mm) for the linear cranial measurements obtained from samples of *Callithrix* species and hybrids. Measurements are identified in the left column, and species/hybrids are identified in the header. (PL) Prosthion to lambda, (EE) euryon to euryon, (iFO) inside frontomalare orbitale to frontomalare orbitale, (BB) bicondylar breadth, (CM) mesial surface of the left upper canine to distal surface of the left second upper molar, (oFO) outside frontomalare orbitale to frontomalare orbitale, (MS) mandibular symphysis height, (lFM) foramen magnum length, (bFM) foramen magnum breadth, (ZA) zygomatic arch breadth.

	Callithrix penicillata	*Callithrix jacchus*	**Hybrids**
PL	44.6 ± 0.86	44.32 ± 1.01	45.26 ± 1.27
EE	25.24 ± 0.65	25.30 ± 0.86	27.43 ± 0.78
iFO	22.77 ± 0.65	23.43 ± 0.90	23.89 ± 0.51
oFO	24.32 ± 0.57	25.12 ± 0.95	25.14 ± 0.59
BB	24.38 ± 0.85	24.54 ± 1.36	25.39 ± 0.67
MS	8.79 ± 0.95	8.08 ± 0.49	8.61 ± 0.61
CM	10.98 ± 0.48	10.72 ± 0.42	11.03 ± 0.29
iFM	10.98 ± 0.48	6.05 ± 0.67	6.73 ± 0.62
bFM	6.66 ± 0.17	6.20 ± 0.49	7.13 ± 0.27
ZA	10.57 ± 0.54	11.05 ± 0.43	11.13 ± 0.56

species (as shown by Student's t-test and one-way ANOVA): iFO, oFO, and MS. In the hybrids, two of these characters (iFO and oFO) were more similar to those of *C. penicillata* and one (CM) was more similar to that of *C. jacchus*. The comparison between the hybrids and each species separately (Student's t-test and one-way ANOVA) showed that EE and CM were significantly different between the hybrids and *C. penicillata*, whereas EE, iFO, oFO, and MS were significantly different between the hybrids and *C. jacchus* (Table 3).

The first three components of the PCA (MS, iFM, and bFM) accounted for most of the observed skull variation (PC1 = 39.9%, PC2 = 26.4%, and PC3 = 13.6%) (Figs 5–6). MS contributed positively and IFM and bFM contributed negatively to PC1; MS and bFM contributed negatively and IFM contributed slightly positively to PC2.

The DFA analysis confirmed the a priori classification, revealing highly significant inter-sample variation (Wilk's lambda = 0.065998, approximate F = 6.0744, p < 0.0001). The scatter plot showed three distinct groups of points (Figs 7–8), each group representing either one of the species or the hybrids. Measurements oFO and ZA contributed most to the first discriminant function while IFM, iFO, and EE contributed most to the second discriminant function (Figs 7–8).

Molecular analyses

The cytochrome b gene, comprising 1140 bp, was sequenced for all specimens. Only the hybrids shared haplotypes. The 13 hybrid sequences had six haplotypes, two of which were shared by more than one specimen. Analysis of sequences from all hybrids revealed 21 variable sites (18 transitions and three transversions),

Table 3. One-way ANOVA and Student's t-test statistical analyses of linear cranial measurements obtained from *Callithrix* species and hybrids.

	PL	EE	iFO	oFO	BB	MS	CM	iFM	bFM	ZA
C. penicillata x *C. jacchus*	ns	ns	*	*	ns	*	ns	ns	ns	ns
C. penicillata x Hybrids	ns	*	ns	ns	ns	*	ns	ns	ns	ns
C. jacchus x Hybrids	ns	*	*	*	ns	*	ns	ns	ns	ns

ns = not significant and * = significant at p < 0.05. (PL) prosthion to lambda; (EE) euryon to euryon; (iFO) inside frontomalare orbitale to frontomalare orbitale; (BB) bicondylar breadth; (CM) mesial surface of the left upper canine to distal surface of the left second upper molar; (oFO) outside frontomalare orbitale to frontomalare orbitale; (MS) mandibular symphysis height; (lFM) foramen magnum length; (bFM) foramen magnum breadth; (ZA) zygomatic arch breadth.

with estimates of genetic distance ranging from 0.001 to 0.01%.

The phylogenetic analyses resulted in a monophyletic *Callithrix* genus (Fig. 9), with *C. aurita* as the sister taxon to the remaining species. The hybrids were grouped into three distinct clades within the *C. penicillata* lineage.

The median-joining network analysis was focused on the relationship between *C. penicillata* and hybrids from the coastal lowlands (CL) and from the metropolitan region (MR) of Rio de Janeiro. It revealed relationships between the hybrid haplotypes of the two regions (Fig. 10). Each of the seven haplotypes in the network is separated by at least one variable site (from a total of 21 variable sites). Haplotype diversity (Hd) was 0.7582 and the nucleotide diversity (Pi) was 0.00537. Seven specimens, including CL and MR hybrids, shared haplotype 1 (H1). Each of haplotypes 2, 3, and 4 (H2, H3 and H4) included one CL individual. Haplotype 5 (H5) was shared by two CL specimens. Haplotype, H6, included two pure *C. penicillata* specimens and a single MR individual. Haplotype 7 (H7) included one pure *C. penicillata* specimen. Results from the network analysis were similar to those described for the phylogenetic analysis.

DISCUSSION

Hybridization has been consistently documented in primates. However, the effects of natural and anthropogenic hybridization on biodiversity remain unclear. Differentiating between these two types of hybridization is a challenge in evolution and conservation studies (Malukiewicz et al. 2014). Previous reports showed that hybrids have intermediate characters between the parental species (Hershkovitz 1977, Alonso et al. 1987, Fuzessy et al .2014, a finding that is partially supported by results of the current study. One possible explanation for such finding is the fact that the parental species are phylogenetically very close. (Tagliaro et al. 1997, Buckner et al. 2014. Mallet (2005) argued that hybridization occurs in approximately 10 percent of the mammalian species, usually between groups that diverged more recently (1 to 2 million years). *C. jacchus* and *C. penicillata* are species that diverge very recently (Malukiewicz et al. 2015).

We observed that the auricular tufts (one of the main characters used to identify marmosets) of the hybrids do not match the description of any species of the genus. In the hybrids the tufts

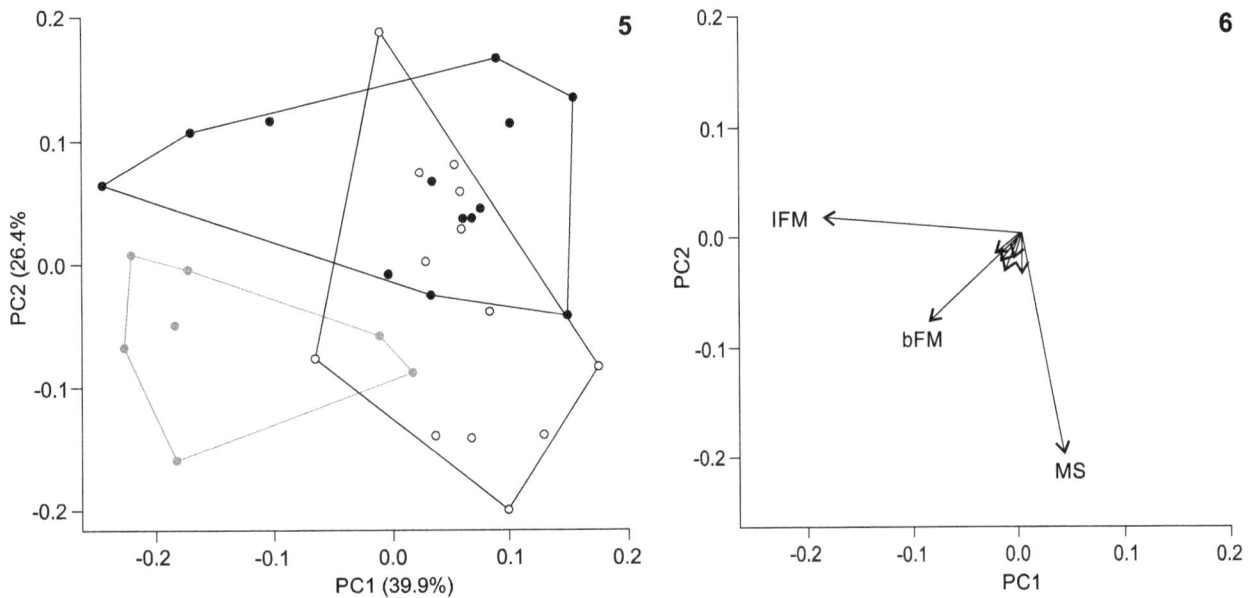

Figures 5–6. PCA analysis results. (5) Scatter plot of scores for principal component 1 x 2. Black circles represent *Callithrix penicillata*, white circles represent *C. jacchus* and grey circles represent hybrids. (6) Contribution of morphometric variables to the principal components. Vectors indicate the loadings of the scores for each variable on the first two principal components.

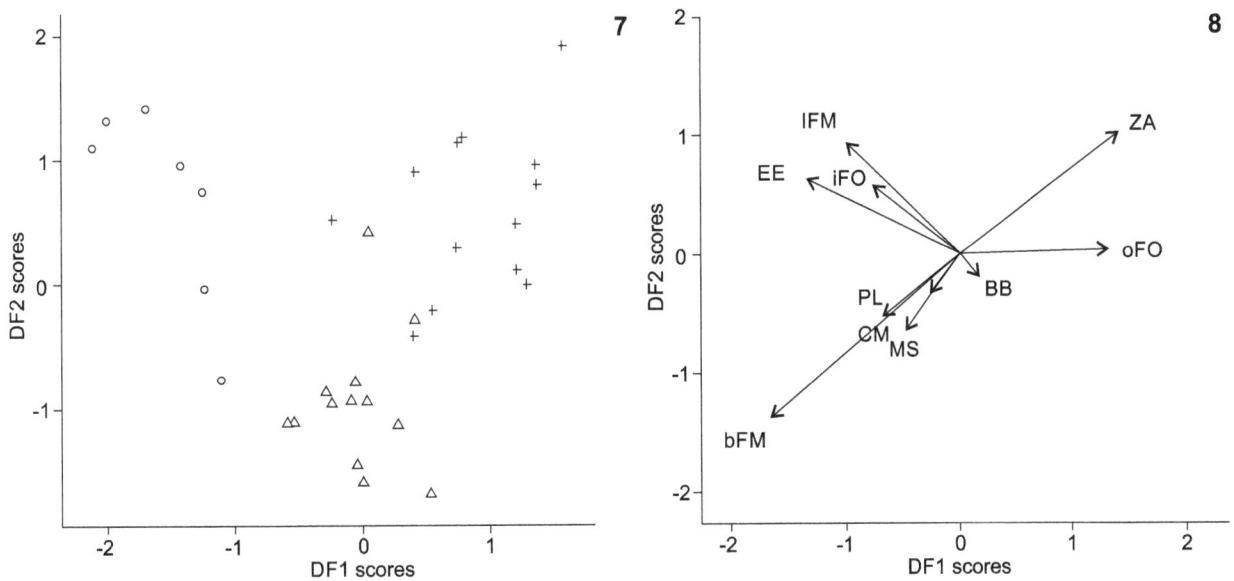

Figures 7–8. DFA analysis results. (7) Scatter plot of scores for discriminant function 1 x 2. Three distinguishable groups characterize the species and hybrids analyzed: (+) *Callithrix penicillata*, (Δ) *C. jacchus*, and (○) hybrids. (8) Contribution of morphometric variables to the discriminant functions. Vectors indicate the loadings of the scores for each variable on the first two discriminant functions.

were white or gray, arranged anterior and laterally to the ear and were broken in some portions, a mosaic that may result from several generations of hybridization. In the study of Alonso et al. (1987),

involving hybrids between *C. jacchus* and *C. penicillata* in natural hybrid zones, the recorded color pattern and shape of the auricular tufts suggest that there is a reproductive isolation mechanism be-

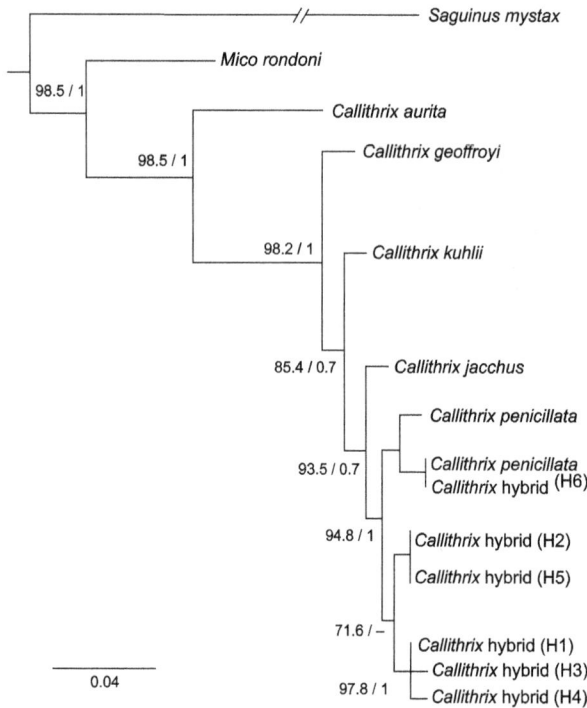

Figure 9. The Bayesian and Maximum Likelihood analyses for MT-CYB of *Callithrix*, rooted by *Saguinus*. Numbers close to branches are boostrap values and posterior probability, respectively.

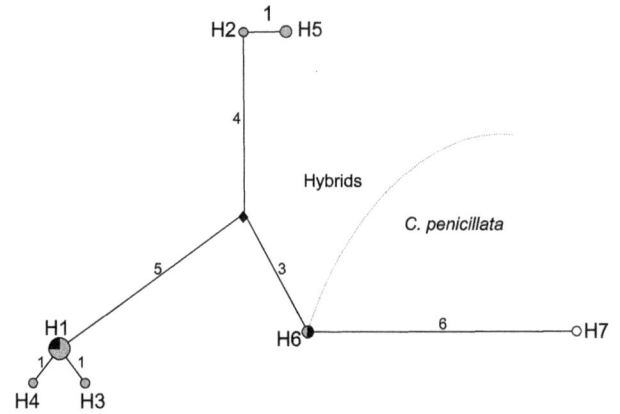

Figure 10. Haplotype network of MT-CYB sequences for *Callithrix penicillata* and *Callithrix* hybrids. Circles represent distinct haplotypes. White circles represent *C. penicillata*, black circles represent hybrids from the MR, and gray circles represent hybrids from the CL. The size of each circle is proportional to the number of individuals per haplotype, with the smallest circle corresponding to n = 1. The lozenge represents the medium-vector. Numbers near lines between haplotypes represent the number of mutations.

tween these species, since there was little penetration of *C. jacchus* characters in the *C. penicillata* population and vice-versa. Individuals with pure parental phenotypes were absent in the hybrid groups of that study, as was observed in the present one.

Hybrids of mixed ancestry between two marmosets species (*C. penicillata* and *C. geoffroyi*) had greater morphological variation compared with individuals of pure ancestry (Fuzessy et al. 2014), which could possibly explain the results found for the hybrid individuals in our study. Malukiewicz et al. (2014) agreed that the current situation of coastal lowland marmosets is the result of multiple introductions and that there some new genetic variations caused by new introductions. These multiple introductions may not only be from pure parental individuals but also from hybrid individuals, resulting in crosses between pure and mixed ancestries that give rise to highly variable phenotypes. The hybrids analyzed herein showed a large variation in body color, with a predominance of light brown regions throughout the body (detected even in the tail). This has not been previously observed in either of the two studied species and may be related a transgressive segregation. Studies on hybrid populations have occasionally reported the presence of phenotypes that are extreme relative to those of either parental line (Rieseberg and Ellstrand 1993, Cosse et al. 1995). The generation of these extreme phenotypes is referred to as transgressive segregation, a phenomenon specific to segregating hybrid generations and refers

to the fraction of individuals that exceed parental phenotypic values in either a negative or positive direction (Rieseberg et al. 1999).

Only one qualitative character differed between *C. jacchus* and *C. penicillata*: the presence/absence of a prolongation of the maxilla, after the last molar. Hybrids had the space, being therefore similar to *C. jacchus*. This is an unprecedented observation in the taxonomic literature, since authors report the skulls of *C. jacchus* and *C. penicillata* as being very similar (Garbino 2015, Natori 1994).

Three measurements explained most of the variation observed in the PCA. When specimens are plotted along PC1 and PC2, the distribution is somewhat scattered. However, there are some trends according to the species and hybrids. An overlap exists between *C. penicillata* and *C. jacchus*, which may be explained by the great similarity between their skulls. Similarly, there is an overlap between hybrids and *C. jacchus*, indicating that hybrids are more related with *C. jacchus* than with *C. penicillata*. This result differs from that obtained by the univariate analysis.

The results of the DFA show that Wilk's lambda was relatively small (0.065998) and the approximate F value was high (6.0744, p < 0.0001), corroborating the a priori classification.

Results of the phylogenetic analysis supported the monophyly of *Callithrix*, as expected, based on previous phylogenies (Tagliaro et al. 1997, Perelman et al. 2011, Garbino 2015). In the current study, *C. jacchus* is the sister group of *C. penicillata*, and these two are the most recent split in the genus. Hybrids were grouped in a clade with *C. penicillata*, suggesting that this species is the one involved in the maternal lineage of the individuals, being a direct or indirect parental species. Where *C. penicillata* was not a direct parent, crosses between lineages of hybrids with *C. penicillata* maternal origin may have occurred. Both cases suggest that males preferentially mate with females from the *C. penicillata* lineage.

Mate choice by males is primarily associated with mate availability and with variation in female quality. If males engage in paternal care, as is the case in *Callithrix*, the average number of available females is likely to be high compared to the capacity of males to mate with them. In this situation, males are less likely to be able to mate with all available females, rejecting some of them. If the benefit of mating with specific females exceeds the cost of assessing them, mate choice can evolve (Edward and Chapman 2011). Although fertile hybrids will be generated between the two species, some matings may result in less viable hybrids than others, influencing the process of mate choice (Coimbra-Filho et al. 1993). The relationship found between the hybrids and *C. penicillata* in this study suggests that hybrid males prefer *C. penicillata* females. This may be because hybrids that result from mating with *C. penicillata* females are more viable than those that result from mating with *C. jacchus* females.

Primates recognize potential mates (members of the same species or not) based on visual (mainly on the face), acoustic, olfactory, and other sensorial cues. Different patterns of facial color and auricular hairs have diagnostic value for each species in the taxonomy of *Callithrix*. One conspicuous morphologic character that distinguishes these four species is the coloration of the auricular tufts (Cavalcanti and Langguth 2008).

Cavalcanti and Langguth (2008) suggested that there is an isolation mechanism based on head color, since when two different species are together, one responds to the other with significantly less frequency than to its own species. However, this is not always the case, suggesting that, in spite of the conspicuous differences in facial coloration patterns, the evaluator species recognizes individuals of the other, "cue-bearing" species as potential sexual competitors. During speciation events in *Callithrix*, reproductive isolation mechanisms did not necessarily appear simultaneously in all the presently recognized species. Populations of each species have a different history. *C. jacchus* and *C. penicillata* are very closely-related and have split recently from a common ancestor. Thus, it is possible that reproductive isolation is not complete in this case. Although the primate literature is relatively rich in studies of sexual selection and mate preference in hybrid zones (Shurtliff 2013), little attention has been given to these topics in recent decades. A study on howler monkeys (*Alouatta palliata* and *A. pigra*) showed that hybridization and subsequent backcrossing are directionally biased, probably producing only fertile hybrid females and inviable or infertile males. This suggests that a process of mate choice may occur for Neotropical primates in hybrid zones (Cortés-Ortiz et al. 2007).

The low genetic distance between hybrids revealed that they are genetically very close, corroborating the hypothesis of multiple introductions suggested by the morphological results and the short passage of new genetic variations.

Haplotype H1 was found in most of the samples, containing both CL and MR individuals and precluding the existence of a geographical structure. A possible explanation is that introductions of marmosets into the coastal lowlands occurred recently with animals from the metropolitan region. De Morais et al. (2008) argued that the history of the introduction of marmosets in the coastal lowlands of Rio de Janeiro is uncertain and emphasized the occurrence of two major releases of marmosets, seized by regulatory agencies (> 60 animals), between 1983-1987, in regions close to the study area. Each of the haplotypes, H2, H3, and H4 was found in only one CL hybrid, and haplotype H5 was found in two CL individuals. This diversity in a single region may be the result of multiple introductions, with the arrival of new individuals from different localities. Haplotype H6 included one *C. penicillata* and one MR individual, but the fact that haplotype H1 also included an MR individual indicates that H6 is not characteristic only of the MR.

The network analysis showed close a relationship between the hybrids and *C. penicillata*, confirming the result from the ML and BI analyses. At least one variable site separates each haplotype, which may reflect differences due to multiple introductions or a polymorphism in the population. Twenty-one variable sites were observed, with eighteen transitions and three transversions. Haplotype diversity among hybrids and *C. penicillata* was high (0.7582), whereas nucleotide diversity was low (0.00537). Low nucleotide diversity may be explained by the founder effect, whereby an introduction is followed by population growth. The several haplotypes also had low diversity, possibly due to multiple anthropogenic introductions that continue to occur. The low level of genetic diversity both in the CL and between the CL and the MR suggests a recent history of population expansion, most likely associated with introductions.

These issues reflect the need to perform more detailed studies concerning hybridization and the development of morphological characters in order to obtain a better understanding of the evolution of *Callithrix*.

ACKNOWLEDGMENTS

We would like to thank the financing agency Conselho Nacional de Desenvolvimento Científico e Tecnológico (CNPq) for providing a scientific initiation scholarship to Adrielle M. Cezar and research fellowships to Leila M. Pessôa (process 308505/2016-6). We also thank João A. de Oliveira and Pablo R. Gonçalves for their help with analyzing the specimens at the Museu Nacional da Universidade Federal do Rio de Janeiro (MN-UFRJ) and the Núcleo em Ecologia e Desenvolvimento Sócio-Ambiental de Macaé (NU-PEM/UFRJ). We are grateful to the members of the monograph examination commission, Alcides Pissinatti, Daniel F. da Silva, Maria Lucia Lorini, and Héctor Seuánez, for their availability and contributions; to the Centro de Primatologia do Rio de Janeiro; to the São Paulo Zoo and to the Dresden Zoo for granting permission to access the materials used in this study. We are grateful to all our laboratory colleagues who helped us during this project. The English spelling and grammar of a previous version of this manuscript were edited by Publicase, PPGBBE, UFRJ financed the

English editing work.

REFERENCES

Akaike H (1973) Information theory and an extension of the maximum likelihood principle. In: Petrov BN, Csáki F (Eds) 2nd International Symposium on Information Theory, Tsahkadsor (Armenia), September 2–8, 1971. Akadémiai Kiadó, Budapest, 267–281.

Alonso C, de Faria DS, Langguth A, Santee DF (1987) Variação da pelagem na área de intergradação entre *Callithrix jacchus* e *Callithrix penicillata*. Revista Brasileira de Biologia 47: 465–470.

Bandelt HJ, Forster P, Röhl A (1999) Median-joining networks for inferring intraspecific phylogenies. Molecular Biology Evolution 16: 37–48. https://doi.org/10.1093/oxfordjournals.molbev.a026036

Bonvicino CR, Lemos B, Seuánez HN (2001) Molecular phylogenetics of howler monkeys (Alouatta, Platyrrhini): A comparison with karyotypic data. Chromosoma 110: 241–246. https://doi.org/10.1007/s004120000128

Buckner JC, Alfaro JL, Rylands AB, Alfaro ME (2014) Biogeography of the marmosets and tamarins (Callitrichidae). Molecular Phylogenetic Evolution 82: 413–425. https://doi.org/10.1016/j.ympev.2014.04.031

Casado F, Bonvicino CR, Nagle C, Comas B, Manzur TD, Lahoz MM, Seuánez HN (2010) Mitochondrial Divergence Between 2 Populations of the Hooded Capuchin, *Cebus (Sapajus) cay* (Platyrrhini, Primates). Journal of Heredity 101: 261–269. https://doi.org/10.1093/jhered/esp119

Cassens I, Vicario S, Waddell VG, Balchowsky H, van Belle D, Ding W, Fan C, Lal Mohan RS, Simões-Lopes PC, Bastida R, Meyer A, Stanhope MJ, Milinkovitch MC (2000) Independent adaptation to riverine habitats allowed survival of ancient cetacean lineages. Proceedings of National Academy of Sciences of the United States of America 97: 11343–11347. https://doi.org/10.1073/pnas.97.21.11343

Cavalcanti GC, Langguth A (2008) Recognition of mate and speciation in marmoset genus Callithrix (Primates, Cebidae, Callithriquinae [sic]). Revista Nordestina de Biologia 19: 59–73.

Coimbra-Filho AF, Pissinatti A, Rylands AB (1993) Experimental multiple hybridism among *Callithrix* species from eastern Brazil. In: Rylands AB (Ed.) Marmosets and tamarins: systematics, ecology and behavior. Oxford University Press, Oxford, 95–120.

Cortés-Ortiz L, Duda Jr TF, Canales-Espinosa D, García-Orduña F, Rodríguez-Luna E, Bermingham E (2007) Hybridization in large-bodied new world primates. Genetics 176: 2421–2425. https://doi.org/10.1534/genetics.107.074278

Cosse AA, Campbell MG, Glover TJ, Linn Jr CE, Todd JL, Baker TC, Roelofs WL (1995) Pheromone behavioral responses in unusual male European corn borer hybrid progeny not correlated to electrophysiological phenotypes of their pheromone-specific antennal neurons. Experientia 51: 809–816. https://doi.org/10.1007/BF01922435

De Morais Jr MM, Ruiz-Miranda CR, Grativol AD, Andrade CC, Lima CA (2008) Os sagüis, *Callithrix jacchus* e *C. penicillata*, como espécies invasoras na região de ocorrência do mico-leão dourado. In: Oliveira PP, Grativol AD, Ruiz-Miranda CR (Eds) Conservação do mico-leão-dourado: enfrentando os desafios de uma paisagem fragmentada. Campos dos Goytacazes, Universidade Estadual do Norte Fluminense, vol. 1, 86–117.

De Vivo M (1991) Taxonomia de *Callithrix* Erxleben, 1777 (Callitrichidae, Primates). Belo Horizonte, Fundação Biodiversitas.

Edward DA, Chapman T (2011) The evolution and significance of male mate choice. Trends in Ecology and Evolution 26: 647–654. https://doi.org/10.1016/j.tree.2011.07.012

Fuzessy LF, de Oliveira SI, Malukiewicz J, Silva FFR, do Carmo PM, Boere V, Ackermann RR (2014) Morphological Variation in Wild Marmosets (*Callithrix penicillata* and *C. geoffroyi*) and Their Hybrids. Evolutionary Biology 41: 480–493. https://doi.org/10.1007/s11692-014-9284-5

Garbino GST (2015) How many marmoset (Primates: Cebidae: Callitrichinae) genera are there? A phylogenetic analysis based on multiple morphological systems. Cladistics 31: 652–678. https://doi.org/10.1111/cla.12106

Guindon S, Dufayard JF, Lefort V, Anisimova M, Hordijk W, Gascuel O (2010) New Algorithms and Methods to Estimate Maximum-Likelihood Phylogenies: Assessing the Performance of PhyML 3.0. Systematic Biology 59: 307–21. https://doi.org/10.1093/sysbio/syq010

Hershkovitz P (1977) Living New World Monkeys (Platyrrhini). With an Introduction to Primates. The University of Chicago Press, Chicago, vol. 1.

Huelsenbeck JP, Ronquist F (2001) MrBayes: Bayesian inference of phylogenetics trees. Bioinformatics 17: 754–755. https://doi.org/10.1093/bioinformatics/17.8.754

Irwin DM, Kocher TD, Wilson AC (1991) Evolution of the cytochrome b gene of mammals. Journal of Molecular Evolution 32: 128–144. https://doi.org/10.1007/BF02515385

Keane TM, Creevey CJ, Pentony MM, Naughton TJ, Mcinerney JO (2006) Assessment of methods for amino acid matrix selection and their use on empirical data shows that ad hoc assumptions for choice of matrix are not justified. BMC Evolution Biology 6: 29. https://doi.org/10.1186/1471-2148-6-29

Librado P, Rozas J (2009) DnaSP v5: A software for comprehensive analysis of DNA polymorphism data. Bioinformatics 25: 1451–1452. https://doi.org/10.1093/bioinformatics/btp187

McCarthy C (1998) Chromas. Queensland, School of Health Science, Griffith University, version 1.45.

Mallet J (2005) Hybridization as an invasion of the genome. Trends in Ecology and Evolution 20: 229–237. https://doi.org/10.1016/j.tree.2005.02.010

Malukiewicz J, Boere V, Fuzessy LF, Grativol AD, French JA, Silva I, Pereira LCM, Ruiz-Miranda CR, Valença Y, Stone AC (2014) Hybridization effects and genetic diversity of the common and black-tufted marmoset (*Callithrix jacchus* and *Callithrix penicillata*) mitochondrial control region. American Journal of Physical and Anthropology 155: 522–536. https://doi.org/10.1002/ajpa.22605

Malukiewicz J, Boere V, Fuzessy LF, Grativol AD, de Oliveira e Silva I, Pereira LCM, Ruiz-Miranda CR, Valença Y, Stone AC (2015) Natural and Anthropogenic Hybridization in Two Species of Eastern Brazilian Marmosets (*Callithrix jacchus* and *C. penicillata*). PLoS ONE 10(6): e0127268. https://doi.org/10.1371/journal.pone.0127268

Menezes AN, Bonvicino CR, Seuánez HN (2010) Identification, classification and evolution of owl monkeys (*Aotus*, Illiger 1811). BMC Evolution Biology 10: 248. https://doi.org/10.1186/1471-2148-10-248

Natori M (1994) Craniometrical variations among eastern Brazilian marmosets and their systematic relationships. Primates 35: 167–176. https://doi.org/10.1007/BF02382052

Perelman P, Johnson WE, Roos C, Seuánez HN, Horvath JE, Moreira MA, Kessing B, Pontius J, Roelke M, Rumpler Y, Schneider MP, Silva A, O'Brien SJ, Pecon-Slattery J (2011) A Molecular Phylogeny of Living Primates. PLoS Genetic 7: e1001342. https://doi.org/10.1371/journal.pgen.1001342

Rieseberg LH, Archer MA, Wayne RK (1999) Transgressive segregation, adaptation and speciation. Heredity 83: 363–372. https://doi.org/10.1038/sj.hdy.6886170

Rieseberg LH, Ellstrand NC (1993) What can morphological and molecular markers tell us about plant hybridization? Critical Review Plant Science 12: 213–241. https://doi.org/10.1080/07352689309701902

Ruiz-Miranda CR, Affonso AG, Martins A, Beck BB (2000) Distribuição do sagui (*Callithrix jacchus*) nas áreas de ocorrência do mico leão dourado no Estado de Rio de Janeiro. Neotropical Primates 8: 98–101.

Rylands AB, Coimbra-Filho AF, Mittermeier RA (1993) Systematics, geographic distribution and some notes on the conservation status of the Callitrichidae. In: Rylands AB (Ed.) Marmosets and tamarins: systematics, ecology and behaviour. Oxford Science Publications, Oxford, 11–77.

Rylands AB, da Fonseca GAB, Leite YLR, Mittermeier RA (1996) Primates of the Atlantic forest: origin, distributions, endemism, and communities. In: Norconk MA, Rosenberger AL, Garber PA (Eds) Adaptive radiations of neotropical primates. Plenum, New York, 21–51. https://doi.org/10.1007/978-1-4419-8770-9_2

Rylands AB, Coimbra-Filho AF, Mittermeier RA (2009) The systematics and distributions of the marmosets (*Callithrix, Callibel-*

la, Cebuella, and *Mico*) and *Callimico* (*Callimico*) (Callitrichidae, Primates. In: Ford SM, Porter LM, Davis LC (Eds) The smallest anthropoids: The marmoset/*Callimico* radiation. Springer, New York, 25–61. https://doi.org/10.1007/978-1-4419-0293-1_2

Sambrook J, Russel DW (2001) Molecular cloning. CSHL Press, Cold Spring Harbor, 3rd ed.

Shurtliff QR (2013) Mammalian hybrid zones: a review. Mammal Review 43: 1–21. https://doi.org/10.1111/j.1365-2907.2011.00205.x

Tamura K, Nei M (1993) Estimation of the number of nucleotide substitutions in the control region of mitochondrial DNA in humans and chimpanzees. Molecular Biology and Evolution 10: 512–526.

Tamura K, Peterson D, Peterson N, Stecher G, Nei M, Kumar S (2011) MEGA5: molecular evolutionary genetics analysis using maximum likelihood, evolutionary distance, and maximum parsimony methods. Molecular Biology and Evolution 28: 2731–2739. https://doi.org/10.1093/molbev/msr121

Tagliaro CH, Schneider MPC, Schneider H, Sampaio I, Stanhope M (1997) Marmoset phylogenetics, conservation perspectives, and evolution of the mtDNA control region. Molecular Biology and Evolution 14: 674–684. https://doi.org/10.1093/oxford-journals.molbev.a025807

APPENDIX 1

Specimens provenance are summarized below.

The *C. penicillata* and *C. jacchus* specimens are deposited in the National Museum, Federal University of Rio de Janeiro. Their voucher numbers are: *Callithrix penicillata* (MN4260-4262, MN4264-4266, MN4268-4270, MN10681, MN11334, MN23798, MN23800 and MN30549) and *C. jacchus* (MN3953, MN5521, MN5528, MN5535, MN5546, MN5551, MN5573, MN17274-17276, MN17291, MN23772, MN23774, MN30544, MN30548).

The hybrid (*Callithrix* sp.) specimens are deposited at the Center for Ecology and Socio-Environmental Development of Macaé, Federal University of Rio de Janeiro (PRG1415-1417, PRG1454, PRG1456, PRG1702, PRG1703, PRG1706, PRG1708, TXD005).

Author Contributions: AMC, LMP and CRB participated equally in the preparation of this article.

19

Minaselates, a new genus and new species of Epiphragmophoridae from Brazil (Gastropoda: Stylommatophora: Helicoidea)

Maria Gabriela Cuezzo[1], Meire Silva Pena[2]

[1]*Instituto de Biodiversidad Neotropical, CONICET-UNT, Facultad de Ciencias Naturales, Universidad Nacional de Tucumán. Miguel Lillo 205, 4000 Tucumán, Argentina.*
[2]*Laboratório de Malacologia, Museu de Ciências Naturais, Pontifícia Universidade Católica de Minas Gerais. Avenida Dom José Gaspar 500, Coração Eucarístico, Belo Horizonte, MG, Brazil.*
Corresponding author: Maria Gabriela Cuezzo (gcuezzo@webmail.unt.edu.ar)

http://zoobank.org/1B7B6395-EE91-46AA-9774-89832FE0F47A

ABSTRACT. We describe a new genus and a new species in the family Epiphragmophoridae, *Minaselates paradoxa* **sp. n.** The new species was found at the National Park Cavernas do Peruaçu, in northern portion of the state of Minas Gerais, Brazil. *Minaselates paradoxa* **sp. n.** is classified in Epiphragmophoridae based on the fact that it shares the following diagnostic features of the family: a dart apparatus with a single dart sac, and two unequal mucous glands at the terminal genitalia. *Minaselates* **gen. n.** differs from *Epiphragmophora* Doering, 1874 by having a granulose protoconch, shell spire with blunt apex, complex microsculpture on the teleoconch and closed umbilicus fused with the shell wall. Also, significant differences between the two genera are the presence of a long and thin kidney that extends more than half the length of the pulmonary cavity, the presence of a flagellar caecum, and a smooth jaw in *Minaselates* **gen. n.** The finding of this new species and genus is particularly significant to refine the definition of the family, since Epiphragmophoridae has been traditionally diagnosed using the same characters of *Epiphragmophora*. *Dinotropis* Pilsbry & Cockerell, 1937, the other valid genus in the family, is monospecific and is only known by the morphology of the shell. In many ways it is similar to *Epiphragmophora*. A cladistics analysis was made in the present study which supports *Minaselates* **gen. n.** as a different entity and as sister group of the *Epiphragmophora* within Epiphragmophoridae.

KEY WORDS. Cerrado, Pleurodontidae, Pulmonata, South America, Taxonomy.

INTRODUCTION

Epiphragmophoridae is a Pulmonate land snail family exclusively distributed in South America. It is composed of the genera *Epiphragmophora* Doering, 1874 and *Dinotropis* Pilsbry & Cockerell, 1937. *Epiphragmophora* is currently composed of 63 species distributed in Peru, Bolivia, Argentina, Paraguay and southern Brazil with a single extra occurrence in Colombia (Linares and Vega 2011). The species are classified into five subgenera (*Epiphragmophora* s.s., *Angrandiella* Ancey, 1886, *Doeringiana* Ihering, 1929, *Karlchmidtia* Hass, 1955 and *Pilsbrya* Ancey, 1887) (Zilch 1959, Richardson 1982, Cuezzo 2006). *Dinotropis* is a monotypic genus, known only by *D. harringtoni* Pilsbry & Cockerel, 1937 from Bolivia.

Epiphragmophoridae is currently diagnosed by the same synapomorphies of *Epiphragmophora* because *Dinotropis* is only known by its original description, which is entirely based on shell characters. Based on a cladistic hypothesis (Cuezzo 2006), *Epiphragmophora* is characterized by the following synapomorphies i) malleated shell body whorl surface with diagonal ribs, ii) umbilicus overlapping, but not fused to the body whorl, iii) thick, widely reflexed peristome, iv) presence of a dart sac apparatus inserting in the vagina, or directly into the atrium, v) mucous glands unequal in size and shape, vi) insertion of mucous glands ducts in middle portion of dart sac, and vii) penial retractor muscle inserting in epiphallus medial portion. The short duct of the bursa copulatrix, a character that traditionally had been used to define the genus, is characteristic only of a small group of species.

Epiphragmophora was classified by Pilsbry (1894) in the Helicidae, tribe Belogona Euadenia. The members of the latter are distinguished by having mucous glands of typically glandular structure, in contrast to the tube-like glands of the Belogona Siphonadenia. Thiele (1929) later classified *Epiphragmophora* in Fruticicolidae: Epiphragmophorinae. Pilsbry (1939) assembled all the American dart-bearing helices in Helminthoglyptidae and restricted *Epiphragmophora* to the South American Epiphragmophorinae Hoffman, 1928 from the same family. Nordsieck (1987) maintained Epiphragmophorinae Hoffman, 1928, but moving it into Xanthonychidae, while stating that its reproductive system resembles that of the Cepoliinae: diverticulum usually missing, one dart sac, dart glands unequal, one elongate and the other compact, inserting on the dart sac or on its base. Finally, Schileyko (1991) elevated Epiphragmophorinae to family within Helicoidea and this classification was followed by Bouchet and Rocroi (2005) in the last gastropod family nomenclator. Cuezzo (2006) studied species of *Epiphragmophora* from Argentina and part of Bolivia, providing the first phylogenetic analysis of the genus. Species from Peru and Paraguay have been scarcely investigated and are mostly known by their original descriptions. Specimens with their entire body preserved are rare in malacological collections globally. Consequently, comprehensive anatomical or molecular studies are not feasible in most cases. In Brazil there is a single species, *Epiphragmophora oresigena bernardius* Ihering in Pilsbry, 1900, known from Serra da Bocaina and Campos do Jordão in São Paulo state and from state of Rio Grande do Sul.

During a field trip to the National Park Cavernas do Peruaçu in northern Minas Gerais, Brazil, to collect gastropods, a striking group of land snails was found. Analysis of the specimens collected revealed that they represent a new species of the family Epiphragmophoridae. The objective of the present work is to describe the new species in a new genus of Epiphragmophoridae and to discuss its position among the South American Helicoidea.

MATERIAL AND METHODS

Snails were collected at the National Park Cavernas do Peruaçu (14°56'S, 44°36'W) located in the state of Minas Gerais, Brazil. This conservation unit was created in 1999 with 143,353.84 ha (http://www.icmbio.gov.br), to protect limestone caverns. The calcareous massif is covered by rare and typical deciduous (Caatinga) or semi-deciduous forests called Seasonal Dry Tropical Forest (SDTF) (Pennington et al. 2000, Prado 2000). Open shrub savannah (Cerrado) also occurs on the top of the mountains, and in poorly drained areas of the Park where the Peruaçu River originates. Shrub density is lower at the watersheds, and there the vegetation is totally herbaceous (Campo Limpo and Veredas), with localized Buriti palm-trees. The Caatinga, the Cerrado and the Chaco dry areas are extensive, open biomes, and form the dry diagonal in South America, which is a natural phytogeographic unit (Pennington et al. 2000, Prado

2000). The Cerrado spreads across 2,031,990 km² of the central Brazilian Plateau and is the second largest of Brazil's major biomes, after the Amazon. This biodiversity hotspot actually receives abundant rainfall (between 1,100 and 1,600 mm per year), although this rainfall is concentrated in a six to seven month period between October and April. The rest of the year is characterized by a pronounced dry season.

Live specimens and dry shells of *Minaselates paradoxa* **sp. n.** were collected from rocky outcrops in dry deciduous forests of the National Park. Dry shells adhered to rocks or to leaves were abundant but live snails were scarce. The collected specimens were drowned in water for relaxation previous to fixation in ethanol 96%. Their shells were then photographed and measured using the software ImageJ 1.49 (Fig. 1). Anatomical information was obtained by dissecting specimens and studying them under a Leica MZ6 stereoscope, illustrations of the dissected parts where made with the aid of a camera lucida. Photographs of the different organ systems were taken using a Nikon 5000 camera attached to the stereoscope. They were enhanced and finalized using the software Corel Draw version X3. The terminology for the anatomical descriptions follows Tompa (1984). The terms proximal and distal refer to the position of an organ or part of an organ in relation to the gamete flow from the ovotestis (proximal) to the genital pore (distal), as in previous studies (Cuezzo 1997, Cuezzo 2006). The limit between the epiphallus and penis is based on the internal sculpture of their inner walls. The radula and the jaw of specimens were observed and photographed with a Jeol Scanning Electron Microscopy 35CF at the Integral Center of Electron Microscopy of the National University of Tucumán, Argentina. Shell microphotographs were obtained with a DSM 950ZEISS SEM at "Centro de Aquisição e Processamento de Imagens" of the Federal University of Minas Gerais, Brazil.

Intitutional abbreviations used in the text: IBN, Instituto de Biodiversidad Neotropical, Tucumán, Argentina; MLP-Ma, Museo de La Plata, Buenos Aires, Argentina; MNRJ, Museu Nacional Rio de Janeiro, Universidade Federal do Rio de Janeiro, Brazil; MCN, Museu Ciências Naturais, Pontifícia Universidade de Minas Gerais, Belo Horizonte, Brazil.

For the cladistic analysis, a matrix of 35 characters from the general anatomy plus shell morphology was generated for 24 terminal taxa (Appendix 1), following characters and codifications of Cuezzo (2003) for *Pleurodonte* Fischer, 1807 and *Labyrinthus* Beck, 1837 and Cuezzo (2006) for species of *Epiphragmophora*, with modifications. The characters used in the analysis are listed in Appendix 2. In the text, characters and state characters numbers are located between parenthesis. Only species for which the anatomy had been described were included in the character matrix. The data matrix was built in Winclada, v. 1.00.08 (Nixon 2002). Non-applicable data were coded as "–" and multistate characters were treated as additive. Cladistic analyses were performed with TNT, version 1.5 (Goloboff et al. 2008) with Character Weighting and Traditional Search basing

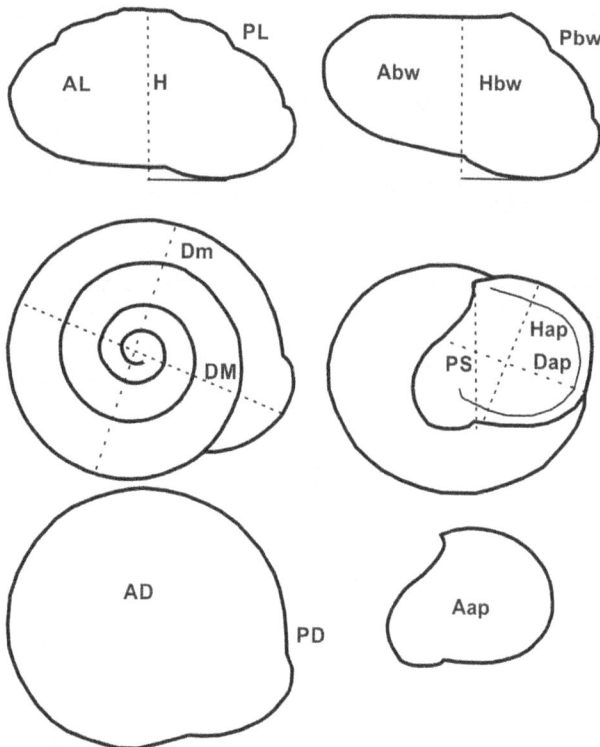

Figure 1. Shell measurements. Abbreviations (AD) dorsal view area, (Aap) apertural area, (Abw) body whorl area, (AL) lateral view area, (Dap) apertural diameter, (DM) major diameter, (Dm) minor diameter, (H) total shell height, (Hap) height of aperture, (Hbw) body whorl height, (Pbw) body whorl perimeter, (PD) dorsal view shell perimeter, (PL) lateral view shell perimeter, (PS) parietal space.

the strategy on RAS + TBR (random addition sequences plus swap by tree bisection and reconnection), with 1,000 replications and 100 trees saved per replication. The default concavity (K) value was used in all analyses (K = 3000). In the analysis, trees were rooted in *Pleurodonte* based on Wade et al. (2007) hypothesis on the phylogenetic relationships of the Helicoidea. Clade support was estimated using symmetric resampling (Goloboff et al. 2003), because the resulting values obtained under this procedure are not distorted by character weighting. Additionally, support for the obtained clades was calculated using the Jackknife resampling method.

TAXONOMY

Supra superfamily classification follows Bouchet and Rocroi (2005) and Ponder and Lindberg (1997). This classification tries to integrate the results of recent cladistics work by using the unranked "clade" above the rank of superfamily while still using the traditional Linnaean ranks for superfamilies and all taxa below the rank of superfamily.

Class Gasteropoda Cuvier, 1795
Clade Heterobranchia Burmeister, 1837
Clade Stylommatophora Schmidt, 1855
Superfamily Helicoidea Rafinesque, 1815
Epiphragmophoridae Hoffmann, 1928

Minaselates gen. n.

http://zoobank.org/48537C28-29CA-4488-97FF-1C570990D9F9

Diagnosis. *Minaselates* gen. n. is distinguished by the following characters: 1) shell globose with blunt apex; 2) protoconch sculptured with granules; 3) teleoconch sculptured with complex microstructures; 4) umbilicus imperforate, parietal wall fused with columellar zone of peristome; 5) wavy spiral lines below the periphery and over ventral teleoconch surface; 6) genitalia with a dart apparatus composed by a single dart sac and two unequal mucous glands, one globose and the other oval; 7) presence of a flagellar caecum; 8) bursa copulatrix duct short, no longer than the sac.

Type species. *Minaselates paradoxa* sp. n. by original designation.

Description. Shell globose, with 4 to 5 convex whorls. Spire conic with blunt apex. Protoconch granulose. Teleoconch sculptured. Wavy spiral grooves at the ventral teleoconch surface. Aperture subcircular with thin peristome. Umbilicus closed. Presence of spiral brownish bands more pronounced in the body whorl. Kidney long and thin, more than half the lung roof length. Genitalia with a dart apparatus and two unequal mucous glands.

Etymology. *Minaselates* is a compound name formed by *Minas* in honor to the Brazilian state where the species was found, and *selates*, a noun in the genitive singular, that derives from the Greek meaning "snail" (Brown 1979).

Remarks. *Minaselates* gen. n. is classified in Epiphragmophoridae because it has a dart apparatus and two unequal mucous glands at the terminal genitalia. These structures are diagnostic of Epiphragmophoridae (Helicoidea) and their morphology serve to differentiate this family from the remaining helicoidean groups. *Dinotropis* differs from *Minaselates* in its depressed shell with an acute peripheral keel and open umbilicus. *Minaselates* differs from *Epiphragmophora* in its general shell shape with blunt apex, granulose protoconch and complex sculpture of the teleoconch surface. The wavy spiral grooves at the ventral teleoconch surface in *Minaselates* are lacking in both, *Epiphragmophora* and *Dinotropis*. The presence of a long and thin kidney in *Minaselates* is very different to the kidney shape in *Epiphragmophora*, which is triangular and shorter.

Minaselates paradoxa sp. n.

http://zoobank.org/9AACED00-8AFB-4736-85DC-14A3399A04CB
Figs 1–26

Diagnosis. Shell globular, with three spiral continuous pigmented bands, the middle, equatorial band thinner. Protoconch granulose. Dorsal side of teleoconch with axial lines bearing triangular lamellae, ventral teleoconch with wavy, concentric, spiral grooves. Imperforate umbilicus. Jaw smooth. Kidney triangular, long and thin, of about 60 to 70% the length of the lung roof. Vas deferens insertion in lower portion of flagellar caecum. Strong, short muscular penial retractor inserting at proximal epiphallus.

Etymology. The species name derives from the Greek *paradoxos* meaning "strange, contrary to expectation" (Brown 1979) as this is a species of Epiphragmophoridae that was not expected to occur in the state of Minas Gerais, Brazil.

Description. **Shell** (Figs 2–13) dextral, helicoidal with 4¾ convex, solid whorls. Coloration pale brown with three spiral darker brown bands more separated from each other in the body whorl. Medial pigmented band thinner than the other bands (Figs 2–7). Suture impressed. Protoconch of 1½ whorls, covered by oval, tightly arranged pustules (Figs 8, 9). Dorsal side of teleoconch with slightly curved axial growth lines that, at higher magnifications, appear as axial thin lines bearing broad-base triangular lamellae. These axial lines are not continuous and are separated by narrow spaces with wrinkles. Each lamellae has an axis from the base to the upper extreme of the triangle (Figs 10–13). The long axes of the lamellae are perpendicular to the shell sutures. This sculpture looks like granules at naked eye. Wavy spiral grooves below the periphery, and in the basal surface area. Peristome thin, reflexed, some specimens showing a basal thickening and or an incipient palatal tooth. Aperture roundish, without angulation, and rounded in the palatal zone, well reflected in the columellar zone covering the umbilicus. Shell imperforate. Shell measurements. Table 1, Fig. 1. **Digestive system.** Radula (Figs 14–17) Central tooth long, unicuspid. Lateral teeth long, with incipient lateral cusps, of about 54–56 µm (n = 10). Marginal teeth bicuspid, similar to laterals in shape and size. Jaw (Figs 18, 19) Horseshoe slightly arched, translucent, with no division. Surface almost smooth with thin, transverse grooves visible with high magnification. **Pallial system** (Figs 20, 21) Kidney triangular, long and thin, of about 60 to 70% the length of the lung roof. Main pulmonary vein thick, splitting into two secondary branches before reaching mantle collar; pulmonary roof dark grey in color, furrowed by well marked, but lesser minor transverse veins. Secondary ureter runs parallel to rectum, completely closed until reaching mantle collar. Ureteric interramus triangular in shape, deeply excavated. **Genital System** (Figs 22–26) Terminal genitalia with dart apparatus and two mucous glands (Figs 22–24). Right ommatophore retractor crosses the distal genital system between penis and vagina. Vagina long. Penis and vagina entering side by side in atrium. Bursa copulatrix with a globular sac and short, thick duct. Bursa copulatrix slightly longer than free oviduct. Single dart sac muscular, with medial constriction, ending in the atrium. Upper dart sac inverted pear shaped, lower portion of dart sac bellow constriction cylindrical, thicker (Fig. 24). Two unequal mucous glands with their respective thin efferent ducts inserting independently above the dart sac constriction (Fig. 25). One of the glands bean shaped with medial

Table 1. Shell dimensions in mm or mm² (n = 14). DM major diameter, Dm minor diameter, AD dorsal area, PD dorsal perimenter, H total height, HBw body whorl height, AL lateral area, PL lateral perimeter, ABw body whorl area, PBw body whorl perimeter, Dap diameter of the aperture, Hap height of the aperture, EP length parietal space, AAp apertural area, Pap apertural perimeter, DP penultimate whorl diameter, DPr protoconch diameter (see Fig. 1).

Character	Mean	SD	Min	Max	Holotype
DM	28.526	0.765	27.103	29.714	28.934
Dm	25.582	0.669	24.182	26.374	26.058
AD	557.304	30.211	502.938	596.790	566.908
PD	85.289	2.375	80.699	88.531	86.135
H	18.663	1.005	17.183	20.370	19.375
HBw	15.068	0.878	13.904	16.351	15.904
AL	375.374	30.818	324.118	411.659	386.807
PL	74.999	3.084	69.055	78.120	76.548
ABw	343.840	28.940	299.433	379.724	355.669
PBw	74.081	2.727	69.154	77.544	74.445
Dap	14.886	1.091	12.997	16.888	15.004
Hap	14.937	0.598	13.647	15.763	15.158
PS	11.382	0.610	10.245	12.335	11.158
AAp	190.094	14.988	158.663	208.360	208.360
PAp	52.650	2.173	49.201	56.690	53.222
DP	16.605	0.874	14.900	17.782	17.782
DPr	3.050	0.264	2.651	3.550	2.873

duct and the other oval with one end bulkier than the other. Vas deferens is a long, narrow duct that passes between one of the mucous glands and the dart sac, over the penial sheath, going down it adheres to penis-vagina at angle with connective tissue and then rises parallel to the penis complex to insert bellow the flagellar caecum. Vas deferens insertion marks the limit between epiphallus and flagellum. Penial complex tubular without external differentiation between penis and epiphallus. Short distal muscular penis sheath. Penial retractor muscle thick inserting in proximal epiphallus. Penial cavity occupied by several thin wavy pilasters and an oval verge sculptured with overlapping lamellae (Fig. 26). Penis longer than epiphallus, limits differentiated through their particular inner sculpture. Penial sheath short of about 1/5 of penial length. Epiphallus inner cavity with three thick pilasters. Proximal epiphallus with a rounded caecum where the vas deferens inserts. Flagellum tubular, longer than epiphallus. Spermoviduct long with uterus zone plicated transversal to the longitudinal axis. Albumen gland bean shaped with a prominent fertilization pouch-spermathecal complex (FPSC). Atrium short.

Type locality. Brazil, *Minas Gerais*: Itacarambi, National Park Cavernas do Peruaçu, Vale dos Sonhos (523m, X = 0599645, Y = 8343426), M.S. Pena, A. Suhett, D.C. Souza leg., December 2010, (MNRJ 34.580), Holotype (ethanol preserved specimen).

Other material examined. Brazil, *Minas Gerais*: Itacarambi, National Park Cavernas do Peruaçu, Nossa Senhora Aparecida Farm (532 m, X = 0589284, Y = 8328970), M.S. Pena, A. Suhett,

Figures 2–7. Shell morphology: (2) Dorsal view of Holotype, MNRJ 34.580; (3) lateral position of holotype shell and soft body; (4) dorsal, (5) ventral, and (6) lateral view of paratype, IBN 861; (7) live snail. Scale bar: 3–5 = 10 mm.

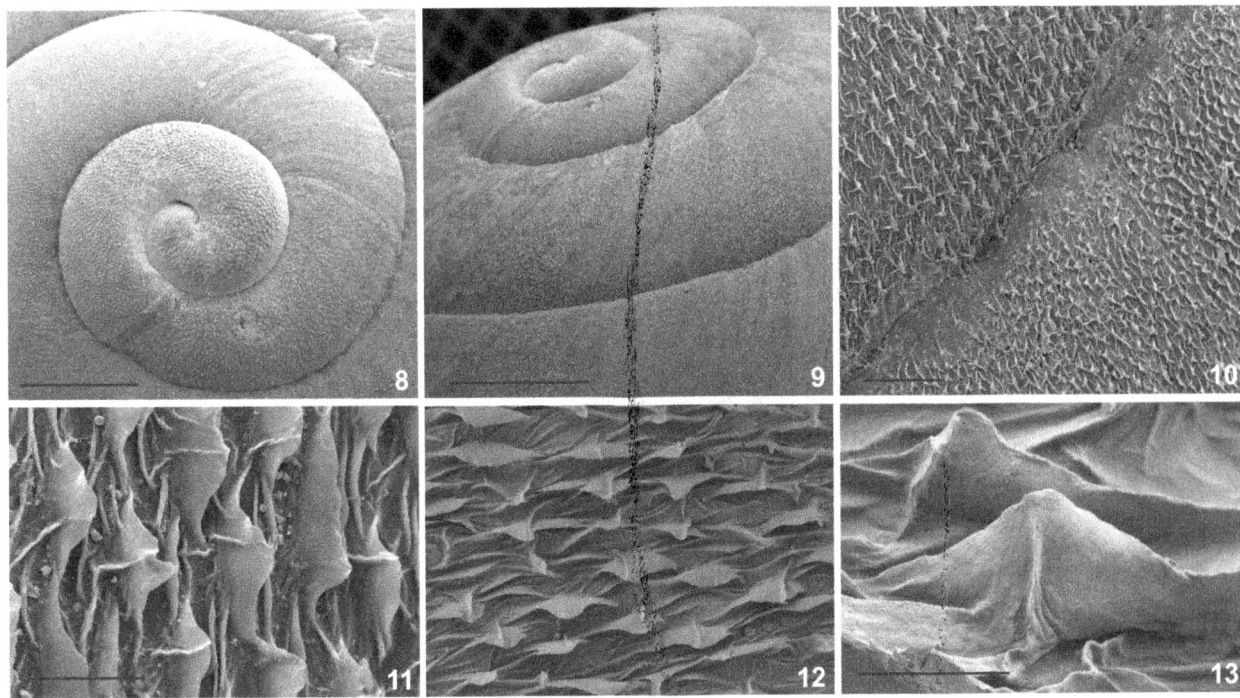

Figures 8–13. Shell morphology: (8) Protoconch in dorsal view, scale bar = 4mm; (9) lateral view of the protoconch and first whorls, scale bar = 4mm; (10) general view of the teleoconch microsculpture, scale bar = 100µ; (11) body whorl microsculpture consisting on axial rows of triangular lamella separated by wrinkles; (12) perpendicular view of the lamella, scale bar = 5µ; (13) detail of a triangular lamellae showing its central axis, scale bar = 2 µm.

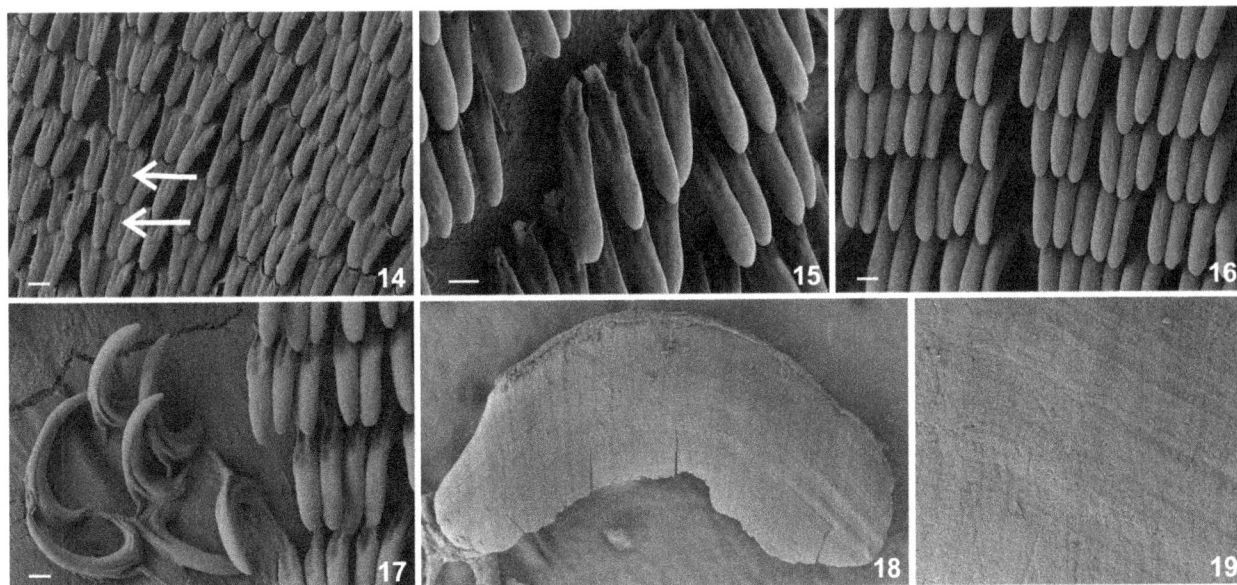

Figures 14–19. Digestive system: (14) Dorso-lateral view of the radula, arrows points to central tooth in two transverse rows of teeth; (15) detail on a dorso-lateral view of the lateral teeth;(16) detail of lateral teeth close to margin in dorsal view (17) margin of the radula showing curve shaped marginal teeth; (18) dorsal view of the smooth, fragile jaw; (19) detail of the dorsal surface of the jaw showing transversal shallow grooves. Scale bars = 10 µm.

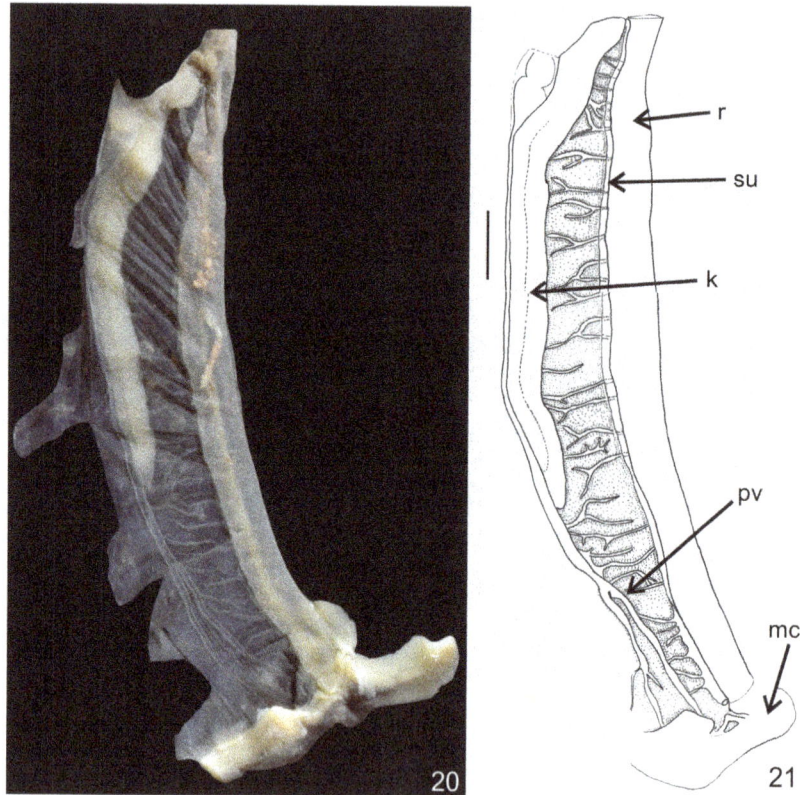

Figures 20–21. Pallial system: (20) photograph of the ventral zone of the pulmonary roof, note the long kidney, with respect to the total length of the lung; (21) line drawing of the same region detailing the limits between kidney and ureter and the shallow veins crossing the pulmonary roof. Scale bar: 21 = 5mm. Abbreviations (k) kidney, (mc) mantle collar, (pv) pericardic vein, (r) rectum, (su) secondary ureter.

D.C. Souza leg., December 2010, (IBN 21-S, MLP-Ma 14216), dry shells, (IBN 861), ethanol preserved specimens. Paratypes. Brazil, *Minas Gerais*: Itacarambi, National Park Cavernas do Peruaçu, Brejal, Peruaçu River side (663 m, X = 0579404, Y = 8332170), (MCN 192), dry shells. Brazil, *Minas Gerais*: Itacarambi, Natiomal Park Cavernas do Peruaçu, Janelão Cave (714 m, X = 0581514, Y = 8329046) M.S. Pena, A. Suhett, D.C. Souza leg., December 2010, (MCN 208).

Distribution. Thus far known only from National Park Cavernas do Peruaçu, northern region of Minas Gerais, Brazil.

Remarks. *Minaselates* differs from all known species of *Epiphragmophora* by having a granulose protoconch, the shell spire with a blunt apex, and by the wavy, spiral grooves in the ventral region of the shell. The fused, imperforate umbilicus on the shells of this new species is not typical of *Epiphragmophora*. Most of species of the genus present an open, perspective umbilicus. The exceptions are *E. argentina* (Holmberg, 1909) and some specimens of *E. variegata* Hylton Scott, 1962. The presence of wavy spiral grooves at the base of the shell of *M. paradoxa* sp. n. is noteworthy, sharply contrasting with the condition found in all other species of *Epiphragmophora*, where it is absent, except for *E.(Pylsbrya) farrisi* (Pfeiffer, 1859), which has shallow spiral lines. *Minaselates paradoxa* sp. n. also differs from the species classified

in *Epiphragmophora* in the shape and length of the kidney, and the smooth jaw. In the species of *Epiphragmophora* for which the anatomy has been studied, the kidney is no more than half the length of the pulmonary roof, while the jaw is ribbed. A noteworthy character present in *Minaselates paradoxa* sp. n. is the presence of a flagellar caecum, a structure not found in *Epiphragmophora*. In this new species, the vas deferens inserts in the lower portion of the caecum. The penial retractor muscle in *Epiphragmophora* is mostly long and thin, inserting in the epiphallus, while in *Minaselates paradoxa* sp. n. this muscle is stronger, inserting in the caecum. *M. paradoxa* sp. n. is isolated from the area of distribution of *Epiphragmophora*, whose area of highest species richness is in the western portion of South America.

Minaselates paradoxa sp. n. resembles some species of Pleurodontidae by the presence of complex structures in the terminal genitalia of the male, such as the flagellar caecum. It is also similar to some Pleurodontidae in its long and thin kidney, the crowded granules in the surface of the shell protoconch, the globular general shape of the shell, the complex microsculpture on the teleoconch and in its smooth jaw. The presence of a flagellar caecum is noteworthy in *Minaselates*, this structure being absent in Epiphragmophoridae, while it is characteristic

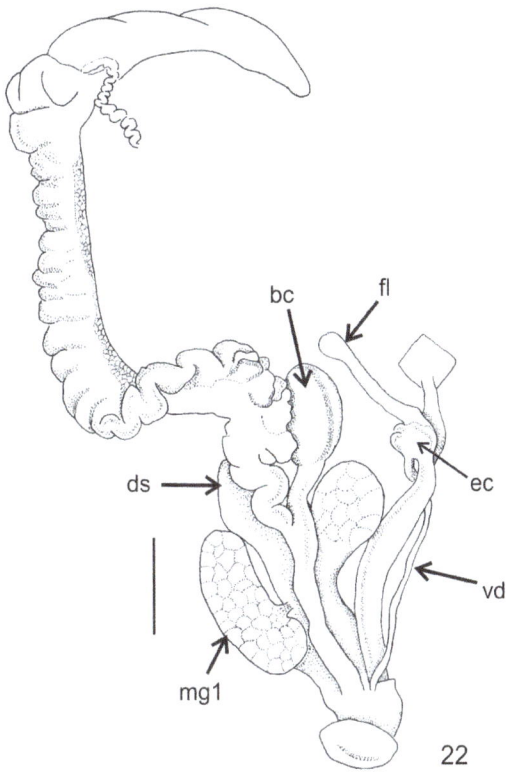

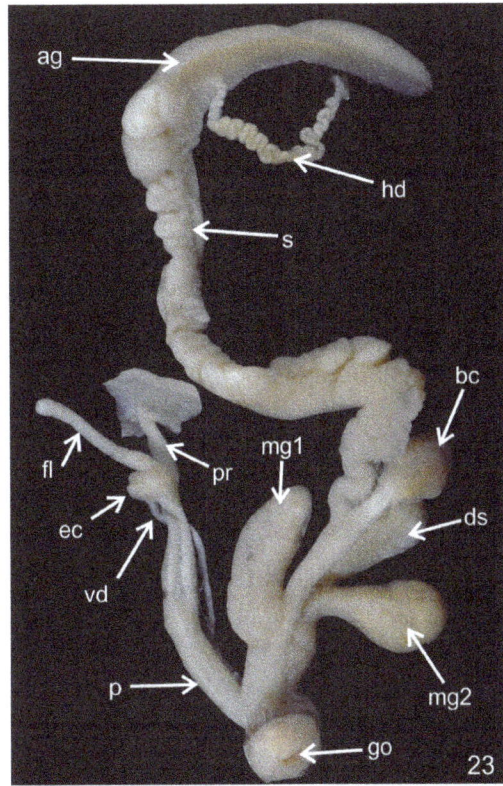

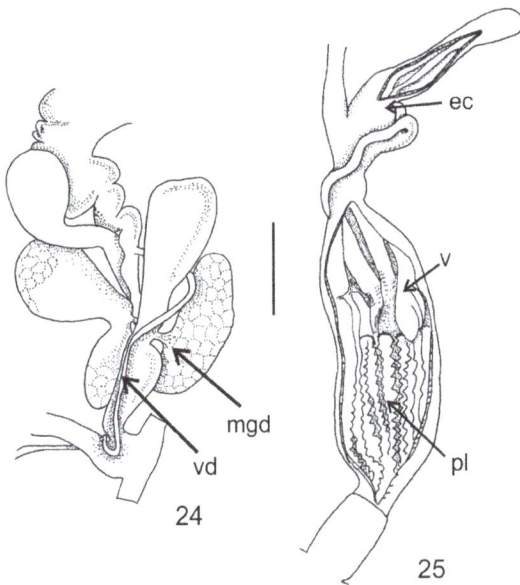

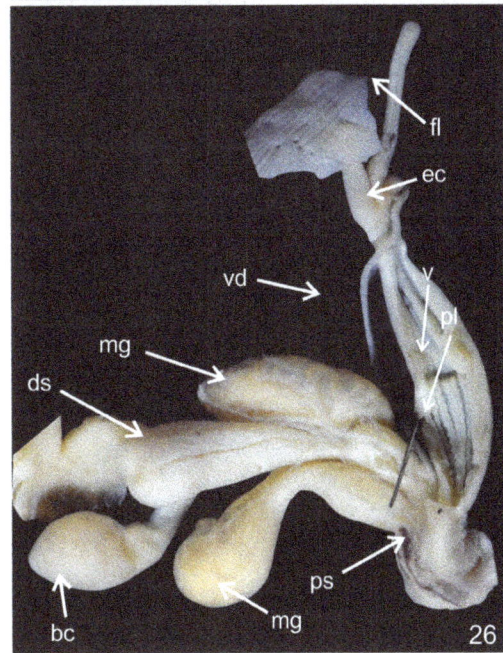

Figures 22–26. Genital system: (22, 23) line drawing and photograph of the complete dissected out reproductive system (ag) albumen gland, (bc) bursa copulatrix, (ds) dart sac, (ec) flagellar caecum, (fl) flagellum, (go) genital opening, (hd) hermaphroditic duct, (mg1) mucous gland 1, (mg2) mucous gland 2, (p) penis, (pr) penial retractor muscle, (s) spermoviduct, (vd) vas deferens; (24) detail of the terminal genitalia (pl) penial plates, (ps) penial sheath, (v) verge; (25) vas deferens and insertion of mucous gland ducts into dart sac (mgd) mucous gland duct; (26) inner sculpture of the penial complex showing the verge inside the penis. Scale bars: 22 and 25 = 5 mm.

of some Pleurodontidae such as *Polydontes* Montfort, 1810 and *Pleurodonte incerta* (Férussac, 1823) (Cuezzo 2003).

In *Minaselates*, however, the vas deferens inserts in the flagellar pouch while in *Polydontes* the vas deferens inserts above this caecum. At first glance, the shell of this new species is very similar to some species of *Pleurodonte* (Pleurodontidae), except for the presence of concentric sculpture below the periphery, and in the basal area. It also differs in the morphology of the terminal genitalia that has a dart complex and a different flagellum shape. In *Pleurodonte*, the vas deferens is twisted around the epiphallus, descending to the peni-oviducal angle, while in *M. paradoxa* **sp. n.** the vas deferens runs straight, parallel to the penis complex, without looping around the epiphallus. Most of the Helicoidean families have a ribbed jaw or 'odontognath jaw', except for the Sagdidae with a 'stegognath jaw' and the Sphicterochilidae and Cepoliinae with a smooth jaw ('oxygnath jaw') (Cuezzo 2003). In *M. paradoxa* sp. n. the jaw was found to be smooth, only having fine transverse lines visible with electron microscopy, similar to jaws in *Caracolus* Montfort, 1810 and *Labyrinthus*.

Minaselates paradoxa sp. n. is distributed within a National Park area where Cerrado and Caatinga are the dominant biomes. These are considered high diversity hotspot areas. Within these hotspots areas specimens were collected in typical deciduous or semi-deciduous forests called Seasonally Dry Tropical Forests (SDTF) (Pennington et al. 2000, Prado 2000). These types of forest are drought-adapted, tree-dominated ecosystems with a more or less continuous canopy. Grasses are not a significant

element in these forests, which are currently scattered in eastern South America. A hypothesis has been advanced predicting that SDTFs were more widespread during drier glacial climates, and that their fragmentation during the current wet interglacial period has contributed to speciation and the distribution patters of species observed today (Pennington et al. 2000, Prado 2000). This hypothesis is for the most part based on the assessment of floristic links among species assemblages in the SDTF areas (Sarkinen et al. 2011). Land snails inhabiting this type of dry forest, from the Cerrado and Caatinga hotspots areas, have been scarcely studied and consequently their conservation status is unknown. Most vertebrates and plants have been more extensively documented in the Cerrado and Caatinga hotspots areas (Overbeck et al. 2015), but this is not the case with many invertebrate groups.

CLADISTIC ANALYSIS

The main goal of the analysis performed was to evaluate and support the description of a new genus. For this, only the species of *Epiphragmophora* with known complete anatomical and shell morphology information were used. The morphological data set analyzed under the implied weights approach resulted in two most parsimonious trees. The resulting strict consensus tree is illustrated (Fig. 27). Synapomorphies that support *Minaselates* gen. n. as a new genus, different from *Epiphragmophora* are: body whorl surface covered with triangular lamellae

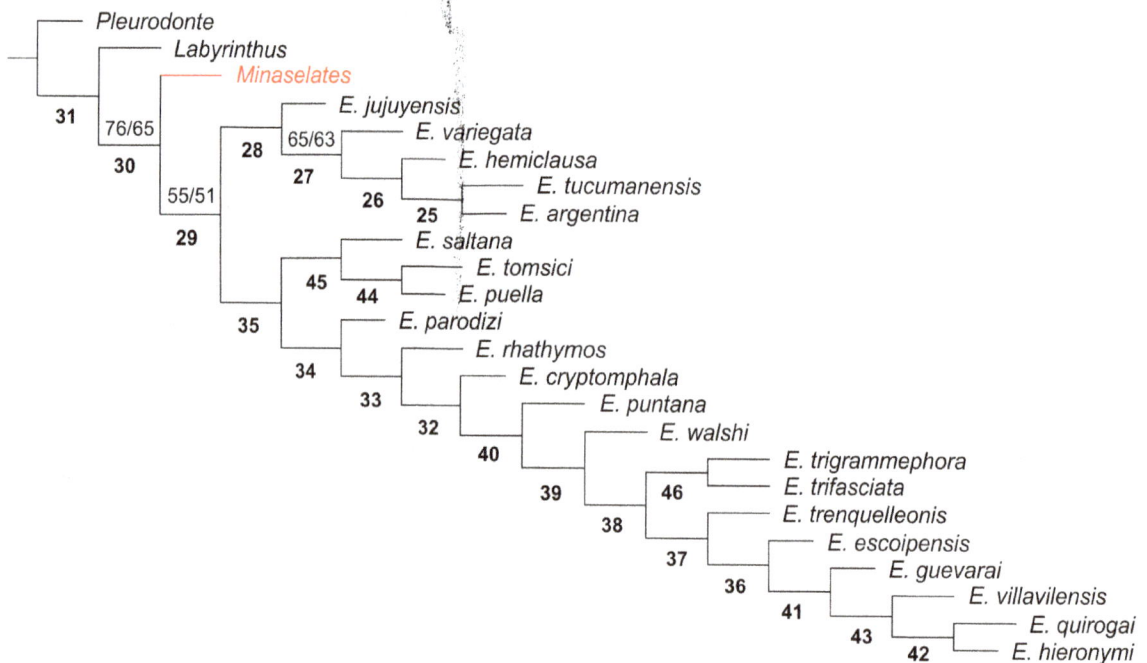

Figure 27. Consensus tree obtained from two trees under implied weight approach highlighting the position of *Minaselates* gen. n. Only taxa with described anatomy were included in the matrix. Numbers below branches are node numbers. Symmetric resampling values (left number) and jackknife values (right number) are located above branches. Only values above 50% are illustrated.

(0: 5); duct of bursa copulatrix extremely short, not longer than sac (22:0); presence of a flagellar caecum (32:1).

Epiphragmophora (node 29) was also recovered as a monophyletic genus in all optimal trees, supported by the following synapomorphies: Body whorl surface with axial ribs regularly distributed (0:3); umbilicus overlapped but not solded to body whorl (1: 1); protoconch smooth (8:0), and Penial retractor muscle inserting in medial zone of epiphallus (21:1). Within *Epiphragmophora*, the clade [*E. variegata* [*E. hemiclausa* [*E. tucumanensis*+*E. argentina*]]] (node 27) has the highest SR and jackknife support and resulted monophyletic in all trees obtained.

Minaselates gen. n. is the sister group of *Epiphragmophora* in both optimal trees. This relationship is supported by the following synapomorphies: Presence of mucous glands in terminal genitalia (10: 1); presence of a dart apparatus (11:1); medium to short length of the bursa copulatrix duct (22:01) and a finger-like short to medium flagellum (25:1).

ACKNOWLEDGEMENTS

Thanks to the Instituto Chico Mendes de Conservação da Biodiversidade for the working permits and support (SIS-BIO-ICMBIO 19133-1 de 08/04/2009) and Fundo de Incentivo à Pesquisa (FIP), Pontifícia Universidade Católica de Minas Gerais for financial support to MSP. MGC is a researcher of the Argentine National Council for Scientific Research (CONICET). Financial support to MGC has been received through PIP 0055 (CONICET). We also thank the anonymous reviewers and the editor for their valuable comments.

REFERENCES

Bouchet P, Rocroi JP (2005) Classification and Nomenclator of Gastropod Families. Malacologia 47: 1–397.

Brown RW (1979) Composition of scientific words. A manual of methods and a lexicon of materials for the practice of logotechnics. Washington, DC, Smithsonian Institution Press.

Cuezzo MG (1997) Comparative anatomy of three species of *Epiphragmophora* Doering, 1874 (Pulmonata: Xanthonychidae) from Argentina. The Veliger 40: 216–227.

Cuezzo MG (2003) Phylogenetic analysis of the Camaenidae with special emphasis on The American taxa. Zoological Journal of the Linnean Society 138: 449–476. https://doi.org/10.1046/j.1096-3642.2003.00061.x

Cuezzo MG (2006) Systematic revision and cladistic analysis of *Epiphragmophora* Doering from Argentina and Southern Bolivia (Gastropoda: Stylommatophora: Xanthonychidae). Malacologia 49: 121–188. https://doi.org/10.4002/1543-8120-49.1.121

Goloboff P, Farris J, Kallersjo M, Oxelmann B, Ramirez M, Szumik C (2003) Improvements to resampling measures of group support. Cladistics 19: 324–332. https://doi.org/10.1111/j.1096-0031.2003.tb00376.x

Goloboff P, Farris JS, Nixon KC (2008) TNT, a free program for phylogenetic analysis. Cladistics 24: 774–786. https://doi.org/10.1111/j.1096-0031.2008.00217.x

Linares EL, Vega ML (2011) Catalogo preliminar de los Moluscos continentales de Colombia. Bogotá, Biblioteca José Gerónimo Triana 22, Instituto de Ciencias Naturales, Facultad de Ciencias, Universidad Nacional de Colombia.

Nixon KC (2002) Winclada Computer program. Ithaca, Published by the author, ver. 1.00.08.

Nordsieck H (1987) Revision des Systems der Helicoidea (Gastropoda: Stylommatophora). Archiv für Molluskenkunde 118: 9–50.

Overbeck GE, Velez-Martin E, Scarano FR, Lewinsohn TM, Fonseca CR, Meyer ST, Müller SC, Ceotto P, Dadalt L, Durigan G, Ganade G, Gossner MM, Guadagnin DL, Lorenzen K, Jacobi CM, Weisser WW, Pillar VD (2015) Conservation in Brazil needs to include non-forest ecosystems. Diversity and Distributions 21: 1455–1460. https://doi.org/10.1111/ddi.12380

Pennington RT, Prado DE, Pendry CA (2000) Neotropical seasonally dry forests and Quaternary vegetation changes. Journal of Biogeography 27: 261–273. https://doi.org/10.1046/j.1365-2699.2000.00397.x

Pilsbry HA (1894) Guide to the study of Helices. In: Manual of Conchology. Philadelphia, Academy of Natural Sciences, vol. 7. 1–366

Pilsbry HA (1939) Land Mollusca of North America (north of Mexico). Academy of Natural Sciences Monographs 1: 216–227.

Ponder WF, Lindberg DR (1997) Towards a phylogeny of gastropod molluscs: an analysis using morphological characters. Zoological Journal of the Linnean Society 119: 83–265. https://doi.org/10.1111/j.1096-3642.1997.tb00137

Prado D (2000) Seasonally dry forests of tropical South America: from forgotten ecosystems to a new phytogeographic unit. Edinburgh Journal of Botany 57: 437–461. https://doi.org/10.1017/S096042860000041X

Richardson L (1982) Helminthoglyptidae: Catalog of species. Tryonia 6: 1–117.

Sarkinen T, Iganci JR, Linares-Palomino R, Simon M, Prado D (2011) Forgotten forests – issues and prospects in biome mapping using Seasonally Dry Tropical Forests as a case study. BMC Ecology 11: 1–16. https://doi.org/10.1186/1472-6785-11-27

Schileyko AA (1991) Taxonomic status phylogenetic relations and system of the Helicoidea sensu lato. Archiv fur Molluskenkunde 120: 187–236.

Thiele J (1929–1935) Handbook of Systematic Malacology 1. Berlin, Jena.

Tompa AS (1984) Land Snails (Stylommatophora). In: Tompa AS, Verdonk H H. Van Den Biggelar JA (Eds) The Mollusca. New York, Academic Press, 47–140. https://doi.org/10.1016/B978-0-08-092659-9.50009-0

Wade CM, Hudelot C, Davison A, Mordan PB (2007) Molecular phylogeny of the helicoid land snails (Pulmonata: Stylommatophora: Helicoidea), with special emphasis on the Camaenidae. Journal of Molluscan Studies 73: 411–415. https://doi.org/10.1093/mollus/eym030

Zilch A (1959–1960) Gastropoda. Euthyneura. In: Handbuch der Paläozoologie 6. Berlin, Borntraeger, 1–825.

APPENDIX 1

Character matrix used in cladistic analysis. Only taxa with described anatomy were included in the matrix. Character codification according to Cuezzo (2003, 2006) with some modifications. "–"are non-applicable characters, "?" are missing characters.

		1	2	3	4	5	6	7	8	9	10	1	2	3	4	5	6	7	8	9	20	1	2	3	4	5	6	7	8	9	30	1	2	3	4
Pleurodonte	4	0	0	1	0	0	0	1	[01]	1	0	0	–	–	–	–	–	–	–	–	0	1	2	2	1	0	2	0	0	[01]	–	0	0	0	0
E. argentina	2	[03]	0	1	0	1	0	1	0	0	1	1	1	0	1	0	1	0	0	1	0	1	2	1	1	0	1	0	1	1	0	0	0	0	0
E. cryptomphala	2	[01]	[12]	1	0	1	0	1	0	0	1	1	0	1	0	1	0	0	1	0	1	1	1	0	1	1	0	0	0	1	0	0	0	0	0
E. escoipensis	0	1	1	0	0	[01]	0	1	0	2	1	1	0	1	1	1	1	1	1	0	1	1	1	0	1	1	0	0	0	0	1	0	0	0	0
E. hemiclausa	2	1	0	1	0	1	0	1	0	2	1	1	1	0	1	0	0	0	0	0	1	1	2	1	0	1	1	0	1	1	0	0	0	0	0
E. guevarai	0	2	0	0	0	1	0	1	0	2	1	1	0	1	1	1	0	0	1	2	1	1	0	0	0	1	0	0	0	0	1	0	0	0	0
E. hieronymi	0	2	0	0	0	[01]	0	0	0	0	1	1	0	1	1	1	0	1	1	1	2	0	0	1	0	0	0	0	1	0	0	0	0	0	0
E. jujuyensis	2	1	1	1	1	1	0	1	0	0	1	1	0	1	1	1	0	1	0	0	2	1	1	0	0	1	1	0	0	0	1	0	0	0	0
E. parodizi	3	1	0	1	0	1	0	1	0	0	1	1	0	1	1	1	0	1	1	2	0	1	0	0	1	0	0	1	1	1	1	0	0	0	0
E. puella	4	2	0	2	0	0	1	0	0	0	1	1	0	?	?	?	0	1	?	?	?	?	0	?	?	1	?	?	0	0	1	?	0	0	0
E. puntana	3	1	1	0	0	1	0	1	0	0	1	1	0	1	1	1	1	1	1	1	2	1	1	0	0	1	0	0	0	0	0	0	0	0	0
E. quirogai	0	2	0	0	0	[01]	0	1	0	2	1	1	0	1	1	1	0	1	1	1	0	1	2	0	0	1	1	0	0	0	1	1	1	0	0
E. rhathymos	2	3	0	1	0	1	0	1	0	0	1	1	0	1	1	1	0	1	1	1	0	1	1	0	0	1	0	0	0	0	1	0	0	0	0
E. saltana	3	1	1	1	0	1	0	1	0	0	1	1	0	1	1	1	0	1	0	1	0	1	1	0	0	1	0	1	1	0	0	0	0	0	0
E. tomsici	2	1	1	1	0	0	0	1	0	0	1	1	0	1	1	1	0	1	0	1	0	0	1	0	0	0	0	1	0	0	1	0	0	0	0
E. trenquelleonis	0	1	0	0	0	1	0	1	0	0	1	1	0	1	1	1	0	1	1	2	1	1	0	2	1	1	0	0	[01]	0	1	0	0	0	0
E. trifasciata	1	2	0	0	0	1	0	1	0	0	1	1	0	1	1	1	0	0	1	2	1	1	0	0	1	1	0	0	0	1	0	0	0	0	0

APPENDIX 2

Character list used in the cladistics analysis of species of *Epiphragmophora* with known anatomy plus the new genus and species, *Minaselates paradoxa* sp. n.

0. Body whorl surface: with thin growth lines = 0; with thick growth ridges = 1; malleated with diagonal ribs = 2; with axial ribs regularly distributed = 3; pustules/granules to wrinkles = 4; triangular lamella in axial rows = 5. [additive].
1. Umbilicus: Fused with basal lip of peristome = 0; overlapped but not fused to body whorl = 1; perspective wide not overlapped by peristomal lip = 2; perspective narrow slightly overlapped = 3; wide partly overlapped = 4.
2. Shape of the aperture: sub circular = 0; oval horizontal = 1; sub quadrangular = 2.
3. Peristome: Thin expanded slightly reflexed = 0; thick wide reflexed = 1; thin highly expanded = 2.
4. Basal callus in peristome: absent = 0; present = 1.
5. Peripheral bands: absent = 0; present = 1.
6. Body whorl periphery: convex = 0; equatorially subcarinated = 1; supraequatorially subcarinated = 2; carinated = 3. [additive].
7. Aperture respect to body whorl: not descending = 0; descending = 1.
8. Protoconch sculpture: smooth = 0; granulose = 1.
9. Spire apex: pointed = 0; dull = 1; not evident = 2.
10. Mucous glands in terminal genitalia: absent = 0; present = 1.
11. Dart apparatus: absent = 0; present = 1.
12. Shape of dart sac: long finger-like usually with constriction = 0; short, cylindrical, no constriction = 1.
13. Dart sac insertion: in vagina = 0; in atrium = 1.
14. Dart sac papillae: absent = 0; present = 1.
15. Relation between ducts of both mucous glands: separated = 0; distally fused or contiguous = 1.
16. Position of left mucous gland duct: distal respect to the body of the gland = 0; equatorial respect to the body of the gland = 1.
17. Shape of right mucous gland: not sac-like = 0; sac-Like = 1.
18. Right mucous gland: not fused with atrium wall = 0; distally fused with atrium wall = 1.
19. Penis length respect to epiphallus: half epiphallus length = 0; as long as epiphallus = 1; longer than epiphallus length = 2.
20. Penial papillae (= verge): absent = 0; present = 1.
21. Penial retractor muscle: inserts in distal epiphallus = 0; inserts in medial zone of epiphallus = 1; inserts in proximal epiphallus = 2; inserts in proximal penis = 3.
22. Duct of bursa copulatrix: extremely short not longer than sac = 0; medium = 1; long = 2. [additive].
23. Vagina: short = 0; medium to long = 1; extremely long = 2. [additive].
24. Atrium: short = 0; medium to long = 1.
25. Flagellum: thin, long = 0; finger-like, short to medium = 1; *Pleurodonte*-like = 2; *Labyrinthus*-like = 3. [additive].
26. Penial muscular band: absent = 0; present = 1.
27. Penial sheath (penial tunica): simple = 0; double or multi-

layer = 1.

28. Microhabitat associated to: rocks = 0; tree trunks = 1.

29. Vas deferens: surrounding dart sac = 0; not surrounding dart sac = 1.

30. Epiphallus proximal portion: Not widen at point entrance vas deferens = 0; Widen at point of entrance of vas deferens = 1.

31. Penial retractor: not = 0; forming a loop around vas deferens before insertion in epiphallus = 1.

32. Flagellar caecum: absent = 0; present = 1.

33. Kidney length respect to pulmonary roof: not exceeding half of pulmonary roof length = 0; more than half the pulmonary roof = 1.

34. Jaw: ribbed = 0; smooth = 1.

Author Contributions: MGC conceived and designed the experiments, analyzed the data, wrote the paper, prepared figures and tables, reviewed drafts of the paper. MSP collect live specimens in the field, analyzed the data, prepared figures, reviewed drafts of the paper.

Competing Interests: The authors have declared that no competing interests exist.

20

Owenia caissara sp. n. (Annelida, Oweniidae) from Southern Brazil: addressing an identity crisis

Luiz Silva[1], Paulo Lana[1]

[1]*Centro de Estudos do Mar, Universidade Federal do Paraná. Avenida Beira Mar, 83255-976 Pontal do Paraná, PR, Brazil*
Corresponding author: Paulo Lana (paulolana@gmail.com)

http://zoobank.org/77080203-3319-4EFB-87A8-7BB838B22634

ABSTRACT. We re-assess the taxonomic status of *Owenia* Delle Chiaje, 1841 from Southern Brazil based on estuarine specimens from Paranaguá Bay (Paraná) and Babitonga Bay (Santa Catarina), and literature records. *Owenia caissara* **sp. n.** is diagnosed by a branchial crown with five pairs of tentacles, branched close to the base of the crown, rectilinear collar with a pronounced lateral slit, two ventrolateral ocelli partially covered by the collar, up to 23 hooks on a single row in the first abdominal segment, regularly curved nuchal shape, regularly moderate teeth curvature, and long and thin scales with oval transition. The description of *Owenia caissara* **sp. n.** reinforces the idea that *Owenia fusiformis* sensu lato is a complex of closely related species that can be distinguished on the basis of both macro- and micro- morphological traits.

KEY WORDS. Estuarine bottoms, Polychaeta, subtidal

INTRODUCTION

Species of *Owenia* Delle Chiaje, 1841 are found from the intertidal zone to 2,000 m deep (Dauvin and Thiébaut 1994). The presumed cosmopolitanism of *Owenia fusiformis* Delle Chiaje, 1844, Oweniidae, originally described from Sicily, Mediterranean Sea, was strongly advocated by Hartman (1959), in her catalogue of the polychaetes of the world. Hartman suggested that the Ammocharidae (a junior synonym of Oweniidae) *Ammochares tegula* Kinberg, 1867, *A. brasiliensis* Hansen, 1882, and *A. sundevalli* Kinberg, 1867, from South America, among many other oweniid taxa from disjoint geographical areas, should be referred to *O. fusiformis*. Subsequent worldwide records (Imajima and Hartman 1964, Plante 1967, Ibanez-Aguirre and Solis-Weiss 1986, Gillet 1988, Dauvin and Gillet 1991) and a biogeographic analysis by Dauvin and Thiébaut (1994) reinforced the notion that *O. fusiformis* is cosmopolitan, based on the presumed high dispersal potential of its larva and the species' capacity to reproduce under variable temperature regimes (Mcnulty and López 1969, Bhaud 1982).

More recently, the cosmopolitan distribution of *O. fusiformis* has been questioned and rejected by many authors based on re-evaluations of the dispersal potential of the mitraria larvae and on more detailed analyses of morphological traits (Blake 2000, Koh and Bhaud 2001, Koh et al. 2003, Guizien et al. 2006,

Martin et al. 2006, Ford and Hutchings 2010). The mitraria larva of oweniids can remain in the plankton for up to 30 days (Wilson 1932, Thiébaut et al. 1992, 1994). Although this might suggest a high potential for dispersion, factual data on dispersal potential are still scarce. Dispersion models tested in Banyuls Bay (NW Mediterranean France) suggested that the dispersion ability of mitraria larvae is in fact very limited and could not explain or substantiate a cosmopolitan distribution (Guizien et al. 2006, Verdier-Bonnet and Carlotti 1997). More detailed analyses of morphological traits with potential diagnostic value, previously underestimated in the literature, also showed that *O. fusiformis* has, in fact, a restricted distribution (Koh and Bhaud 2001, 2003, Koh et al. 2003, Martin et al. 2006, Ford and Hutchings 2010). Blake (2000) included novel morphological characters in his partial revision of *Owenia*. After comparing specimens from California and locations near the type locality of *O. fusiformis*, he revalidated *O. collaris* Hartman, 1955 and described a new species, *O. johnsoni* Blake, 2000. He also suggested that conventional diagnostic characters should be supplemented with analysis of the neuropodial rings.

Based on such novel morphological traits, Koh and Bhaud (2001) described *O. gomsoni* from the Yellow Sea in Southern Korea. Koh and Bhaud (2003) also established a new set of traits with forty-eight morphological characters for the identification of *Owenia* species. They used measurements of the thorax, capillary notochaetae, and hooks as novel diagnostic features of species. They confirmed the validity of *O. collaris*, *O. johnsoni*, and described four new species, *O. polaris*, *O. borealis*, *O. petersenae*, and *Owenia* sp. n. not formally named at that time, but later described as *O. persica* Martin, Koh, Bhaud, Dutrieux & Gil, 2006. More recently, five new species were recorded from Australia: *O. australis* Ford & Hutchings, 2010; *O. bassensis* Ford and Hutchings, 2010; *O. mirrawa* Ford & Hutchings, 2010, *O. dichotoma* Parapar & Moreira, 2015 and *O. picta* Parapar & Moreira, 2015.

The taxonomical knowledge of the genus in Brazil is unsatisfactory, although specimens of *Owenia* are often found, and are often numerically dominant in estuarine or shallow shelf benthic assemblages. Ecological surveys, in particular, tend to cluster all species of the genus under the name *O. fusiformis*. *A. sundevalli* and *A. brasiliensis*, both collected in shallow continental shelf bottoms off Brazil in the second half of the nineteen century, were later referred to *O. fusiformis* by Augener (1934) and Hartman (1959). This treatment is still followed in the World Register of Marine Species, which keeps both species as subjective synonyms of *O. fusiformis*. It is unlikely that they are indeed synonyms of *O. fusiformis*; however, since the original descriptions are succinct, and the type series are severly damaged or in a bad state of conservation (Gustavo Sene-Silva, pers. obs. in an unpublished MSc thesis), the revaluation of their actual taxonomical status is difficult.

Following the recent trend of taxonomic reassessments of the genus and hoping to address a taxonomical identity crisis, we began to re-evaluate *Owenia* from Southern Brazil, describing a new species based on the morphological analysis of estuarine populations collected from the Paranaguá Bay (state of Paraná) and Babitonga Bay (state of Santa Catarina).

MATERIAL AND METHODS

Oweniid specimens were collected from shallow subtidal locations of the Paranaguá Bay (Paraná, Brazil) and Babitonga Bay (Santa Catarina, Brazil). Samplings in Paranaguá Bay were carried out from December 2013 to June 2014 near the mouth of the Baguaçu River (25°33′S, 48°23′W). Subtidal samples were taken with a Petit Ponar grab or shovels manually operated during scuba diving. In Babitonga Bay, samples were taken between April and August 2014 in Paulas Beach (26°13′S, 48°37′W), with a Petersen grab.

The characteristic tubes of *Owenia* were manually separated from the sediment still in the field, stored in plastic jars with water from the collection site, and then taken to the Centro de Estudos do Mar (CEM) at the Universidade Federal do Paraná (UFPR). For morphological descriptions, 16 individuals from Paranaguá Bay and 21 from Babitonga Bay were evaluated (including type-material listed in the corresponding section, and non-deposited individuals); they were removed from tubes and kept in Petri dishes with sea water and 8% magnesium chloride for one hour. At least ten individuals from each site were ob-

served under a stereoscopic microscope for the description of in vivo coloration. The animals were photographed with a Sony NEX3 digital camera. The length of individuals was measured with the aid of a scale built into the stereoscopic microscope.

After fixation, mucus and sediment particles were removed from the body; hooks and chaetae were extracted from four individuals from each site. Fragments of the epidermis with notochaetal bundles on the first abdominal segment and uncini bundles were dipped three times in distilled water for thirty minutes to remove the remaining attached particles. After this, the material was preserved in 70% alcohol. This material was ran through a graded ethanol series to reach the critical point and coated with gold, and examined and photographed in a Zeiss EVO LS15-100 scanning electron microscope (SEM) at the Electron Microscopy Center (CME) at CEM. The terminology and measurements for the descriptions (Figs 1–4) followed the scheme of Koh and Bhaud (2003). Measurements of hard parts, hooks, and notochaetae were based on ten hooks and six notochaetae from the first abdominal segment.

Methyl green colour patterns were assessed by staining five individuals for five minutes with a solution of 0.05 g of methyl green powder in 10 ml of distilled water. Excess was removed by washing in 70% alcohol under visual control in dorsal and ventral thoracic sections were photographed (Martin et al. 2006).

Type-material was deposited at the Zoology Museum of Campinas University, ZUEC (state of São Paulo, Brazil).

TAXONOMY

Owenia caissara sp. n.

http://zoobank.org/D36477B5-63A6-41B4-9D4B-92EA85C330E1

Diagnosis. Crown with five pairs of tentacles. Tentacular branches beginning near collar base, numerous near crown base and apex. Collar rectilinear, with pronounced lateral slit. Two ventrolateral ocelli partially covered by collar. First abdominal segment with rows of up to 23 hooks. Hooks with 0° to 90° angles in relation to anteroposterior body axis, and nuchal shape regularly curved; teeth curvature moderate. Notochaeta scales long and thin with oval transition between A and B (Fig. 4).

Description (based on holotype; numbers between brackets refer to average measurements in Babitonga and Paranaguá Bay specimens, except body length which represents the maximum and minimum in both places). Body 21 (15–32) mm long. Width at collar height 1.18 (1.16 and 1.24) mm. Body divided into tentacular crown, thorax, and abdomen (Figs 5, 10). Tentacular crown and thorax separated by a thin membrane forming a collar (Figs 6, 7, 11, 14, 17, 20). Rectilinear collar with a lateral slit in angle 63° (77° and 67°) on average. Short crown (crown/abdomen length ratio = 1:2), with five pairs of tentacles (Fig. 17). Dorsal branches longer than ventral branches (Figs 7, 13, 17, 20). Tentacular branches 0.83 (1 and 0.89) mm long. Crown/collar length ratio of 0.70 (0.89 and 0.70). Thorax/collar length

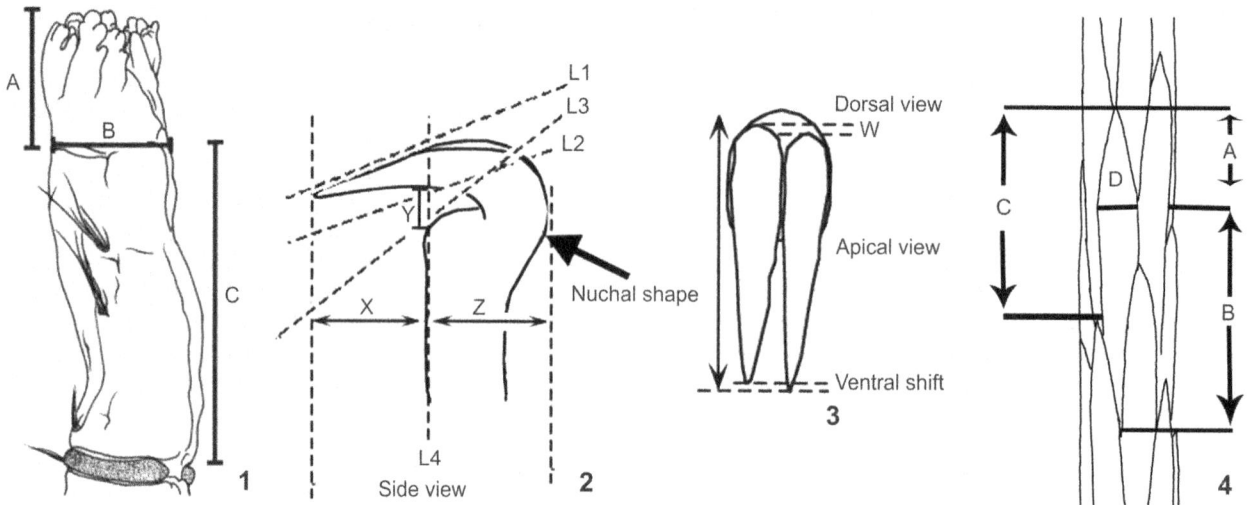

Figures 1–4. Measurements of morphological traits, maximum length of tentacular crown (A), Collar length (B), Thorax length (C). Redrawn from Martin et al. (2006). (2-3) Hook in lateral and apical view, distance between the tip of shaft and teeth ventral margin (X), distance starting from the hook base to the ventral margin of the teeth ends (Y), distance between the manubrium ventral face and teeth distal dorsal face (Z), dorsal and ventral shift of the teeth (W). L1: tangent from the teeth top edge. L2: tangent from the teeth bottom margin. L3: tangent from the shoulder ventral margin. L4: manubrium direction observed in the ventral position. Redrawn modified from Martin et al. (2006). (4) Median section of a capillary notochaeta indicating the meaning of measures A, B, C, and D taken on scales. A: Distance from the free end until the widest scale part. B: Distance from the widest scale part to where it is completely overlapped by two adjacent scales. C: Longest distance between the ends of two successive scales. D: Maximum scale width at which point A and B intersect. Redrawn from Koh and Bhaud (2003).

ratio of 1.95 (1.97 and 1.84). Thorax with three segments with capillary notochaetae in lateral bundles on first two segments, dorso-lateral on third segment. Thorax 1.6 (2.12 and 2.22) mm long on average. Abdomen with thirteen to eighteen biramous segments, each one with one pair of capillary notochaetae bundles and one neuropodial ring almost encircling the body. Dorsal ridges of the fifth segment with clavate glandular fractures, curved and expanded, almost touching along the middorsal body line. Posterior abdominal region without a dorsal groove. Neuropodial ring with rows of minuscule bidentate hooks.

First abdominal segment with rows of up to 23 hooks (Figs 18, 21), in 0° to 90° angles in relation to antero-posterior body axis. Teeth on hooks with a space in between, in an inverted V-shape (Figs 23, 24, 26, 27). Hooks with rectilinear shoulder and regularly curved nuchal shape (Figs 16, 19, 22). Average teeth protrusion of 2.02 μm (Figs 4A, 19, 22). Distance between head of shaft and lower part of the teeth (length of opening) 0.83 μm (moderate), (Fig. 2:Y). Hooks were not measured in the holotype to avoid damage to the individual, and following figures are measurements of five individuals from Paranaguá and Babitonga bays, respectively. Maximum hook width (X + Z) of 4.79 and 5.01 μm (Fig. 2) and X/Z ratio of 0.73 and 0.67 (Fig. 2). Moderate teeth curvature, with average angle formed by meeting of L2/L4 tangents from 54° and 63° (Figs 19, 22). Long and thin scales, total length of notochaetal scales on first abdominal segment (A + B) of 4.87 and 4.29 μm (Figs 4, 25, 28). A + B/D ratio of 9.9 (Figs 4, 25, 28). Average length of scale's free part (C) of 2.5 μm and 2.32 μm (Fig. 4). Oval transition area between A and B (Figs 4, 15, 25, 28).

Living specimens with dark brown coloration at the base and terminal region of tentacular branches (Figs 11, 13). Red tinged thorax and beginning of abdomen due to body transparency, which highlights vascularization; remaining abdomen pinkish (Figs 5, 10). Color absent in alcohol – preserved animals, except one pair of reddish ocelli at ventrolateral base of tentacular branches, partially covered by collar (Figs 6, 9), and brown spots basally on tentacular branches and on terminal regions (Figs 6, 7).

Methyl green staining pattern characterized by tentacular branches unreceptive to staining, dorsal side of the collar and two longitudinal dorso-lateral lines strongly stained. On the ventral side, the two V shaped lines were unreceptive to methyl green but the border of these lines and the collar were strongly stained (Fig. 29, Table 1).

Tubes with medium and coarse particles (481-586 μm), coalesced by mucus in an imbricated pattern. In cross-section, smaller particles near lumen and larger on tube edge. Quartz particles dominates (99%) followed by magnetite (0.6%), biotite (0.37%), and shell and echinoderm fragments (0.03%). Tubes from 26 to 57 mm long (n = 70).

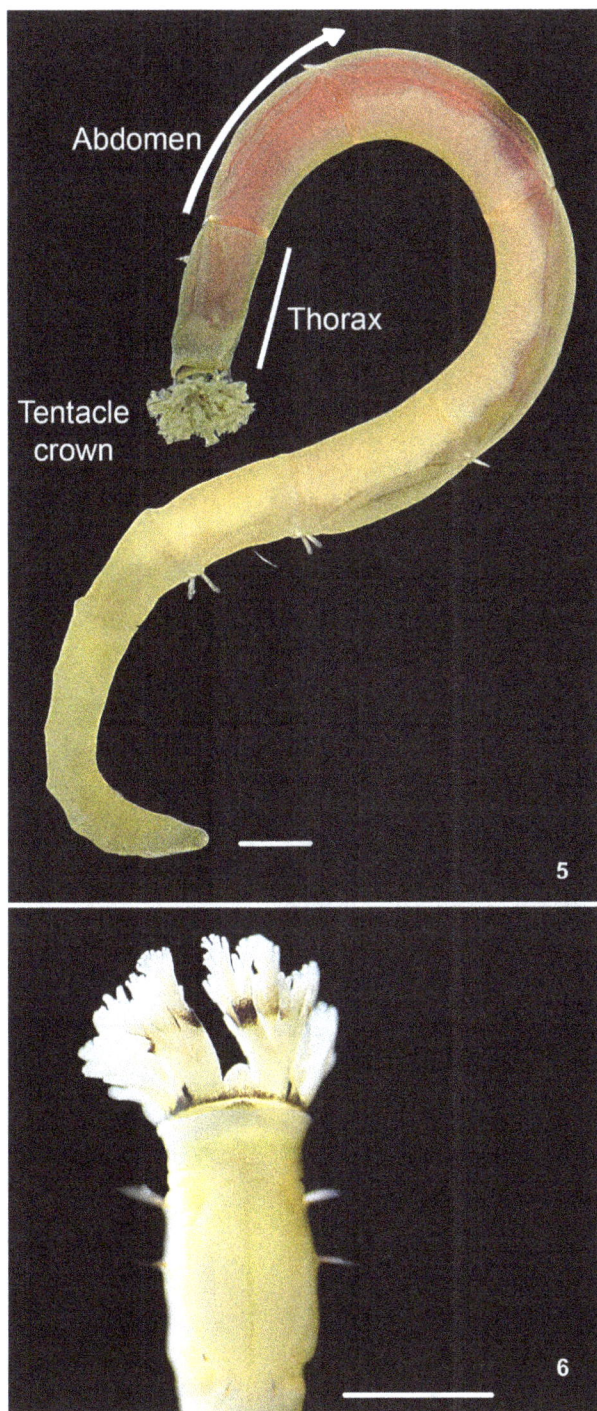

Figures 5–6. *Owenia caissara* sp. n. from Paranaguá Bay, Polychaeta 17525; lateral view showing the three body regions: tentacular crown, thorax, and abdomen (5), cephalic region in dorsal view showing dark brown pigmentation at the base and near the apex of tentacles in one specimen fixed in 70% alcohol (6). Scale bars: 1 mm (5), 1.4 mm (6).

Material examined. Holotype: ZUEC Polychaeta 17486, 21 mm, Santa Catarina, Babitonga Bay, Paulas Beach, 15/Jun./2014. Paratypes: ZUEC Polychaeta 17517-17522, Santa Catarina, Babitonga Bay, Paulas Beach, 22/Aug./2014, 6 specimens; Polychaeta 17523-17525, Paraná State, Paranaguá Bay, Cotinga Channel, 4/Jun./2014, 3 specimens; ZUEC Polychaeta 17487-17516, Santa Catarina, Babitonga Bay, Paulas Beach, 3/Oct./2014, 29 specimens.

Type locality. Paulas Beach, Babitonga Bay, Santa Catarina State, 26°13'S, 48°37'W.

Distribution. Currently known only from estuarine habitats along the coasts of the states of Paraná and Santa Catarina (Brazil).

Etymology. The species name honors fisherfolk from traditional communities still found along the southern and southeastern Brazilian coasts. We prefer the archaic spelling "caissara" to the modern "caiçara" to avoid the usage of the cedilla diacritical mark in the taxonomic literature.

Habitat. Populations of *Owenia caissara* sp. n. are frequent in shallow subtidal bottoms with a predominance of medium sand, at 0.5 to 5 m depth at Babitonga and Paranaguá Bays.

Remarks. *Owenia caissara* sp. n. has five pairs of tentacles (four in *O. fusiformis*), a tentacular branching close to the collar base (clearly more terminal in *O. fusiformis*), a collar with a pronounced slit (absent or inconspicuous in *O. fusiformis*), ventrolateral ocelli partially covered by the collar (completely exposed in *O. fusiformis*), hooks of the first abdominal segment in 0° to 90° angles (varying from 0° to 5° in *O. fusiformis*) and the transition between A and B (Fig. 4) on scales is oval (curved in *O. fusiformis*, Koh and Bhaud 2003), tentacular branches unreceptive to staining (strongly receptive in *O. fusiformis* tentacular branches) (Table 1).

The Californian *O. johnsoni* and *O. collaris* differ from *O. caissara* sp. n. in having four tentacular branches with few dichotomies. In *O. johnsoni*, the crown is long and hooks are in a 45° angle on the first abdominal segment. In *O. collaris*, the angle of the first abdominal segment ranges from 30° to 45°.

The Korean species *O. gomsoni* differs from *O. caissara* sp. n. in having five or more pairs of tentacles (always five in *O. caissara* sp. n.), a curved collar (straight in *O. caissara* sp. n.) and a transition area of notochaeta scale angular (oval in *O. caissara* sp. n.), tentacular branches strongly receptive to staining (unreceptive in *O. caissara* sp. n. tentacular branches), V shaped lines strongly receptive to methyl green (unreceptive in *O. caissara* sp. n.) (Table 1).

Owenia *borealis* and *O. polaris* differ from *O. caissara* sp. n. in having only four pairs of tentacular branches, an angular collar (straight in *O. caissara* sp. n.) and an angular transition area in notochaeta scale (oval in *O. caissara* sp. n.). In addition, the dorsal and ventral tentacles do not differ in size in *O. polaris*, and the tentacular branches are receptive to staining (Table 1). *Owenia persica* and *O. petersenae* differ from *O. caissara* sp. n. in having only four pairs of tentacular branches. Moreover, *O. persica* has the dorsal and ventral tentacle branches equal in

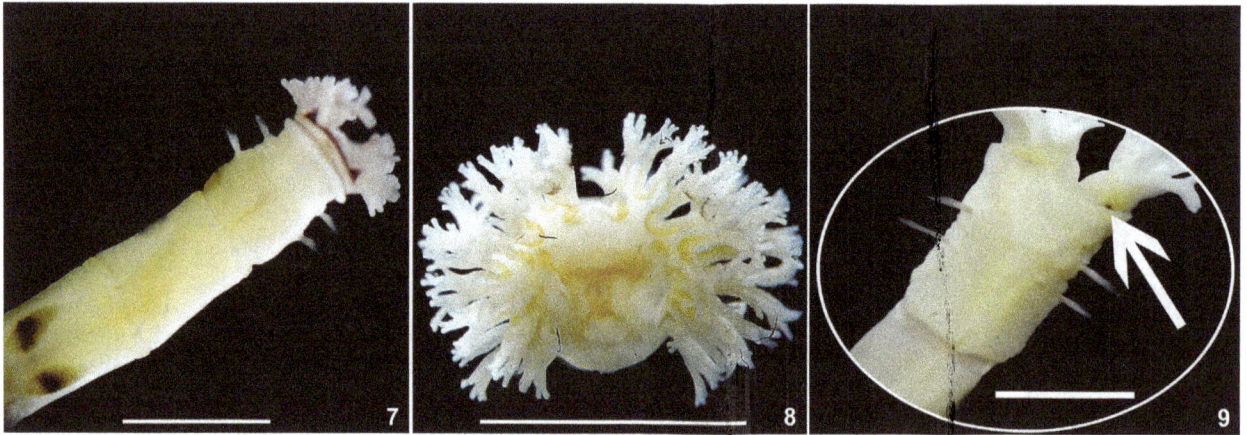

Figures 7–9. *Owenia caissara* sp. n. from Paranaguá Bay, Polychaeta 17524; in dorsal view (7), in frontal view with tentacular branches in detail (8), in ventral view with ventrolateral ocelli partially covered by the collar (9). Scale bars: 3 mm (7), 2 mm (8), and 1.5 mm (9).

Figures 10–13. *Owenia caissara* sp. n. from Babitonga Bay, Polychaeta 17520; lateral view showing the tentacular crown, thorax and abdomen (10), anterior region in dorsal view (11), anterior region in ventral view (12), anterior region in lateral view (13). Scale bars: 1 mm (10), 0.2 mm (11, 12 and 13).

size, the first tentacular branches are far away from the collar, which is angular, and the hooks lack shoulders. In *O. petersenae* dichotomies of tentacles are only observed at the distal end, the collar is curved (convex), and the tentacular branches are receptive to staining (Table 1).

The Australian *O. australis*, *O. mirrawa*, *O. dichotoma* and *O. picta* differ from *O. caissara* sp. n. in having four pairs of tentacular branches. *Owenia dichotoma* has a shorter tentacle crown with fewer ramifications than *O. caissara* sp. n. and *O.*

picta has fewer tentacle crown ramifications than *O. caissara* sp. n., and a bilobed structure between the tentacles of the left and right sides in ventral view, which is lacking in *O. caissara* sp. n. (Table 1).

The specimens collected in Ubatuba (Northern coast of São Paulo, SE Brazil) differ from *O. caissara* sp. n. in having four pairs of tentacular branches with dorsal and ventral branches of equivalent length and the angles of the hooks ranging from 0° to 45° in the first abdominal segment (Table 1).

Table 1. Main characters used to distinguish *Owenia* species (modified from Koh and Bhaud 2003).

Caracteres		*O. johnsoni* - California	*O. collaris* - California	Ubatuba - Brazil	*O. caissara* - Paranaguá Bay - Brazil	*O. caissara* - Babitonga Bay - Brazil	*O. borealis* - Iceland	*O. polaris* - Norwegian Sea	Portugal	Seine Bay	*O. fusiformis* - Banyuls Bay	*O. persica* - Persian Gulf	Madagascar Tulear	*O. gomsoni* - Yellow Sea	Japan Sea	West of Australia	*O. mirrawa* - Australia	*O. bossensis* - Australia	*O. australis* - Australia	*O. dichotoma* - Australia	*O. picta* - Australia	*O. petersene* - New Zealand
Size	Branchial lenth (1)	L	M	S	M	M	S	S	M	M	M	L	M	M	S	M					S	M	L
	Thorax length (2)	L	M	M	M	M	M	M	L	M	M	L	M	M	M	M	S	S	S				M
	Body width (3)	M	M	L	M	M	M	M	M	M	M	S	L	L	M	S							M
	Number of segments (range)	19–20	20		13–21	16–21	22	18–22		19–24	21–30	8–23		22–24			14	17	19		22		
Tentacles	Number of trunks	4	4	4	5	5	4	4		4	4	4	4	5	4	4	4	3	4	4	4	4	4
	Dorsal and Ventral length	≠	≠	=	≠	≠	≠	=		≠	≠	≠	=	≠	≠	=	≠		≠	=		=	≠
Collar	Curved									*				*	*		*						*
	Straight	*	*	*	*	*				*		*			*			*	*	*	*	*	
	With angle						*	*				*											
	Slit length: Short, Middle or long	S	M	M			M	M	S	S	M	M	S	M	M	M	M						L
Thorax	Line of 3 thoracic bundles: / or _/	/	_/	/	_/	_/	/	_/	/	_/	_/	_/	_/	/	_/	_/	_/	/					_/
	Direction (°) of hooks on abdominal segment 1 (range)	0	30 / -45	0 / -45	0 / -90	0 / -90	45	0 / -10	45	0 / -90	0 / -5	45	70 / -90	0 / -30	10 / -90	45							30 / -90
Hooks	Length of tooth (X) (4)	M	L	M	M	M	S	M	M	M	M	S	L	M	M	S							M
	Length of opening (Y) (5)	M	M	M	M	M	M	M	S	M	M	L	M	M	M	M							S
	Ratio X/Z (6)	M	M	M	M	M	S	M	M	M	M	M	M	M	M	M							M
	Angle of tooth (7)	S	B	M	S	S	B	B	M	M	M	M	M	M	S	M	B						S
	Dorsal shift present (pr), or absent (ab)	pr	pr	ab	pr	pr	pr	pr	ab	pr	ab	pr	pr	ab	ab	pr							pr
	Ventral shift present (pr), absent (ab)	pr+	pr+	pr+	pr	pr	pr	ab	pr	pr	pr	pr	ab	pr	pr	pr							pr
Scales on chaetae	Length (a + b)/d (8)	M	M	L	L	L	S	M	L	M	M	M	S	M	M	M							L
	Sharpness (a/d) (9)	A	A	S	M	M	A	M	S	S	M	M	A	S	M	S							S
	Length of free part (c) (10)	L	M	L	M	M	S	M	L	M	M	L	S	S	M	M							L
	Transition area: < > or ( )	<>	<>	()	()	()	<>	<>	()	<>	<>	()	()	()	<>	()	()	()	()				()
Methyl Green Staining	Tentacles			–	–	–	+				*	–			*	–							*
	Dorsal collar			*	+	+	+				*		+		+								*
	2 ventral lines of thorax			–	–	–	–				–				+	–							–
	V-shaped area from mouth to setiger 2			⊽	∪		⊽	∪			⌣	∨	∪		∪	∪							∨

(1) A/C (Branchial length/Collar length, Fig. 1) ≥ 1.27 : L (long), ≤ 0.68 : S (short), 0.68 < M < 1.27; (2) B/C (Collar length/Thoracic length, Fig. 1) ≥ 2.62: L (long) ≤ 1.41 : S (short), 1.41 < M < 2.62; (3) B (Collar length, Fig. 1) ≥ 1.36 mm : L (long), ≤ 0.73 mm : S (short), 0.73 < M < 1.36; (4) X (See Fig. 2) ≥ 2.93 μm : L (long), ≤ 1.58 μm : S (short), 1.58 < M < 2.93 ; (5) Y (See Fig. 2) ≥ 1.37 μm : L (long), ≤ 0.74 μm : S (short), 0.74 < M < 1.37

(6) X/Z (See Fig. 2) ≥ 0.98 : L (long), ≤ 0.53 : S (short), 0.53 < M < 0.98; (7) L2/L4(°) (See Fig. 2) ≥ 80 : B (big), ≤ 70° : S (small), 70° < M < 80°; (8) (A+B)/D (See Fig. 4) ≥ 9.01 : L (long), ≤ 4.85 : S (short), 4.85 < M < 9.01; (9) A/D (See Fig. 4) ≥ 3.12 : S (sharp), ≤ 1.68 : A (acute), 1.68 < M < 3.12; (10) C (See Fig. 4) ≥ 4.30 μm : L (longo), ≤ 2.32 μm : S (curto), 2.32 < M < 4.30; (11) *: strong coloration, +: coloration, – : without coloration.

DISCUSSION

Although direct comparison with type-material from other species is an almost mandatory practice to describe a new species, most of this material was not available for loans to Brazil. We are convinced that *Owenia fusiformis* and related taxa are morphologically close but not cryptic, and we believe that morphological characters provide good evidence to recognize and treat them as separate evolutionary lineages. Very good and detailed descriptions of some of the closest species to *O. caissara* sp. n. are currently available and were extensively used in our study (see Table 1). Due to the loss or bad state of conservation of their type series (communication by Gustavo Sene-Silva, a former student of PCL), we decided not to detail the taxonomic affinities

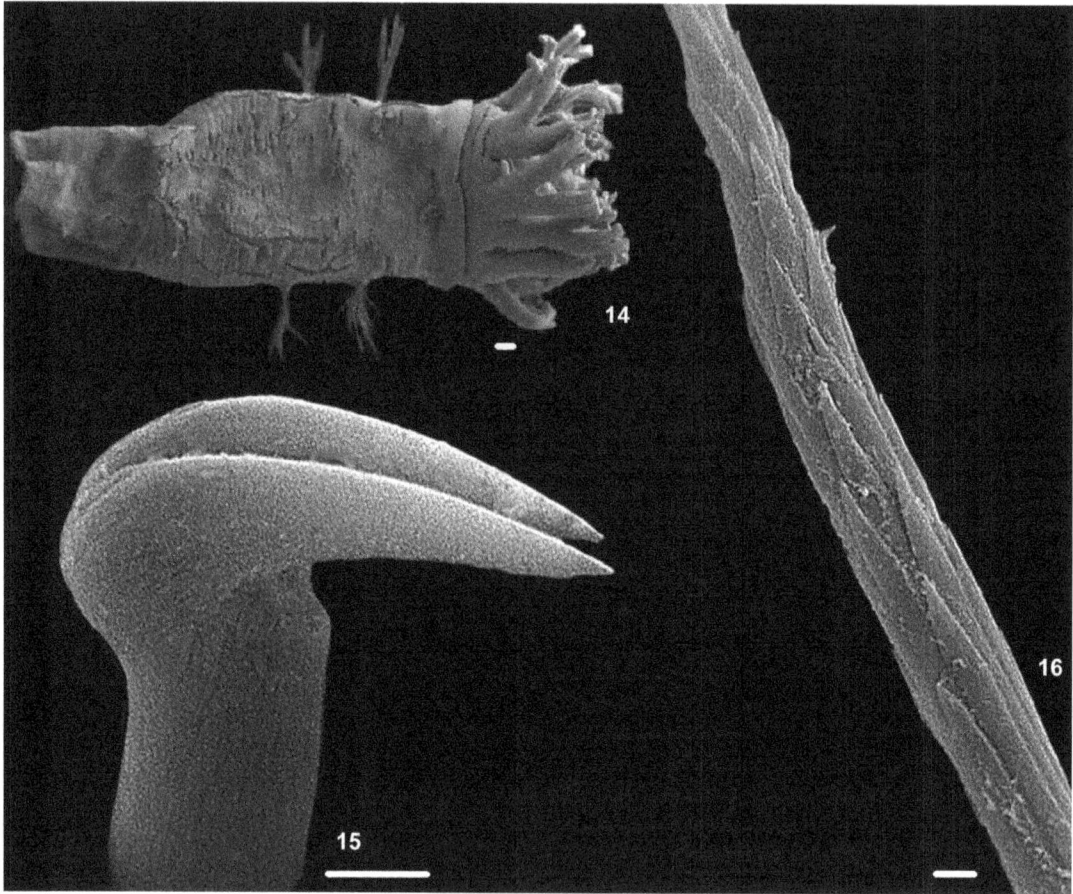

Figures 14–16. *Owenia caissara* sp. n. from Babitonga Bay, Polychaeta 17522. Anterior region in lateral view showing tentacular crown, thorax, and first neuropodial ring (14), scales on notochaeta from the first abdominal segment (15), lateral view of a hook from the first neuropodial ring (16). Scales bars: 100 μm (14), 1 μm (15 and 16).

of *O. sundevalli* (Kinberg, 1867), *O. tegula* (Kinberg, 1866) and *O. brasiliensis* (Hansen, 1882), from shallow continental shelf bottoms off the southwestern Atlantic. However, even if succinct, their original descriptions strongly indicate that they differ from *O. fusiformis* and *O. caissara* sp. n., so that re-description, revalidation, and neotype designation are much needed. Since geographic range and habitat preferences are also good criteria to separate species of *Owenia*, we thus believe that the current delineation of *O. caissara* sp. n. is well justified and rely on multiple lines of morphological and biogeographical evidence.

By describing a new species from southern Brazil, previously referred to as *O. fusiformis*, we reinforce the growing understanding of the large worldwide diversity of *Owenia*, supporting that the existing environmental barriers effectively limit larval dispersal (Norris and Hull 2012). Taxonomically robust morphological characters allowed for the unambiguous recognition of a new taxon, contributing to mitigate a true identity crisis still persistent in the regional literature.

We emphatically anticipate the need for a taxonomic revision of the material so far recorded along the southwestern Atlantic, by combining both modern morphological criteria and molecular data. Even in the absence of such revision, we do not recommend keeping *O. sundevalli*, *O. tegula*, and *O. brasiliensis* as synonyms of *O. fusiformis*, which is still the case in the World Register of Marine Organisms (Read 2015). Traits such as the number of body segments and length of the tentacular crown are not diagnostic at the species level, and should be considered unreliable or inconsistent for the diagnosis or synonymies among the species of *Owenia* (Koh and Bhaud 2003, Ford and Hutchings 2010). Trying to address this issue, Blake (2000) and Koh and Bhaud (2001, 2003) used additional or novel macro- and microscopic morphological features based on hard structures. Moreover, since measurements of soft parts are prone to errors due to tissue contraction after fixation, they suggested the usage of relative proportions, as the tentacular crown vs. thorax length ratio. Ford and Hutchings (2010) showed, however, that even

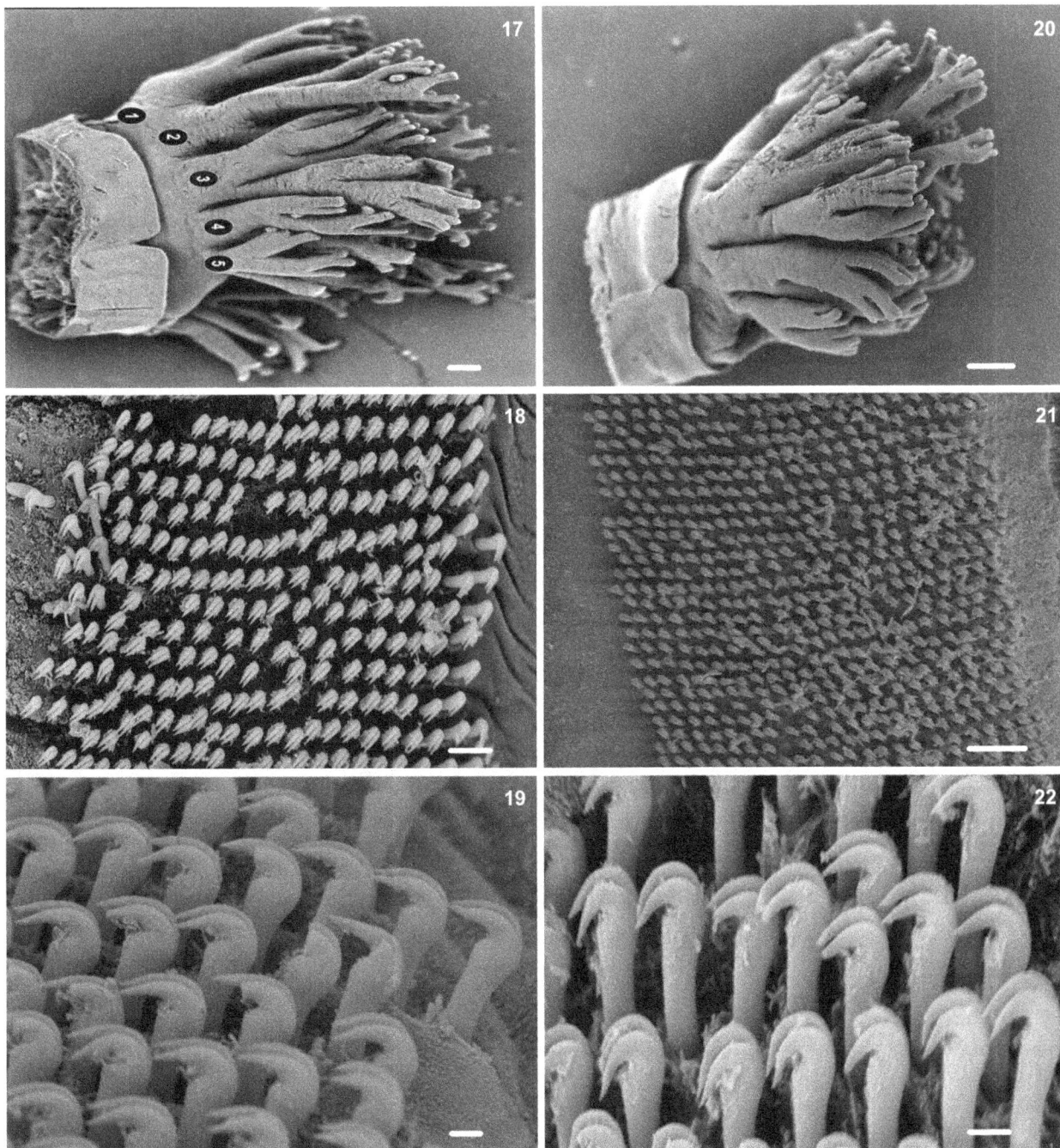

Figures 17–22. SEM Images. 17-19 (first column), *Owenia caissara* sp. n. from Babitonga Bay Polychaeta 17518, 17519. 20-22 (second column), *Owenia caissara* sp. n. from Paranaguá Bay. 17 and 20, tentacular crown and thorax portion; 18 and 21 band of neuropodial hooks on the first abdominal segment (band median portion of band); 19 and 22, lateral view of hooks on the first abdominal segment. Scale bars: 100 µm (17 and 20), 10 µm (18), 2 µm (19), 20 µm (21), 3 µm (22).

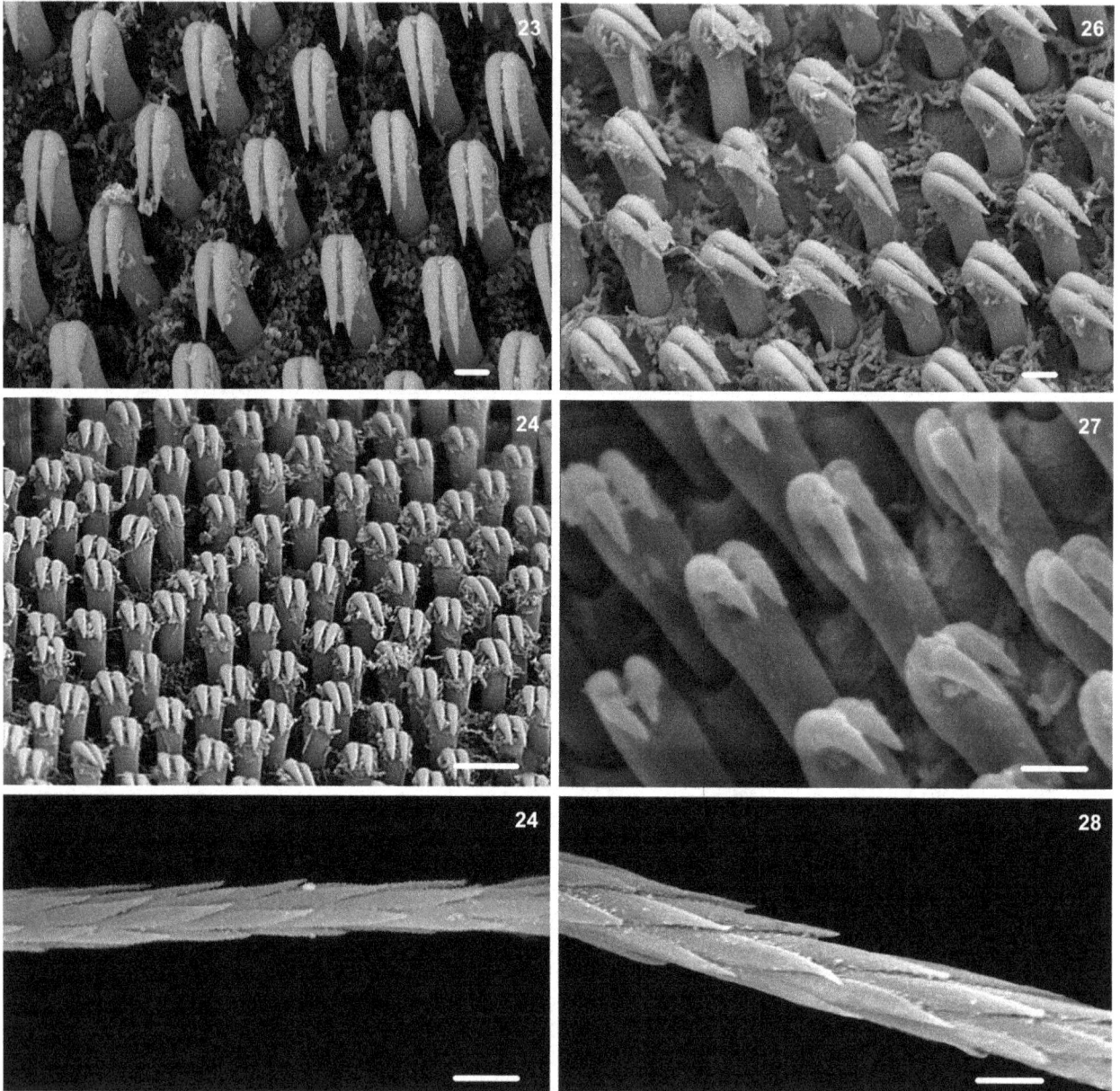

Figures 23–28. SEM Images. 23–25 (first column), *Owenia caissara* sp. n. from Babitonga Bay, Polychaeta 17517, 17518. 26–28 (second column), *Owenia caissara* sp. n. from Paranaguá Bay, Polychaeta 17523. 23 and 26, apical view of hooks on the first abdominal segment; 24 and 27, frontal view of hooks on the first abdominal segment; 25 and 28, median portion of notochaeta from first abdominal segment, showing scales. Scale bars: 2 μm (23), 5 μm (24), 2 μm (26, 28), 4 μm (25), 1 μm (27).

the relative proportions may vary depending on the animal's age due to allometric growth.

Chitinous structures, such as hooks and other chaetae, do not suffer alterations or deformation after fixation. Therefore, the use of morphometric proportions and measurements of hard 'structures would also allow for a better assessment of intra- and interspecific variability. However, there is still no consensus that morphometric data are sufficient for the unequivocal recognition of new species. For example, the significant variability between chaetal scales in one single individual may hinder the usefulness of this character to diagnose species. The same holds true for the number of hooks in neuropodial rows, which is also influenced by animal development.

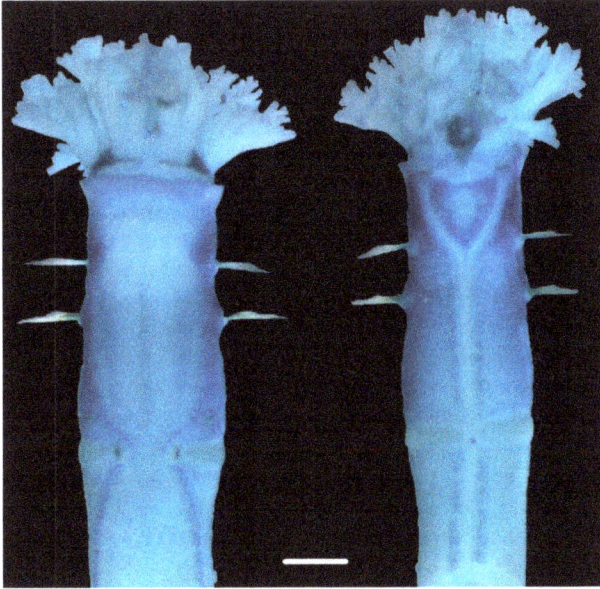

Figure 29. Methyl green colour pattern of anterior end of the body in dorsal (left) and ventral (right) views. Scale bar: 5 μm.

Koh and Bhaud (2003) suggested that the high morphological variability found in *Owenia* should be evaluated with caution since it can only reflect inter-population phenotypic plasticity. Unfortunately, the few available studies addressing growth and development are restricted to the Mediterranean *O. fusiformis* (Gentil et al. 1990, Dauvin and Gillet 1991).

The question about using *Owenia* tubes as useful diagnostic features for species recognition remains open. Koh and Bhaud (2001, 2003) noticed that the greater or lesser prevalence of shells, quartz, and heavy minerals, could reflect interspecific variability. Experiments conducted with populations from the Mediterranean and Yellow Seas indeed demonstrated distinct preferences for certain particles for tube building. Mediterranean animals preferred particles of quartz and carbonate while those from the Yellow Sea preferred only quartz particles (Koh and Bhaud 2001). Conversely, Ford and Hutchings (2010) suggested that this apparent preference would only indicate greater or lesser availability of these materials in the sediment, making it difficult or impossible to use this feature to distinguish between species. Therefore, further studies on the ability to select particles and its possible diagnostic and taxonomic implications are needed. Strategies and tube-building strategies by *O. caissara* sp. n. will be presented elsewhere.

The difficulty in effectively establishing diagnostic morphological characteristics for the recognition of *Owenia* species still remains. Molecular data will likely help to address the still prevalent identity crisis of the genus along the southwestern Atlantic. Unfortunately, the availability of molecular data is still incipient, and the sequences available on GenBank have not yet been used for phylogenetic studies or species differentiation.

ACKNOWLEDGEMENTS

We are grateful to all people who helped us during field work, especially Tamara Aparecida Carlini and her family, who housed us in São Francisco, and Gabriela Truppel. Juliana Ferreira helped with the MEV pictures. The Lab of Minerals and Rocks (LAMIR/UFPR) supported us in the identification of tube particles. LS was supported by a M Sc grant from CAPES (Coordenação de Aperfeiçoamento de Pessoal de Nível Superior).

REFERENCES

Augener H (1934) Polychaeten aus den Zoologischen Museen von Leiden und Amsterdam. IV Schluss. Zoologische Mededeelingen s'Rijks 17: 67–160.

Bhaud M (1982) Relations entre stratégies de reproduction et aire de répartition chez les annélides polychètes. Oceanologica Acta 5: 465–472.

Blake JA (2000) Family Oweniidae Rioja. In: Blake JA, Hilbig B, Scott PV (Eds) Taxonomic Atlas of the Benthic Fauna of the Santa Maria Basin and Western Santa Barbara Chanel. Santa Barbara, Santa Barbara Museum of Natural History, vol. 7, 97–127

Dauvin JC, Gillet P (1991) Spatio-temporal variability in population structure of Owenia fusiformis Delle Chiaje (Annelida: Polychaeta) from the Bay of Seine (Eastern English Channel). Journal of Experimental Marine Biology and Ecology 152: 105–122. https://doi.org/10.1016/0022-0981(91)90138-M

Dauvin J-C, Thiébaut E (1994) Is Owenia fusiformis a cosmopolitan species? Mémoires du Muséum National d'Histoire Naturelle Paris 162: 383–404.

Ford E, Hutchings P (2010) An analysis of morphological characters of Owenia useful to distinguish species: description of three new species of Owenia (Oweniidae: Polychaeta) from Australian waters. Marine Ecology 26: 181–196. https://doi.org/10.1111/j.1439-0485.2005.00062.x

Gentil F, Dauvin J-C, Ménard F (1990) Reproductive biology of the polychaete Owenia fusiformis Delle Chiaje in the Bay of Seine (eastern English Channel). Journal of Experimental Marine Biology and Ecology 142: 13–23. https://doi.org/10.1016/0022-0981(90)90134-X

Gillet P (1988) Structure des peuplements intertidaux d'annélides polychètes de l'estuaire du Bou Regreg (Maroc). Bulletin d'écologie 19: 33–42.

Guizien K, Brochier T, Duchêne J-C, Koh BS, Marsaleix P (2006) Dispersal of Owenia fusiformis larvae by wind-driven currents: turbulence, swimming behaviour and mortality in a three-dimensional stochastic model. Marine Ecology Progress Series 311: 47–66. https://doi.org/10.3354/meps311047

Hartman O (1959) Catalogue of the polychaetous annelids of the world. Los Angeles, Allan Hancock Foundation publications, Occasional paper #23, 197p.

Ibanez-Aguirre AL, Solis-Weiss V (1986) Anélidos poliquetos de las praderias de *Thalassia testudinum* del noroeste de la Laguna de Términos, Campeche, México. Revista de Biologia Tropical 34: 35–47.

Imajima M, Hartman O (1964) The polychaetous annelids of Japan. Part II. Los Angeles, Allan Hancock Foundation Publications, University of Southern California.

Koh B-S, Bhaud MR (2001) Description of *Owenia gomsoni* n. sp (Oweniidae, Annelida Polychaeta) from the Yellow Sea and evidence that *Owenia fusiformis* is not a cosmopolitan species. Vie et Milieu 51: 77–86.

Koh B-S, Bhaud MR (2003) Identification of new criteria for differentiating between populations of *Owenia fusiformis* (Annelida, Polychaeta) from different origins: Rehabilitation of old species and erection of new species. Vie et Milieu 53: 65–95.

Koh B-S, Bhaud MR, Jirkov IA (2003) Two new species of *Owenia* (Annelida: Polychaeta) in the northern part of the North Atlantic Ocean and remarks on previously erected species from the same area. Sarsia 88: 175–188. https://doi.org/10.1080/00364820310001318

Martin D, Koh B-S, Bhaud M, Dutrieux E, Gil J (2006) The genus *Owenia* (Annelida: Polychaeta) in the Persian Gulf, with description of *Owenia persica* sp. n. Organisms Diversity and Evolution 15: 1-21. https://doi.org/10.1016/j.ode.2006.01.001

Mcnulty JK, López NL (1969) Year-round production of ripe gametes by benthic polychaetes in Biscayne Bay, Florida. Bulletin of Marine Science 19: 945-954.

Norris RD, Hull PM (2012) The temporal dimension of marine speciation. Evolutionary Ecology 26: 393–415. https://doi.org/10.1007/s10682-011-9488-4

Plante R (1967) Étude quantitative du benthos de Nosy-Bé: note préliminaire. Cah. O.R.S.T.O.M., Série Océanografie, 5: 95–108.

Read G (2015) *Owenia*. In: Read G, Fauchald K (Ed.) World Polychaeta database. http://www.marinespecies.org/aphia.php?p=taxdetails&id=129427 [Accessed: 15/01/2015]

Thiébaut E, Dauvin J, Lagadeuc Y (1992) *Owenia fusiformis* larvae (Annelida: Polychaeta) in the Bay of Seine. I. Vertical distribution in relation to water column stratification and ontogenic vertical migration. Marine Ecology Progress Series 80: 29–39. https://doi.org/10.3354/meps080029

Thiébaut E, Dauvin J, Lagadeuc Y (1994) Horizontal distribution and retention of *Owenia fusiformis* Larvae (Annelida: Polychaeta) in the Bay of Seine. Journal of the Marine Biological Association of the United Kingdom 74: 129–142. doi: http://dx.doi.org/10.1017/S0025315400035712

Verdier-Bonnet C, Carlotti F (1997) A model of larval dispersion coupling wind-driven currents and vertical larval behaviour: Application to the recruitment of the annelid *Owenia fusiformis* in Banyuls Bay. Marine Ecology Progress Series 160: 217–231. https://doi.org/10.3354/meps160217

Wilson DP (1932) On the mitraria larva of *Owenia fusiformis* Delle Chiaje. Philosophical Transactions of the Royal Society B: Biological Sciences 221: 334. https://doi.org/10.1098/rstb.1932.0004

Author Contributions: LPS and PL designed and conducted the experiments; LPS and PL analyzed the data and wrote the paper.
Competing Interests: The authors have declared that no competing interests exist.

21

Phylogeny of the Neotropical longhorn beetle genus *Ateralphus* (Coleoptera: Cerambycidae: Lamiinae)

Diego de S. Souza[1], Marcela L. Monné[2], Luciane Marinoni[1]

[1]*Departamento de Zoologia, Universidade Federal do Paraná. Caixa Postal 19020, 81531-990 Curitiba, PR, Brazil.*
[2]*Departamento de Entomologia, Museu Nacional, Universidade Federal do Rio de Janeiro. Quinta da Boa Vista, São Cristóvão, 20940-040 Rio de Janeiro, RJ, Brazil.*
Corresponding author: Diego de S. Souza (diegosantanasouza@hotmail.com)

http://zoobank.org/E6B8E187-01A7-4EAB-AD79-4D74DA5EC644

ABSTRACT. *Ateralphus* Restello, Iannuzzi & Marinoni, 2001 is a Neotropical genus of longhorn beetles composed of nine species. This genus was proposed from splitting *Alphus* White, 1855 into other two genera: *Ateralphus* and *Exalphus* Restello, Iannuzzi & Marinoni, 2001. Even though *Ateralphus* (nine species), *Alphus* (four) and *Exalphus* (18) were recently revised, their validity has not been tested using phylogenetic methods. In this study, we carried out a cladistic analysis of *Ateralphus* and its related genera, *Alphus* and *Exalphus*, based on 44 morphological characters of the adults, to test their monophyly and infer the relationships between their species. Our results support the monophyly of the three genera and recovered two clades that corroborate the species-groups previously recognized in *Ateralphus*. A new genus, *Grandateralphus* **gen. n.**, is proposed for one of these clades, which is supported by three synapomorphies: width of upper ocular lobes less than width between the lobe and the coronal suture (character state 6: 0), genae parallel in frontal view (8: 1) and scape gradually expanded toward apex, reaching widest diameter just near apex (9: 2). *Grandateralphus* **gen. n.** includes three new combinations: *G. lacteus* (Galileo & Martins, 2006), **comb. n.**; *G. tumidus* (Souza & Monné, 2013), **comb. n.**; and *G. variegatus* (Mendes, 1938), **comb. n.** Notes on the distribution of *G. variegatus* **comb. n.** and a new record of *E. cicatricornis* Schmid, 2014 for Bolivia (Santa Cruz) are provided.

KEY WORDS. Acanthoderini, cladistics, distribution, new combination, taxonomy.

INTRODUCTION

Ateralphus Restello, Iannuzzi & Marinoni, 2001 is a Neotropical genus of longhorn beetles belonging to Acanthoderini, a tribe of Lamiinae described by Thomson (1860). The tribe Acanthoderini includes 553 species classified into 67 genera, according to Tavakilian and Chevillotte (2016), and is characterized mainly by the piriform scape, shorter than antennomere III, and by the spines or lateral tubercles of the prothorax. Originally, *Ateralphus* species were classified in *Alphus*, a genus described by White (1855) for five species, including *Alphus subsellatus* White, 1855 (subsequently designated as type-species of *Ateralphus* by Restello et al. (2001)). Restello et al. (2001) proposed two additional genera for species previously included in *Alphus* White, 1855: *Ateralphus* and *Exalphus*; and summarized, in a table, the main characteristics to differentiate the three genera from one another. As part of that contribution, 15 species were included

in *Exalphus* (13 new combinations and two new species), six new combinations were proposed for *Ateralphus*, and only three species were kept in *Alphus*.

All species of *Alphus* and *Ateralphus* are distributed in South America. Most of them occur in dense forests, such as the Atlantic and the Amazon forests. There are only a few records from open biomes such as the Cerrado and Chaco for *Ateralphus subsellatus* (White, 1855) and *Ateralphus dejeani* (Lane, 1973). *Exalphus*, in contrast, has a broader distribution, from Guatemala to south of South America, with only one species occurring in Central America (*Exalphus cavifrons* (Bates, 1872)).

Recently, *Alphus* (four species) and *Exalphus* (18) were studied by Souza and Monné (2013a) and Souza and Monné (2014), respectively, who updated the distribution data and provided an identification key for their species. After Souza and Monné (2014), two new species were described in *Exalphus*: *E. docquini* Tavakilian & Néouze, 2013 and *E. cicatricornis* Schmid,

2014. *Ateralphus* (nine species) was revised by Souza and Monné (2013b). They recognized two groups of species in the genus, but failed to name them. In the first group, characterized mainly by the upper ocular lobes separated by three or more times their width, lower ocular lobes narrow and rectangular-shaped, genae parallel to divergent and tibiae with one ring or spot of dark brown setae on sub-apical region, they included *A. lacteus* Galileo & Martins, 2006, *A. tumidus* Souza & Monné, 2013 and *A. variegatus* (Mendes, 1938). In the second group, characterized by the upper ocular lobes separated by less than or equal to twice their width, lower ocular lobes large and rounded, genae convergent and tibiae with dense ring of dark brown setae on apical region, they included the other species of the genus.

In Cerambycidae, new taxa are frequently proposed based only on descriptive taxonomy. In this contribution, we evaluate whether the split of *Ateralphus* and *Exalphus* from *Alphus*, proposed by Restello et al. (2001), can be corroborated through cladistic methods. Our specific goals are: (1) to test the monophyly of *Ateralphus* and its related genera (*Alphus* and *Exalphus*) through phylogenetic analysis based on morphological characters; and (2) to evaluate whether the species groups mentioned by Souza and Monné (2013b) for *Ateralphus* may be corroborated and formalized as autonomous genera. Additionally, notes on the taxonomy and distribution of some species are provided.

MATERIAL AND METHODS

Taxon sampling

We included in the phylogenetic analysis all species currently classified in *Ateralphus*, *Alphus* and 15 of the 18 species placed in *Exalphus*. *Exalphus simplex* (Galileo & Martins, 1998), *E. vicinus* Galileo & Martins, 2003 and *E. docquini* were not included in this study because we were not able to obtain specimens for examination. Since there are no previous phylogenetic studies on the Acanthoderini, we chose our outgroups from species which had been considered close to the ingroup taxa, *Alphus*, *Ateralphus* and *Exalphus*, in previous taxonomic treatments (i.e., Bates (1862), Lacordaire (1872), Martins (1985) and Martins and Galileo (2007)). Additionally, we also selected some representatives of Acanthoderini that bear some morphological resemblance to the species of the ingroup. Therefore, the following species were chosen as outgroups: *Cotyzineus bruchi* (Melzer, 1931), *Myoxinus pictus* (Erichson, 1847), *Nesozineus alphoides* (Lane, 1977) and *Acanthoderes albitarsis* Laporte, 1840, which was used to root the tree in our analysis. In total, 32 terminal taxa were included in the cladistics analysis, comprising 28 ingroup species and four outgroup species (Table 1).

Characters and analysis procedures

The characters were constructed from the external morphology of the adults and male genitalia. Several are modified from previous taxonomic studies on the ingroup species, such

as Bates (1862), Restello et al. (2001) and Souza and Monné (2013b). In total, 44 characters were included in the analysis (27 binary and 17 multistate). The description of the characters and their states follows the structural concepts proposed by Sereno (2007). In the matrix, missing data is indicated by '?'. The characters and their respective states are listed below, and the data matrix is in Table 1. Following the description of each character, in the list of characters, we provide the length (L), consistency index (CI) and retention index (RI) based on the selected most parsimonious topology.

All characters were treated as unordered (or non-additive) and equally weighted. The search for the most parsimonious topologies was conducted in WinClada version 1.00.08 (Nixon 1999–2002), through NONA version 2.0 (Goloboff 1993), using the following commands: number of replications (mult*N) = 1000; starting tree per replication (hold/) = 10; random seed = 1; Multiple TBR + TBR (mult*max*). The character transformations presented on the selected topology are either unambiguous changes or were optimized under fast optimization (ACCTRAN). Branch supports values were calculated in TNT (version 1.1, [August, 2011], Goloboff et al. 2008). Bootstrap values were calculated from an independent analysis using 1000 pseudoreplicates and Bremer support (Bremer 1994) was calculated based on the strict consensus topology using 1000 suboptimal trees up to one step longer.

List of characters

1. Head, posterior margin, shape: (0) rounded (Fig. 1); (1) triangular (Fig. 2). L = 2; CI = 50; RI = 92.
2. Head, coronal suture, relative to posterior margin: (0) not reaching (Fig. 2); (1) reaching (Fig. 1). L = 2; CI = 50; RI = 88.
3. Head, erect setae on base of antennal tubercles: (0) absent; (1) present. L = 1; CI = 100; RI = 100.
4. Head, coarse punctation between upper ocular lobes: (0) absent; (1) present. L = 4; CI = 25; RI = 76.
5. Frons, erect setae near the genal suture: (0) a row reaching the base of the lower ocular lobes (Fig. 6); (1) a row reaching approximately basal half of the antennal tubercles (Fig. 7); (2) a row reaching antennal tubercles (Fig. 8); (3) only one seta at half of the lower ocular lobes (Fig. 9). L = 4; CI = 75; RI = 90.
6. Upper ocular lobes (U), width relative to the width from lobe to coronal suture (W): (0) U less than W; (1) U subequal to W; (2) U larger than W. L = 9; CI = 22; RI = 50.
7. Lower ocular lobes (L), height relative to gena (G): (0) L less than G; (1) L subequal to G; (2) L larger than G. L = 3; CI = 66; RI = 83.
8. Genae, shape, in frontal view: (0) divergent toward apex (Fig. 6); (1) parallel (Fig. 8); (2) convergent toward apex (Fig. 7). L = 4; CI = 50; RI = 83.
9. Antennae, scape, shape: (0) piriform (Fig. 3); (1) gradually expanded toward apex, reaching widest diameter at one third from apex (Fig. 4) (2) gradually expanded toward

Table 1. Data matrix of morphological characters used in the cladistic analysis of *Ateralphus* and related genera, *Alphus* and *Exalphus* (Coleoptera: Cerambycidae: Lamiinae). Outgroups are indicated with an asterisk.

Characters (1–44)

Taxa	1	2	3	4	5	6	7	8	9	10	11	12	13	14	15	16	17	18	19	20	21	22	23	24	25	26	27	28	29	30	31	32	33	34	35	36	37	38	39	40	41	42	43	44
Acanthoderes albitarsis*	1	0	0	0	0	0	0	0	0	0	2	2	0	1	0	0	1	0	0	0	0	1	0	0	0	0	0	0	0	0	0	0	0	0	1	1	0	0	0	1	1	1	0	0
Alphus capixaba	0	0	0	1	1	2	2	2	1	0	0	1	0	0	2	1	0	1	1	0	0	0	1	0	0	1	0	0	0	0	0	1	0	0	0	0	0	0	0	0	?	?	?	?
Alphus marinonii	0	0	0	0	1	2	2	2	1	0	0	0	0	0	2	1	0	1	1	1	0	0	1	0	0	1	0	0	0	0	0	1	0	0	0	0	0	0	0	0	?	?	?	?
Alphus similis	0	0	0	0	1	1	2	2	1	0	0	1	0	0	2	1	0	1	1	1	0	0	1	0	0	1	0	0	0	0	0	1	0	0	0	0	0	0	0	0	1	1	0	1
Alphus tuberosus	0	0	0	0	1	2	2	2	1	0	0	0	0	0	2	1	0	1	1	1	0	0	1	0	0	1	0	0	0	0	0	1	0	0	0	0	0	0	0	0	1	1	0	1
Ateralphus auritarsus	0	1	0	0	1	1	1	2	1	0	1	2	0	1	0	0	1	2	0	0	0	1	1	0	0	0	0	1	0	1	0	0	0	0	0	1	0	1	0	1	?	?	?	?
Ateralphus dejeani	0	1	0	0	1	1	2	2	1	0	1	2	0	1	0	0	1	2	0	0	0	1	1	0	1	0	0	1	0	1	1	0	1	0	0	1	0	1	0	1	1	1	0	0
Ateralphus javariensis	0	1	0	0	1	2	2	2	1	0	1	2	0	1	0	0	1	2	0	0	0	1	1	0	0	0	0	1	0	1	1	0	1	0	0	1	0	1	0	1	?	?	?	?
Ateralphus lacteus	0	1	0	0	1	0	1	1	2	0	1	2	0	1	0	0	1	0	0	0	0	1	1	0	0	0	0	1	0	0	0	0	0	0	0	2	0	1	0	1	1	1	1	0
Ateralphus lucianeae	0	1	0	0	1	1	2	2	1	0	1	2	0	1	0	0	1	2	0	0	0	1	1	0	1	0	0	1	0	1	1	0	1	0	0	1	0	1	0	1	?	?	?	?
Ateralphus senilis	0	1	0	0	1	2	2	2	1	0	1	2	0	1	0	0	1	2	0	0	0	1	1	0	1	0	0	1	0	1	1	0	1	0	0	1	0	1	0	1	1	1	0	0
Ateralphus subsellatus	0	1	0	0	1	1	1	2	1	0	1	2	0	1	0	0	1	2	0	0	0	1	1	0	0	0	0	1	0	1	0	0	0	0	0	1	0	1	0	1	1	1	0	0
Ateralphus tumidus	0	1	0	0	1	0	1	1	2	0	1	2	0	1	0	0	1	0	0	0	0	2	1	0	0	0	0	1	0	0	0	0	0	0	0	2	1	1	1	1	1	1	0	0
Ateralphus variegatus	0	1	0	0	1	0	1	1	2	0	1	2	0	1	0	0	1	0	0	0	0	1	1	0	0	0	0	1	0	0	0	0	0	0	0	2	0	1	1	1	1	1	1	0
Cotyzineus bruchi*	0	0	0	0	3	1	1	2	1	0	0	1	0	1	0	1	1	1	0	0	0	1	1	0	0	0	0	0	0	0	0	0	0	0	1	0	1	0	0	0	0	1	1	0
Exalphus aurivillii	?	?	0	1	2	2	2	2	1	0	0	2	1	1	1	2	1	1	1	0	1	1	0	1	0	0	1	0	1	1	0	2	0	0	1	2	1	1	0	0	?	?	?	?
Exalphus biannulatus	1	0	0	1	2	2	2	1	1	0	0	2	1	1	1	2	1	1	1	1	0	1	0	1	0	0	1	0	1	0	0	2	0	0	1	2	1	1	0	0	1	1	0	0
Exalphus calvifrons	1	0	1	0	2	2	2	1	2	0	0	1	0	1	1	2	1	1	1	1	0	1	0	1	0	0	1	0	1	0	0	1	0	0	1	2	1	1	0	0	2	1	2	0
Exalphus cicatricornis	?	?	1	0	2	2	2	1	2	1	0	1	0	1	1	2	1	1	1	0	1	0	1	0	0	1	0	1	0	0	1	0	0	1	2	1	1	0	?	?	?	?	?	?
Exalphus colasi	1	0	0	1	2	1	2	2	1	0	0	1	1	1	1	2	1	1	1	1	1	0	1	1	0	1	1	1	0	0	0	2	0	0	1	2	1	1	0	0	1	0	0	0
Exalphus confusus	1	0	0	1	1	1	2	2	1	0	0	1	0	1	1	1	1	1	1	0	1	1	0	1	0	0	1	0	0	0	0	2	0	0	1	2	1	1	0	0	1	2	0	0
Exalphus foveatus	1	0	0	1	1	2	2	2	1	0	0	1	0	1	1	1	1	1	1	0	1	1	0	1	0	0	1	0	0	0	0	2	0	0	1	2	1	1	0	0	1	2	0	0
Exalphus gounellei	1	0	0	1	1	2	2	1	1	0	0	1	1	1	1	2	1	1	1	0	0	1	0	1	0	0	1	0	0	1	0	2	0	0	1	2	1	1	0	0	1	2	0	0
Exalphus guaraniticus	1	0	1	1	2	1	2	1	1	0	0	1	1	1	1	2	1	1	1	0	1	0	1	0	0	1	0	0	1	0	0	1	0	0	1	2	1	1	0	0	3	1	0	0
Exalphus leuconotus	1	0	1	1	2	1	2	1	1	0	0	1	1	1	1	2	1	1	1	0	1	0	1	0	0	1	0	0	1	0	0	1	0	0	1	2	1	1	0	0	3	1	0	0
Exalphus lichenophorus	1	0	1	1	2	2	2	1	2	0	0	1	1	1	1	2	1	1	1	0	1	0	1	0	0	1	0	0	1	0	0	1	0	0	1	2	1	1	0	0	?	?	?	?
Exalphus malleri	1	0	1	0	2	2	2	1	2	1	0	1	1	1	1	2	1	1	1	0	1	0	1	0	0	1	0	0	1	0	0	1	0	0	1	2	1	1	0	0	1	1	0	0
Exalphus solangeae	?	?	0	1	1	1	2	1	1	0	?	?	1	1	1	2	1	1	1	1	1	0	1	0	1	0	0	0	0	0	0	2	0	0	1	2	1	1	0	0	?	?	?	?
Exalphus spilonotus	1	0	0	1	1	2	2	2	1	0	0	1	0	1	1	1	1	1	1	0	1	1	0	1	1	0	0	0	0	1	0	2	0	0	1	2	1	1	0	0	3	2	0	0
Exalphus zellibori	1	0	0	1	2	1	2	2	1	0	0	1	1	1	1	2	1	1	1	1	1	0	1	1	0	1	1	1	0	0	0	2	0	0	1	2	1	1	0	0	1	1	0	0
Myoxinus pictus*	1	0	0	1	0	2	1	0	1	0	2	2	1	1	1	1	0	0	0	0	1	0	0	0	0	1	0	1	0	0	0	0	0	0	1	2	0	0	0	1	2	0	0	0
Nesozineus alphoides*	0	1	0	0	1	2	2	2	1	0	0	1	0	0	0	1	1	1	1	0	0	0	1	0	0	0	0	0	0	0	0	0	0	0	0	0	1	0	0	0	0	0	0	0

apex, reaching widest diameter near the apex (Fig. 5). L = 3; CI = 66; RI = 83.

10. Antennae, scape, scar at outer surface of apex: (0) absent; (1) present. L = 2; CI = 50; RI = 0.

11. Antennae, antennomere III, length relative to antennomere IV: (0) III shorter than IV; (1) III subequal to IV; (2) III longer than IV. L = 2; CI = 100; RI = 100.

12. Antennae, antennomere III, length relative to antennomere V: (0) III shorter than V; (1) III subequal to V; (2) III longer than V. L = 4; CI = 50; RI = 84.

13. Antennae, antennomeres III to XI, stain of dark brown setae on half basal region: (0) absent; (1) present. L = 3; CI = 33; RI = 80.

14. Antennae, antennomeres III to XI, stain of dark brown setae on apical region: (0) absent; (1) present. L = 1; CI = 100; RI = 100.

15. Prothorax, coverage pattern of setae: (0) uniformly covered by same color and density setae; (1) with a denser line of white setae at median region; (2) with denser distinct stains of white setae near the lateral tubercle. L = 3; CI = 66; RI = 92.

16. Pronotum, anterior margin, shape: (0) straight (Fig. 10); (1) slightly depressed (Fig. 11); (2) deeply depressed (Fig. 12). L = 3; CI = 66; RI = 94.

17. Pronotum, midline, glabrous tubercle or elevation: (0) absent; (1) present. L = 1; CI = 100; RI = 100.

18. Pronotum, elevation post-lateral to median tubercles: (0) absent; (1) slightly elevated, inconspicuous; (2) distinctly elevated, as a pair of tubercles. L = 2; CI = 100; RI = 100.

19. Pronotum, longitudinal elevation from pair of median tubercles toward posterior margin: (0) absent; (1) present. L = 1; CI = 100; RI = 100.

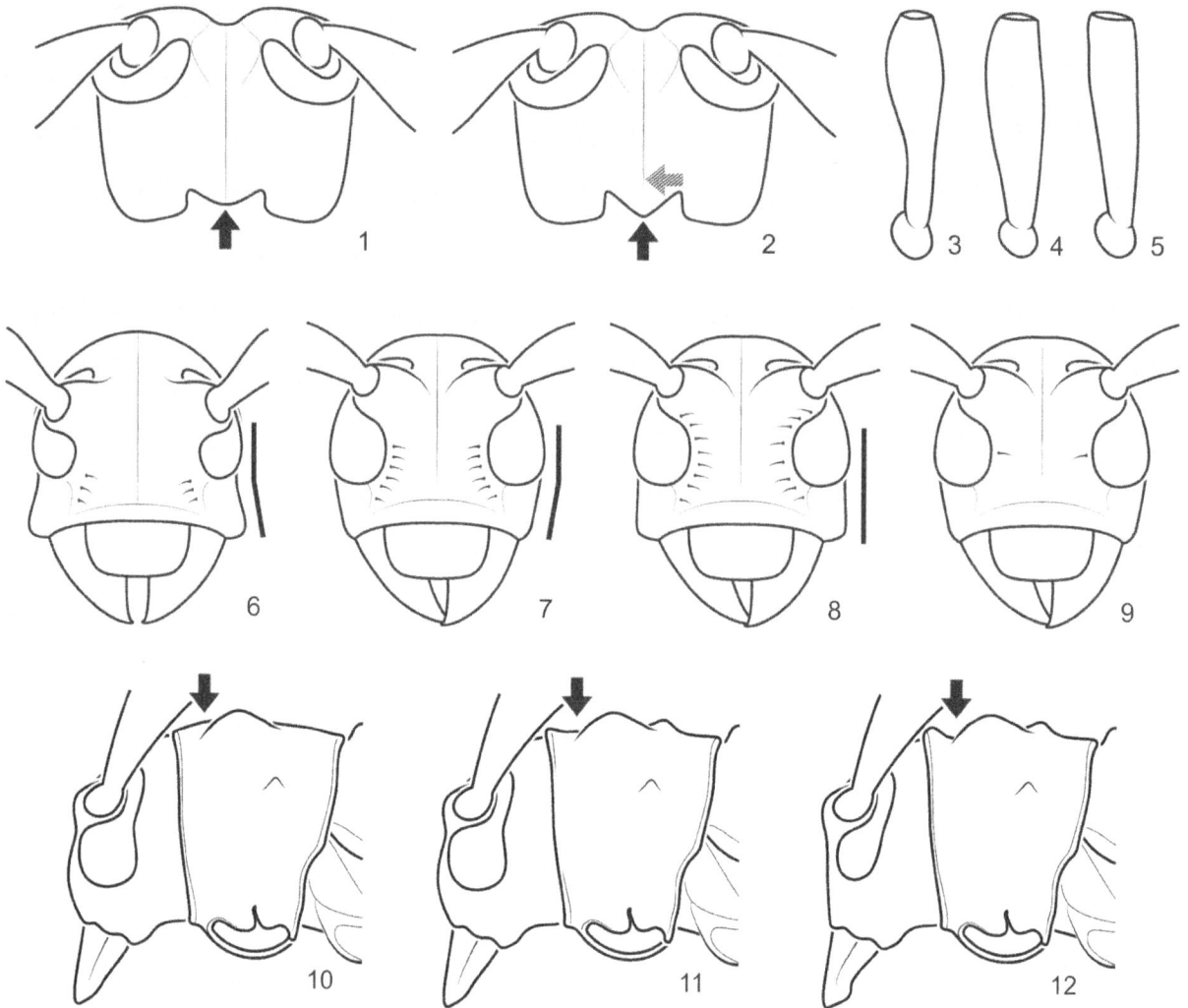

Figures 1–12. Schematic representation of the character states used in the cladistics analysis of *Ateralphus* and related genera, *Alphus* and *Exalphus*. 1–2, head in dorsal view: 1, *Ateralphus subsellatus*, posterior margin rounded (1:0; black arrow) and coronal suture reaching posterior margin (2:1); 2, *Exalphus leuconotus* (Thomson, 1860), posterior margin triangular (1:1; black arrow) and coronal suture reaching posterior margin (2:0; striped arrow). 3–5, shape of scape: C3, *A. albitarsis*, scape piriform (9:0); 4, *A. subsellatus*, scape gradually expanded toward apex, reaching the widest diameter at one third from apex (9:1); 5, *Ateralphus variegatus*, scape gradually expanded toward apex, reaching the widest diameter near the apex (9:2). 6–9, head in frontal view: 6, *A. albitarsis*, row of setae reaching the base of the lower ocular lobes (5:0) and genae divergent toward apex (8:0); 7, *A. subsellatus*, row of setae reaching the basal half of the antennal tubercles (5:1) and genae convergent toward apex (8:2); 8, *E. leuconotus*, row of setae reaching the antennal tubercles (5:2) and genae parallel (8:1); 9, *Cotyzineus bruchi*, only one seta at half of the lower ocular lobes (5:3). 10–12, prothorax in lateral view (character states are indicated by black arrow): 10, *Alphus tuberosus* (Germar, 1824), pronotum straight near anterior margin (16:0); 11, *A. subsellatus*, pronotum slight depressed near anterior margin (16:1); 12, *E. leuconotus*, pronotum deeply depressed near anterior margin (16:0).

20. Prosternal process, lateral margins, shape: (0) straight; (1) with a prominence at post-median region. L = 5; CI = 20; RI = 69.

21. Mesosternal process (Mp), length relative to mesocoxa (Mc): (0) Mp shorter than Mc; (1) Mp subequal to Mc; (2) Mp larger than Mc. L = 2; CI = 50; RI = 80.

22. Mesosternal process, shape: (0) slightly convex; (1) straight; (2) tumescent. L = 2; CI = 100; RI = 100.

23. Mesosternal process, approximated angle at the apical third relative to mesosternum: (0) 90°; (1) 45°; L = 2; CI = 50; RI = 93.

Figures 13–15. Punctures on basal third of the elytra (character 20) observed among the species included in the cladistic analysis. 13–14, Punctures irregularly distributed, represented by *Ateralphus subsellatus* and *Exalphus leuconotus*, respectively; 15, punctures arranged in longitudinal rows, represented by *Alphus tuberosus*.

24. Meso- and metasternum (males), ventral surface, sexual setae: (0) absent; (1) present. L = 1; CI = 100; RI = 100.

25. Scutellum, setae, color pattern: (0) similar to pronotum; (1) covered by dark-brown setae. L = 2; CI = 50; RI = 75.

26. Elytra, basal third, punctures: (0) irregularly distributed (Figs 13, 14); (1) arranged in longitudinal rows (Fig. 15). L = 1; CI = 100; RI = 100.

27. Elytra, basal-crests: (0) absent or slight raised; (1) distinctly raised. L = 2; CI = 50; RI = 92.

28. Elytra, region between basal-crests, setae, color pattern: (0) similar to pronotum; (1) covered by dark-brown setae. L = 2; CI = 50; RI = 90.

29. Elytra, humeral surface: (0) punctate, without raised tubercles; (1) with at least one differentiated and well-developed tubercle at base of humeral carina. L = 3; CI = 33; RI = 83.

30. Elytra, lateral margin, color pattern of setae: (0) with regularly spaced dark-brown stains; (1) similar to the dorsum. L = 1; CI = 100; RI = 100.

31. Femora, basal inner surface, setae, coverage pattern: (0) entirely covered; (1) meso- and/or metafemora glabrous. L = 1; CI = 100; RI = 100.

32. Femora, globose region, setae, coverage pattern: (0) evenly covered with short white setae; (1) mainly covered with sparse white setae, with denser areas forming irregular maculae; (2) mainly covered with sparse white setae, with denser areas forming transverse bands. L = 3; CI = 66; RI = 94.

33. Meso- and metafemora, circular stain of dark brown setae on anterior surface: (0) absent; (1) present. L = 1; CI = 100; RI = 100.

34. Protibiae, shape: (0) straight at base and slight convex toward apex; (1) concave. L = 2; CI = 50; RI = 0.

35. Meso- and metatibiae, half basal region, stain of dark brown setae: (0) absent; (1) present. L = 2; CI = 50; RI = 92.

36. Meso- and metatibiae, apical region, dark brown setae: (0) absent; (1) present. L = 4; CI = 50; RI = 81.

37. Tarsi, tarsomere V, basal third, stain of dark brown setae: (0) absent; (1) present. L = 2; CI = 50; RI = 93.

38. Tarsi, tarsomere V, apical third, stain of dark brown setae: (0) absent; (1) present. L = 2; CI = 50; RI = 85.

39. Abdomen, sternites II-IV, lateral margin, glabrous circular macula: (0) absent; (1) present. L = 1; CI = 100; RI = 100.

40. Sternite V (females), basal region, longitudinal sulcus: (0) absent; (1) present. L = 2; CI = 50; RI = 88.

41. Male genitalia, median lobe, ventral lobe, shape of apex: (0) rounded; (1); truncate; (2) acuminate; (3) with a median salience. L = 4; CI = 75; RI = 50.

42. Male genitalia, median lobe, ventral lobe (V), length relative to dorsal lobe (D): (0) V shorter than D; (1) V subequal to D; (2) V larger than D. L = 6; CI = 33; RI = 20.

43. Male genitalia, tegmen, parameres, distribution pattern of setae: (0) only on apex; (1) on apex and inner margin; (2) on apex and ventral margin. L = 3; CI = 66; RI = 0.

44. Male genitalia, tegmen, basal apophysis, apical region, sclerotization: (0) complete and uniform; (1) with a diagonal line more sclerotized near apex. L = 1; CI = 100; RI = 100.

RESULTS

The cladistic analysis resulted in two most parsimonious cladograms (L = 111, CI = 55 and RI = 86). The cladogram selected (showing unambiguous characters only) to represent the relationships among the taxa is presented in Fig. 16 and a summarized tree with the character states optimized under ACCTRAN optimization is presented in Fig. 17. The monophyly

of *Ateralphus*, *Alphus* and *Exalphus* were corroborated in our analysis, with strong branch support values, supporting Restello et al.'s (2001) hypotheses.

The monophyly of *Alphus* was strongly supported in our results by four unambiguous synapomorphies. Three of them are non-homoplastic (Fig. 16), as follows: prothorax with distinct stains of denser setae near the lateral tubercle (character 15: 2); absence of glabrous tubercle or elevation at midline of the

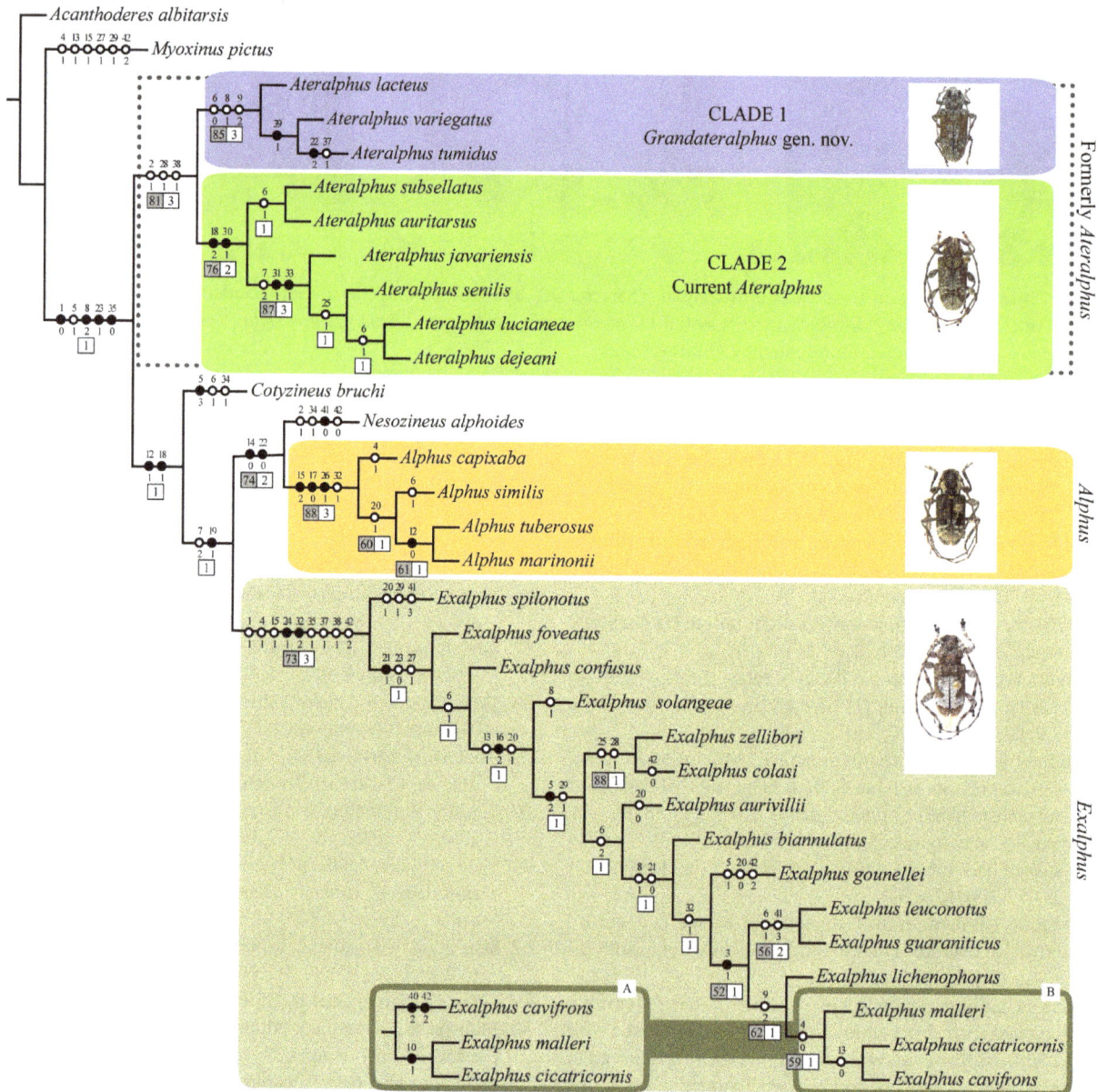

Figure 16. Tree obtained from the cladistics analysis of *Ateralphus* and related genera, *Alphus* and *Exalphus*. Only unambiguous transformations are shown. A and B show the differences between the two most parsimonious topologies. Bootstrap (gray box) and Bremer support (white box) are indicated below each node. Clades are illustrated by the following species: *Alphus similis* Martins, 1985; *Grandateralphus variegatus* comb. n.; *Ateralphus subsellatus*; and *Exalphus leuconotus*.

pronotum (17: 0); and punctures on basal third of the elytra arranged in longitudinal rows (26: 1; Fig. 15). In addition, one homoplastic character also supports this group: the globose region of the femora, mainly covered with sparse white setae, with denser areas forming irregular maculae (32: 1). This state is also present in some apical lineages of *Exalphus*, including *E. calvifrons*, *E. cicatricornis*, *E. gounellei*, *E. guaraniticus*, *E. leuconotus*, *E. lichenophorus* and *E. malleri*. Additionally, one ambiguous, non-homoplastic synapomorphy also corroborates the monophyly of *Alphus* under fast optimization (Fig. 17): basal apophysis with a diagonal line more sclerotized near apex (44: 1).

Exalphus is monophyletic, supported by the following unambiguous synapomorphies (Figs 16, 17): posterior margin of head triangular (1: 1; Fig. 1); presence of coarse punctation between upper ocular lobes (4: 1); prothorax with denser line of white setae at median region (15: 1); males with sexual setae on ventral surface of the meso- and metasternum (24: 1); globose region of femora mainly covered with sparse white setae, with denser areas forming transverse bands (32: 2); meso- and metatibiae with stain of dark brown setae on basal half (35:1); tarsomere V bicolored, with dark brown setae at base and basal

third (37: 1); tarsomere V bicolored, with dark brown setae at base and apical third (38: 1); ventral lobe larger than dorsal lobe (42: 2). The character states 24: 1, 32: 2, 37:1 and 38: 1 were first mentioned by Restello et al. (2001) to diagnose *Exalphus* and are confirmed here as synapomorphies for the genus.

Finally, the monophyly of *Ateralphus* was corroborated by three unambiguous, homoplastic synapomorphies (Fig. 16): coronal suture reaching posterior margin of head (2: 1; Fig. 1B); region between basal-crests covered with dark brown setae (28: 1); and tarsomere V with dark brown setae on apical third (38: 1). Additionally, the following three synapomorphies also support the monophyly of *Ateralphus* under fast optimization (Fig. 17): antennomere III subequal to antennomere IV in length (11: 1); pronotum straight near the anterior margin (16: 0); and sternite V with longitudinal sulcus at basal region of the females (40: 1).

Two main clades were obtained in *Ateralphus*, corroborating the species-groups recognized by Souza and Monné (2013b). Clade 1, which includes *A. lacteus*, *A. tumidus* and *A. variegatus*, was supported by the following homoplastic synapomorphies: width of upper ocular lobes less than width between the lobe and the coronal suture (6: 0); genae parallel in frontal view (8:

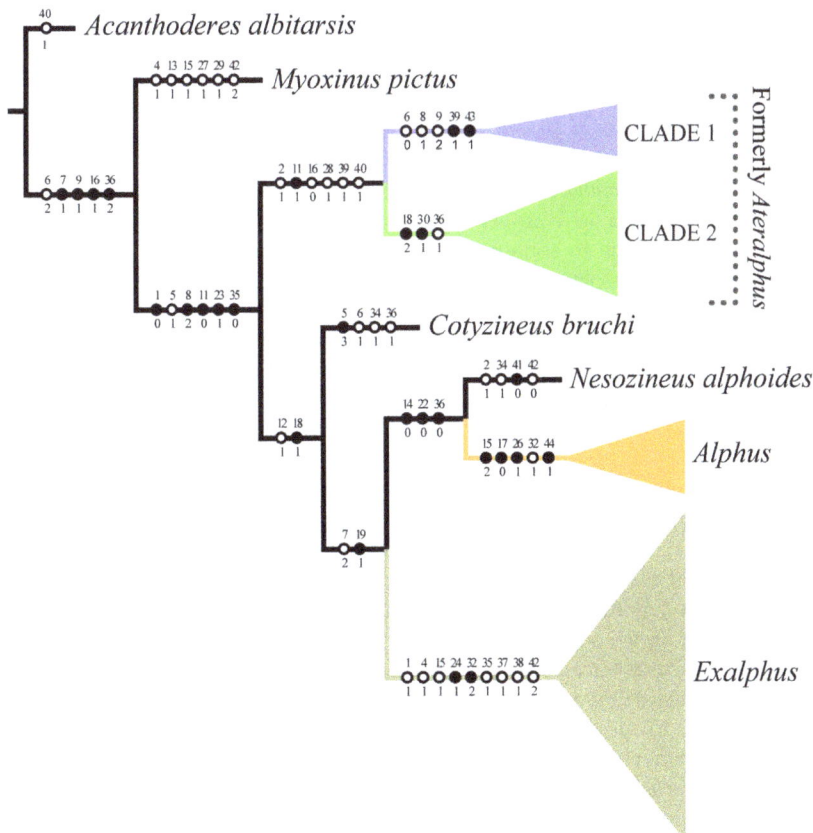

Figure 17. Summarized topology obtained from the cladistics analysis of *Ateralphus* and related genera *Alphus* and *Exalphus*, showing all character state transformations under ACCTRAN optimization. CLADE 1: *Grandateralphus* gen. n.; CLADE 2: Current *Ateralphus*.

1; Fig. 8); and scape gradually expanded toward apex, reaching widest diameter just near apex (9: 2; Fig. 12). Clade 2 includes *A. auritarsus* Souza & Monné, 2013, *A. dejeani*, *A. javariensis* (Lane, 1965), *A. lucianeae* Souza & Monné, 2013, *A. senilis* Bates, 1862 and *A. subsellatus*. This clade was supported by the following character states: post-lateral to median tubercles on the pronotum distinctly raised, appearing as a pair of tubercles (18: 2) and color pattern of the lateral margin of the elytra similar to the rest of the elytra. (30: 1). An additional synapomorphy was obtained under fast optimization for clades 1 and 2, respectively: parameres with setae on apex and inner margin (43: 1); and presence of dark brown setae at apical third (36: 1).

DISCUSSION

Among the genera studied, *Alphus* may be easily recognized by the punctures on the elytra. Restello et al. (2001) cited the arrangement of the punctures on the elytra as one of the most remarkable characteristics that enable the differentiation of *Alphus* from *Ateralphus* and *Exalphus*. In *Alphus*, the elytral punctation is present only on the basal half, arranged in longitudinal rows (Fig. 15), a autapomorphy of the genus in our results. By contrast, in *Ateralphus* and *Exalphus* the elytral punctation is irregularly distributed on the elytra (character 26; Figs 13, 14).

Although the monophyly of *Ateralphus* has been corroborated in this study, we consider that clades 1 and 2 are robust enough (synapomorphic characters, bootstrap and Bremer supports; Fig. 16) to be treated as independent genera. Most character states supporting clade 1 are homoplastic (6: 0; 8: 1 and 9: 2), since they are also observed in *Acanthoderes albitarsis* and in several lineages of *Exalphus*. However, we consider that these character states are important enough to establish the separation of clade 1 from the remaining species of *Ateralphus*. Particularly, we highlight the relevance of characters 6 and 8, which were first used by Souza and Monné (2013b) as key characters to differentiate between the two species groups of *Ateralphus*.

Under ACCTRAN optimization, character state 6: 0 appeared as a plesiomorphic condition, being interpreted here as a reversion in clade 1 (Fig. 17). The transformations of character 8 resulted in a homoplastic synapomorphy for clade 1, with all states widely distributed throughout the topology. The ancestral condition of this character is represented by character state 8: 0 (genae divergent toward apex), whereas character state 8: 1 is an apomorphic condition that appeared independently in several lineages of *Exalphus* and in the representatives of clade 1. As with character 8, character state 9: 2 can also be interpreted as a case of parallelism or convergence since it has two independent origins in our chosen topology. It is a synapomorphy for clade 1 and for the most apical clade of *Exalphus*, which includes *E. lichenophorus*, *E. malleri*, *E. cicatricornis* and *E. calvifrons* (Fig. 16).

In addition to those synapomorphies mentioned above, other characteristics also mentioned by Souza and Monné

(2013b) (e.g., shape of the lower ocular lobe, length of antennae and length of the sternite V relative to sternites II, III and IV), are taxonomically important to separate the species of clade 1 from other *Ateralphus*. Thus, considering the taxonomic significance of the characters supporting the clades 1 and 2, we propose splitting *Ateralphus* in two genera: *Ateralphus* (represented by the clade 2), with six species (*A. subsellatus* (type-species of *Ateralphus*), *A. auritarsus*, *A. javariensis*, *A. senilis*, *A. lucianeae* and *A. dejeani*); and *Grandateralphus* gen. n., allocating the species previously grouped in clade 1 (*A. tumidus*, *A. lacteus* and *A. variegatus*). All systematic changes proposed in this study are summarized in the 'TAXONOMY' section below.

Although the monophyly of *Ateralphus*, *Alphus* and *Exalphus* is strongly supported in our analysis, the phylogenetic relationships between them are not well resolved. A single intergeneric clade with bootstrap over 50% (i.e., *Nesozineus alphoides* + *Alphus*) was recovered. It is supported by the following synapomorphies: absence of stain of dark brown setae at apex of antennomeres III to XI (14: 0); mesosternal process slightly convex (22: 0); and, additionally under fast optimization, absence of a stain of dark brown setae on apex of meso- and metatibiae (36: 0).

Despite the low support values, the distribution of the characters on the topology strongly suggests that *Ateralphus* is the sister group of the lineage (*Cotyzineus bruchi* + ((*Nesozineus alphoides* + *Alphus*) + (*Exalphus*)). Supporting this relationship are the following synapomorphies: posterior margin of head triangular (1: 0; Fig. 1B); presence of a row of erect setae reaching basal half of antennal tubercles (5: 1); genae convergent in frontal view (8: 2); mesosternal process 45° angulated at apical third relative to mesosternum (23: 1); and absence of a stain of dark brown setae on half basal region of the meso- and metatibiae (35: 0)). *Myoxinus pictus* has many morphological resemblances with the representatives of *Exalphus* (which is also expressed by the high number of homoplastic characters exclusively shared between *M. pictus* and the *Exalphus* species, i.e., 13: 1; 15: 1; 27: 1 and 29: 1). However, these genera were not recovered as sister groups. Instead, *M. pictus* resulted as a basal lineage (resulting as sister group of all other species included in the analysis), while *Exalphus* was corroborated as sister group of (*Nesozineus alphoides* + *Alphus*), supported by two synapomorphies: lower ocular lobe larger than gena (7: 2); and absence of elevation post-lateral to the median tubercles of the pronotum (19: 1).

After the description of *Alphus*, *Ateralphus* and *Exalphus* by Restello et al. (2001), no other study has treated the relationships among them or between other genera of Acanthoderini. One exception is the discussion of some characters of these genera by Souza and Monné (2013b). Bates (1862) thought that *Alphus* (which at that date also comprised species currently placed in *Ateralphus* and *Exalphus*) was close to *Myoxinus*, comparing and differentiating them from *Acanthoderes* by the shape of the scape, width of the head (measured through the vertex) and shape of the malar area. In our analysis, such characteristics were adapted and codded in characters 9, 6 and 8, respectively.

Except for *Acanthoderes*, all other species included in our analysis have scape not piriform, slightly shorter than antennomere III (characteristic atypical among the representatives of Acanthoderini). Based mainly on that character, Bates (1862) suggested that *Alphus* was the lineage link between the tribes Acanthoderini and Acanthocinini, the length of the scape being the key character that keeps it within the Acanthoderini, since in Acanthocinini the length of the scape exceeds that of antennomere III. Apart from the shape of the scape, Bates (1862) also pointed out other characteristics that highlight *Alphus* among the Acanthoderini, such as mesosternum narrower toward anterior region, base of coxa angled outward and anterior tarsi of males not dilated.

Considering the insights mentioned by Bates (1862) on the evolution of Acanthoderini and the great number of homoplastic characters in our results due to the morphological plasticity observed among the representatives of the subfamily, we suggest that a more comprehensive phylogenetic study including more representatives of Acanthoderini should be conducted, not only to further explore the relationship between these genera, but also to understand the relationships among the genera of Acanthoderini and to evaluate their tribal classification.

TAXONOMY

Grandateralphus Souza, Monné & Marinoni, gen. n.

http://zoobank.org/82FD03E6-1193-42F5-9A66-D65495D2E0D1

Type-species. *Alphus variegatus* Mendes, 1938.

Description. Frons rectangular; slightly convex, almost flat in lateral view; with a row of long setae near malar area; longitudinal suture well-defined, reaching occiput. Head finely punctate; slightly depressed, with coarse and sparse punctation on vertex; antennal tubercles slightly prominent, obliquely directed. Eyes coarsely faceted. Upper ocular lobes semicircular, bordered at vertex by one row of straight setae; separated by three or more times their width. Lower ocular lobes narrow and rectangular, height less than height of gena. Genae parallel. Labrum covered with dense and short setae; with a transversal row of long setae at median region. Mandibles triangular, symmetrical, apex acuminate; outer margin densely covered with short setae. Scape gradually expanded toward apex, reaching widest diameter near apex; slightly shorter than antennomere III. Pedicel short, gradually expanded toward apex. Antennomeres III–XI with a ring of dark brown setae at apical margin; gradually decreasing in length. Prothorax transverse; coarse and irregularly punctate; covered with short setae; sides with a pointed lateral tubercle. Pronotum straight near anterior margin; disc with a pair of median tubercles and, posterior to these, a small tubercle at midline. Pro-, meso- and metasternum dense and finely punctate. Prosternal process width about 2–3 times narrower than

diameter of one procoxa; longitudinally depressed; posterior margin truncate. Mesosternal process straight or tumescent at posterior half; subequal in length to mesocoxa; lateral margins without tubercles; posterior margin bilobed. Elytra completely covered with setae; with coarse punctation irregularly distributed, denser at basal and lateral areas; slightly convex apically; almost straight at basal third; basal-crests raised, with rounded tubercles slightly elevated; from basal-crests, a sinuous carina extending toward apex; with a diagonal carina from humerus to basal-crest carinae. Humeri rounded, slightly projected anteriorly; with small tubercles. Pro- and mesocoxae globular. Femora and tibiae subequal in length. Femora pedunculate. Tibiae gradually enlarged apically; with a spot of dark brown setae at subapical region. Tarsomeres V bicolored, with dark brown setae on apical third or on base and apical third. Abdomen fine and irregularly punctate. Sternite I as long as sternites II, III and IV together; anterior margin long and acuminate; length about two thirds its total length. Sternites II, III and IV subequal in length. Sternite V wider than long; length about equal or less than length of sternites III and IV together; sternite V of females with a median longitudinal sulcus at basal fourth.

Remarks. *Grandateralphus* gen. n. is closely related to *Ateralphus*. Their sister-group relationship is supported by the following synapomorphies: antennomere III subequal to antennomere IV; pronotum straight near anterior margin; and sternite V with longitudinal sulcus at basal region the female abdomen. *Grandateralphus* gen. n. is supported by the following synapomorphies: width of upper ocular lobes less than width between the lobe and the coronal suture (6: 0), genae parallel in frontal view (8: 1) and scape gradually expanded toward apex, reaching widest diameter just near apex (9: 2). In addition to these synapomorphies, *Grandateralphus* gen. n. can be differentiated from *Ateralphus* by the lower ocular lobes rectangular, lesser in height than gena; meso- and metatibiae with a subapical stain of dark brown setae; and sternite V equal or less in length than sternites III and IV together.

Etymology. *Grandateralphus* gen. n. is a combination of the Latin word *grand* (= large) with *Ateralphus*. It is allusive to the size of the representatives of the new genus, which are usually larger in total length than the representatives of *Ateralphus*.

Grandateralphus gen. n. includes the following species (new combinations proposed in this study):

Grandateralphus lacteus (Galileo & Martins, 2006), comb. n.
Grandateralphus tumidus (Souza & Monné, 2013), comb. n.
Grandateralphus variegatus (Mendes, 1938), comb. n.

Taxonomic notes

Taxonomic notes are provided from examination of material belonging to the American Coleoptera Museum, Texas, United States of America (ACMT). Souza and Monné (2013b) registered *G. variegatus* comb. n. (cited as *A. variegatus*) from Bolivia (Santa Cruz) based on primary records in the literature (Wappes et al. 2006). Recently, we had the opportunity to examine some material from the ACMT and, based on our observations, we exclude the record of *G. variegatus* comb. n. from Bolivia, considering that these records actually correspond to *G. lacteus* comb. n. Also, based on that material, we confirm the literature records of *A. subsellatus* cited by Souza and Monné (2013b) and provide a new country record of *E. cicatricornis* to Bolivia (Santa Cruz).

Material examined. *Grandateralphus lacteus* comb. n. (Galileo & Martins, 2006). Bolivia, Santa Cruz, 4–6 km SSE Buena Vista, F & F Hotel, 1 male, 22–31.x.2002, Wappes and Morris leg.; 1 female, 29–30.x.2003, Robin Clarke leg.; 1 female, 10–15. xi.2003, Robin Clarke leg.; 1 male, 21–24.xi.2003, Wappes, Morris and Nearns leg.; 1 female, 30.ix.2004, Robin Clarke leg.; 1 male, 3–8.x.2004, Wappes and Morris leg.; Reserva Natural Potrrillo del Guenda, Snake Farm, 17°40′26″S, 63°27′43″W, 400 m, 2 males, 2 females, 6–9.x.2006, Wappes, Nearns and Eya leg.; Potrerillo del Guenda, 370 m, 1 female, 16–22.x.2006, Nearns and Eya leg.; Potrerillo del Guenda, Reserva Natural, 40 km Santa Cruz, 17°40′S, 63°27′W, 370 m, 1 male, 16–21.x.2007, F. and J. Romero leg.; Potrerillo del Guenda, Snake Farm, 17°40′S, 63°27′W, 350–400 m, 1 male, 15–22.xi.2011, Bettela, Bonaso and Romero leg.; Potrerillo del Guenda, 17°40′S, 63°27′W, 350–400 m, 1 female, 1.xii.2011, Bettela, Bonaso and Romero leg.; Potrerillo del Guenda, Snake Farm, 17°40′S, 63°27′W, 350–400m, 1 female, 10–18.xii.2011, Bettella, Bonaso and Romero leg.; Potr. Del Guenda, Reserva Natural, Snake Farm, 17°40′15″S, 63°27′26″W, 400 m, 1 male, 1 female, 23–30.x.2013, Wappes and Kuckartz leg.; Potr. Del Guenda, 1 male, 23–30.x.2013, Wappes and Kuckartz leg.; Huaico, Potrerillo, across Guenda fm Potrerillo, 17°40′35″S, 63°26′59″W, 1,270 ft., 1 female, 18.xi.2012, Windsor and Gowin leg.; Huaico, Potrerillo, 17°40′S, 63°26′W, 430m, MV/UV lights, 4 males, 2 females, 27–29.x.2013, Wappes and Kuckartz leg. (ACMT). *Ateralphus subsellatus* (White, 1855). Bolivia, Santa Cruz, Potrerillo del Guenda, Reserva Natural, Snake Farm, 17°40′15″S, 63°27′26″W, 400 m, 1 male, 24–30.x.2012, Betella, Bonaso and Romero leg.; 20 km N Camiri, road to Eyti, 6–8 km E Hwy 9, 19°52′S, 63°29′W, 1250 m, 1 male, 5, 6, 10.xii.2012, Wappes, Bonaso and Skillman leg. (ACMT). *Exalphus cicatricornis* Schmid, 2014. Bolivia, Santa Cruz, Huaico, 17°40′S, 63°24′W, 430 m, 1 female, 21.xi.2013, Skillman and Wappes leg. (ACMT).

ACKNOWLEDGMENTS

We are grateful to James E. Wappes for the loan of specimens for examination; to Tatiana A. Sepúlveda Villa for the valuable considerations to the manuscript; and to Coordenação de Aperfeiçoamento de Pessoal de Nível Superior (CAPES) for the first author's, PhD scholarship. LM and MLM are fellows of the Conselho Nacional de Desenvolvimento Científico e Tecnológico (CNPq) under the following process numbers, respectively: 307732/2015-0 and 304718/2014-9.

REFERENCES

Bates HW (1862) Contributions to an insect fauna of the Amazon Valley. Coleoptera: Longicornes. Annals and Magazine of Natural History 9: 117–124, 396–405, 446–45. https://doi.org/10.5962/bhl.title.110205

Bremer K (1994) Branch support and tree stability. Cladistics 10: 295–304. https://doi.org/10.1111/j.1096-0031.1994.tb00179.x

Goloboff PA (1993) NONA. Noname (a bastard son of Pee-wee). Tucumán, Fundación y Instituto Miguel Lillo, v. 2.0 [computer program distributed by the authors].

Goloboff PA, Farris JS, Nixon K (2008) TNT: a free program for phylogenetic analysis. Cladistics 24: 774–786. https://doi.org/10.1111/j.1096-0031.2008.00217.x

Lacordaire JT (1872) Histoire Naturelle des Insectes. Genera des Coléoptères, ou exposé méthodique et critique de tous les genres proposés jusqu'ici dans cet ordre d'insectes. Librairie Encyclopédique de Roret 9: 411–930. https://doi.org/10.5962/bhl.title.8864

Martins UR (1985) Novos táxons, sinonímias, notas e nova combinação em Cerambycidae (Coleoptera) neotropicais. Revista Brasileira de Entomologia 29: 169–180.

Martins UR, Galileo MHM (2007) Notas e descrições em Acanthoderini (Coleoptera, Cerambycidae, Lamiinae). I. Novos táxons, nova sinonímia e novos registros. Papéis Avulsos de Zoologia 47: 159–164. https://doi.org/10.1590/S0031-10492007001200001

Nixon KC (1999–2002) Winclada. Ithaca, v. 1.00.00 [computer program distributed by the author]

Restello RM, Iannuzzi L, Marinoni RC (2001) Descrição de dois novos gêneros afins a *Alphus* White e duas novas espécies (Cerambycidae, Lamiinae, Acanthoderini). Revista Brasileira de Entomologia 45: 295–303.

Sereno PC (2007) Logical basis for morphological characters in phylogenetics. Cladistics 23: 565–587. https://doi.org/10.1111/j.1096-0031.2007.00161.x

Souza DS, Monné ML (2013a) *Alphus marinonii* sp. n., nova espécie para o Peru e Brasil (Coleoptera, Cerambycidae, Lamiinae). Revista Brasileira de Entomologia 57: 9–11. https://doi.org/10.1590/S0085-56262013000100002

Souza DS, Monné ML (2013b) Revision of the genus *Ateralphus* Restello, Iannuzzi & Marinoni, 2001 (Coleoptera: Cerambycidae: Lamiinae). Zootaxa 3736: 301–337. https://doi.org/10.11646/zootaxa.3736.4.1

Souza DS, Monné ML (2014) Synopsis of the genus *Exalphus* Restello, Iannuzzi & Marinoni (Coleoptera, Cerambycidae, Lamiinae), with description of a new species and new country records. Revista Brasileira de Entomologia 58: 19–24. https://doi.org/10.1590/S0085-56262014000100004

Tavakilian G, Chevillotte H (2016) Titan: base de données internationales sur les Cerambycidae ou Longicornes. Available online at http://titan.gbif.fr/accueil_uk.html [Accessed: 03/05/2017]

Thomson J (1860) Essai d'une classification de la famille des cé-rambycides et matériaux pour servir a une monographie de cette famille. Bouchard-Huzard, Paris, 404 pp. https://doi.org/10.5962/bhl.title.9206

Wappes JE, Morris RF, Nearns EH, Thomas MC (2006) Preliminary checklist of Bolivian Cerambycidae (Coleoptera). Insecta Mundi 20: 1–45.

White A (1855) Catalogue of the coleopterous insects in the collection of the British Museum. 8. Longicornia 2, London, 175–412.

Author Contributions: DSS wrote the paper; LM and MLM meticulously reviewed the text and the characters used in the phylogenetic analysis.

Competing Interests: The authors have declared that no competing interests exist.

PERMISSIONS

All chapters in this book were first published in Zoologia (Curitiba), by Sociedade Brasileira de Zoologia; hereby published with permission under the Creative Commons Attribution License or equivalent. Every chapter published in this book has been scrutinized by our experts. Their significance has been extensively debated. The topics covered herein carry significant findings which will fuel the growth of the discipline. They may even be implemented as practical applications or may be referred to as a beginning point for another development.

The contributors of this book come from diverse backgrounds, making this book a truly international effort. This book will bring forth new frontiers with its revolutionizing research information and detailed analysis of the nascent developments around the world.

We would like to thank all the contributing authors for lending their expertise to make the book truly unique. They have played a crucial role in the development of this book. Without their invaluable contributions this book wouldn't have been possible. They have made vital efforts to compile up to date information on the varied aspects of this subject to make this book a valuable addition to the collection of many professionals and students.

This book was conceptualized with the vision of imparting up-to-date information and advanced data in this field. To ensure the same, a matchless editorial board was set up. Every individual on the board went through rigorous rounds of assessment to prove their worth. After which they invested a large part of their time researching and compiling the most relevant data for our readers.

The editorial board has been involved in producing this book since its inception. They have spent rigorous hours researching and exploring the diverse topics which have resulted in the successful publishing of this book. They have passed on their knowledge of decades through this book. To expedite this challenging task, the publisher supported the team at every step. A small team of assistant editors was also appointed to further simplify the editing procedure and attain best results for the readers.

Apart from the editorial board, the designing team has also invested a significant amount of their time in understanding the subject and creating the most relevant covers. They scrutinized every image to scout for the most suitable representation of the subject and create an appropriate cover for the book.

The publishing team has been an ardent support to the editorial, designing and production team. Their endless efforts to recruit the best for this project, has resulted in the accomplishment of this book. They are a veteran in the field of academics and their pool of knowledge is as vast as their experience in printing. Their expertise and guidance has proved useful at every step. Their uncompromising quality standards have made this book an exceptional effort. Their encouragement from time to time has been an inspiration for everyone.

The publisher and the editorial board hope that this book will prove to be a valuable piece of knowledge for researchers, students, practitioners and scholars across the globe.

LIST OF CONTRIBUTORS

Fernando Ferreira de Pinho and Guilherme Braga Ferreira
Instituto Biotrópicos. Praça JK 25, 39100-000 Diamantina, MG, Brazil

Adriano Pereira Paglia
Programa de Pós-Graduação em Ecologia, Conservação e Manejo da Vida Silvestre, Departamento de Biologia Geral, Universidade Federal de Minas Gerais. Avenida Antonio Carlos 6627, 31270-901 Belo Horizonte, MG, Brazil

Thyara Noely Simões
Universidade Estadual de Santa Gruz, Programa de Pós-graduação em Ecologia e Conservação da Biodiversidade. Rodovia Jorge Amado, km 16, Salobrinho, 45662-900 Ilhéus, BA, Brazil

Arley Candido da Silva
Ecoassociados NGO – Conservação de tartarugas marinhas, baobás e recifes de corais. Rua das Caraúnas, Porto de Galinhas, 55590-000 Ipojuca, PE, Brazil

Carina Carneiro de Melo Moura
Department of Biology, Institute of Pharmacy and Molecular Biotechnology, Heidelberg University. Heidelberg, Im Neuenheimer Feld 364, 69120, Heidelberg, Germany

Daniel F. Perrella
Programa de Pós-graduação em Ecologia e Recursos Naturais, Universidade Federal de São Carlos. Rodovia Washington Luís km 235, 13565-905 São Carlos, SP, Brazil

Paulo V. Davanço, Leonardo S. Oliveira, Livia M.S. Sousa and Mercival R. Francisco
Departamento de Ciências Ambientais, Universidade Federal de São Carlos. Rodovia João Leme dos Santos km 110, 18052-780 Sorocaba, SP, Brazil

Carolina A. Freire, Leonardo de P. Rios, Eloísa P. Giareta and Giovanna C. Castellano
Departmento de Fisiologia, Setor de Ciências Biológicas, Universidade Federal do Paraná. Centro Politécnico, Jardim das Américas, 81531-980 Curitiba, PR, Brazil

Carlos Alberto S. de Lucena, Amanda Bungi Zaluski and Zilda Margarete Seixas de Lucena
Museu de Ciências e Tecnologia, Pontifícia Universidade Católica do Rio Grande do Sul. Avenida Ipiranga 6681, Caixa, 90619- 900 Porto Alegre, RS, Brazil

Henrique Alencar Meira da Silva and Lycia de Brito-Gitirana
Laboratório de Histologia Integrativa, Programa de Pesquisa em Glicobiologia, Instituto de Ciências Biomédicas, Universidade Federal do Rio de Janeiro. Avenida Carlos Chagas Filho 373, Bloco B1-019, 21941-902 Rio de Janeiro, RJ, Brazil

Thiago Silva-Soares
Laboratório de Zoologia, Museu de Biologia Prof. Mello Leitão, Instituto Nacional da Mata Atlântica. Avenida José Ruschi, 29650-000 Santa Teresa, ES, Brazil

Lygia C. Ruas
Programa de Pós-graduação em Aquicultura e Pesca, Instituto de Pesca. Avenida Francisco Matarazzo 455, Parque da Água Branca, 05001-970 São Paulo, SP, Brazil

André M. Vaz-dos-Santos
Laboratório de Esclerocronologia, Departamento de Biodiversidade, Universidade Federal do Paraná. Rua Pioneiro 2153, Jardim Dallas 85950-000 Palotina, PR, Brazil

Raphael Aquino Heleodoro and José Albertino Rafael
Programa de Pós-graduação em Entomologia, Instituto Nacional de Pesquisas da Amazônia. Avenida André Araújo 2936, Petrópolis, 69067-375 Manaus, AM, Brazil

Ricardo Andreazze
Departamento de Microbiologia e Parasitologia, Centro de Biociências, Universidade Federal do Rio Grande do Norte. Campus Universitário Lagoa Nova, Caixa, 59078-900 Natal, RN, Brazil

Laura Gomez-Mesa
Departamento de Ecologia, Universidade do Estado do Rio de Janeiro. Rua São Francisco Xavier 524, PHLC sala 220, Maracanã, 20550-019 Rio de Janeiro, RJ, Brazil
Programa de Biología, Universidad CES-EIA, Calle 10 A, No. 22 – 04, Medellín, Colombia

Juliane Pereira-Ribeiro, Atilla C. Ferreguetti, Marlon Almeida-Santos, Helena G. Bergallo and Carlos F. D. Rocha
Departamento de Ecologia, Universidade do Estado do Rio de Janeiro. Rua São Francisco Xavier 524, PHLC sala 220, Maracanã, 20550-019 Rio de Janeiro, RJ, Brazil

Luciane R. da Silva Mohr and Eduardo Périco
Programa de Pós-graduação em Ambiente e Desenvolvimento, Museu de Ciências Naturais, Setor de Ecologia e Evolução, Centro Universitário Univates. Rua Avelino Tallini 171, 95900-000 Lajeado, RS, Brazil

Vanda S. da Silva Fonseca
Programa de Pós-graduação em Ambiente e Desenvolvimento, Museu de Ciências Naturais, Setor de Ecologia e Evolução, Centro Universitário Univates. Rua Avelino Tallini 171, 95900-000 Lajeado, RS, Brazil
Bioimagens Consultoria Ambiental. Rua Felicíssimo de Azevedo 1352, 90540110 Porto Alegre, RS, Brazil

Alexsandro R. Mohr
Bioimagens Consultoria Ambiental. Rua Felicíssimo de Azevedo 1352, 90540110 Porto Alegre, RS, Brazil
Graduação em Biologia, Universidade de Santa Cruz do Sul. Avenida Independência 2293, 96816-501 Santa Cruz do Sul, RS, Brazil

Igor Souza-Gonçalves
Programa de Pós-Graduação em Ecologia, Departamento de Biologia Geral, Universidade Federal de Viçosa. 36570-900 Viçosa, MG, Brazil
Laboratório de Sistemática e Biologia de Coleoptera, Departamento de Biologia Animal, Universidade Federal de Viçosa. 36570-900 Viçosa, MG, Brazil

Cristiano Lopes-Andrade
Laboratório de Sistemática e Biologia de Coleoptera, Departamento de Biologia Animal, Universidade Federal de Viçosa. 36570-900 Viçosa, MG, Brazil

Renata C. Lima-Gomes
Programa de Pós-Graduação em Biologia de Água Doce e Pesca Interior, Instituto Nacional de Pesquisas da Amazônia. Caixa, 69080- 971 Manaus, AM, Brazil

Jô de Farias Lima
Empresa Brasileira de Pesquisa Agropecuária – Embrapa Amapá. Rodovia Juscelino Kubitschek, km 5, 2600, Caixa, 68906-970 Macapá, AP, Brazil

Célio Magalhães
Coordenação de Biodiversidade, Instituto Nacional de Pesquisas da Amazônia. Caixa, 69080-971 Manaus, AM, Brazil

Somaye Vaissi, Paria Parto and Mozafar Sharifi
Department of Biology, Faculty of Science, Razi University. Baghabrisham 6714967346, Kermanshah, Iran

Lucas L. Lanna, Cristiano S. de Azevedo, Ricardo M. Claudino, Reisla Oliveira and Yasmine Antonini
Instituto de Ciências Exatas e Biológicas, Universidade Federal de Ouro Preto. Campus Morro do Cruzeiro, Bauxita, 35400-000 Ouro Preto, MG, Brazil

Natália Lima Boroni
Departamento de Zoologia, Universidade Federal de Minas Gerais. 31270-901 Belo Horizonte, MG, Brazil
Departamento de Biologia Animal, Universidade Federal de Viçosa. 36570-900 Viçosa, MG, Brazil

Leonardo Souza Lobo
Laboratório de Processamento de Imagem Digital, Museu Nacional, Universidade Federal do Rio de Janeiro. 20940-040 Rio de Janeiro, RJ, Brazil
Departamento de Biologia Animal, Universidade Federal de Viçosa. 36570-900 Viçosa, MG, Brazil

Pedro Seyferth R. Romano and Gisele Lessa
Departamento de Biologia Animal, Universidade Federal de Viçosa. 36570-900 Viçosa, MG, Brazil

Julissa M. Churata-Salcedo and Lúcia M. Almeida
Laboratório de Sistemática e Bioecologia de Coleoptera, Department of Zoology, Universidade Federal do Paraná. Caixa, 81581-980 Curitiba, PR, Brazil

Rodrigo S. Bouzan, João Paulo P. Pena-Barbosa and Antonio D. Brescovit
Laboratório Especial de Coleções Zoológicas, Instituto Butantan. Avenida Vital Brasil 1500, 05503-090 São Paulo, SP, Brazil

Adrielle M. Cezar and Leila M. Pessôa
Laboratório de Mastozoologia, Departamento de Zoologia, Instituto de Biologia, Universidade Federal do Rio de Janeiro. Avenida Brigadeiro Trompowski, Ilha do Fundão, 21941-590 Rio de Janeiro, RJ, Brazil

Cibele R. Bonvicino
Laboratório de Biologia e Parasitologia de Mamíferos Silvestres Reservatórios, Instituto Oswaldo Cruz. Avenida Brasil 4365, Manguinhos, 21040-360 Rio de Janeiro, Brazil
Divisão de Genética, Instituto Nacional do Câncer. Rua André Cavalcanti 37, Centro, 20231-050 Rio de Janeiro, RJ, Brazil

Maria Gabriela Cuezzo
Instituto de Biodiversidad Neotropical, CONICET-UNT, Facultad de Ciencias Naturales, Universidad Nacional de Tucumán. Miguel Lillo 205, 4000 Tucumán, Argentina

Meire Silva Pena
Laboratório de Malacologia, Museu de Ciências Naturais, Pontifícia Universidade Católica de Minas Gerais. Avenida Dom José Gaspar 500, Coração Eucarístico, Belo Horizonte, MG, Brazil

Luiz Silva and Paulo Lana
Centro de Estudos do Mar, Universidade Federal do Paraná. Avenida Beira Mar, 83255-976 Pontal do Paraná, PR, Brazil

Diego de S. Souza and Luciane Marinoni
Departamento de Zoologia, Universidade Federal do Paraná. Caixa, 81531-990 Curitiba, PR, Brazil

Marcela L. Monné
Departamento de Entomologia, Museu Nacional, Universidade Federal do Rio de Janeiro. Quinta da Boa Vista, São Cristóvão, 20940-040 Rio de Janeiro, RJ, Brazil

Index

www.ingramcontent.com/pod-product-compliance
Lightning Source LLC
Chambersburg PA
CBHW082031190326
41458CB00010B/3331